# SOLITONS IN MULTIDIMENSIONS
## INVERSE SPECTRAL TRANSFORM METHOD

# SOLITONS IN MULTIDIMENSIONS

## INVERSE SPECTRAL TRANSFORM METHOD

### B G KONOPELCHENKO
*Budker Institute of Nuclear Physics*
*Novosibirsk, Russia*

**World Scientific**
*Singapore • New Jersey • London • Hong Kong*

*Published by*

World Scientific Publishing Co. Pte. Ltd.
P O Box 128, Farrer Road, Singapore 9128
*USA office:* Suite 1B, 1060 Main Street, River Edge, NJ 07661
*UK office:* 73 Lynton Mead, Totteridge, London N20 8DH

**Library of Congress Cataloging-in-Publication Data**
Konopelchenko, B. G. (Boris Georgievich), 1948–
    Solitons in multidimensions : inverse spectral transform method /
B. G. Konopelchenko.
     p.   cm.
    ISBN 9810213484
    1. Solitons.   2. Inverse scattering transform.   I. Title.
QC174.26.W28K65   1993
530.1'4--dc20                                                     93-16338
                                                                                     CIP

Copyright © 1993 by World Scientific Publishing Co. Pte. Ltd.

*All rights reserved. This book, or parts thereof, may not be reproduced in any form or by any means, electronic or mechanical, including photocopying, recording or any information storage and retrieval system now known or to be invented, without written permission from the Publisher.*

Printed in Singapore by Utopia Press.

# Preface

The soliton phenomena and integrable nonlinear equations represent an important and well established field of modern physics, mathematical physics and applied mathematics. Solitons are found in various areas of physics from hydrodynamics and plasma physics, nonlinear optics and solid state physics, to field theory and gravitation. Nonlinear differential equations which describe soliton phenomena have an universal character. They are of great mathematical interest too.

The inverse scattering (spectral) transform method discovered twenty five years ago provided a mathematical tool for study of solitons. The inverse spectral transform is now a very powerful method in the investigation of nonlinear partial differential equations. Since its discovery this method has been applied to a great variety of nonlinear equations which arise in various fields of physics. The majority of integrable equations treated heretofore are nonlinear differential equations in two independent variables which usually correspond to one spatial and one temporal variables. Methods of solution of such (1+1)-dimensional soliton equations are well established. These methods have been described in a variety of monographs and reviews.

In this monograph we present the principal ideas, methods and results concerning multidimensional soliton equations. The multidimensional inverse spectral transform method has been developed mainly during the last decade. Our aim here is to describe both methods for the construction of multidimensional soliton equations together with the technique for calculation of their exact solutions. We confine ourselves by the 2+1-dimensional (two spatial and one temporal dimensions) integrable equations. A main attention is paid to the $\bar{\partial}$-dressing method. This method is a very powerful and effective tool for the construction of soliton equations and calculation of their exact solutions simultaneously. We consider also the linearization of the initial value problem within the framework of the inverse spectral transform method. The quasi-local $\bar{\partial}$-problem and the nonlocal Riemann—Hilbert problem are the typical corresponding generating problems. The Kadomtsev—Petviashvili, the modified Kadomtsev—Petviashvili, the Davey—Stewartson, the Ishimori and the 2+1-dimensional integrable sine-Gordon equations are our basic examples.

This volume is devoted to the mathematical aspects of 2+1-dimensional solitons. Accordingly, it is addressed mainly to those who are interested in the mathematical methods of the modern multidimensional soliton theory. It is intended primarily for readers who already acquainted with the basic elements of soliton theory. However, while such knowledge is desirable it is not indispensable. The exposition is basically self-contained.

The author is grateful to V.G. Dubrovsky with whom some of the results presented in this monograph have been obtained.

Novosibirsk, July 1992

# Contents

**Preface** v

**Chapter 1. Introduction** 1
  1.1. Inverse spectral transform method in 1+1-dimensions . . . . . . . . . . . 1
  1.2. Basic integrable equations in 2+1-dimensions . . . . . . . . . . . . . . . . 11
  1.3. Solitons and initial value problem in 2+1-dimensions . . . . . . . . . . . 21

**Chapter 2. Dressing method** 29
  2.1. $\bar{\partial}$-equation . . . . . . . . . . . . . . . . . . . . . . . . . . . . . . . . . . . . . . 29
  2.2. Nonlocal $\bar{\partial}$-problem . . . . . . . . . . . . . . . . . . . . . . . . . . . . . . . . . 32
  2.3. $\bar{\partial}$-dressing method . . . . . . . . . . . . . . . . . . . . . . . . . . . . . . . . . 38
  2.4. Examples . . . . . . . . . . . . . . . . . . . . . . . . . . . . . . . . . . . . . . . . . 46
  2.5. $\bar{\partial}$-dressing on nontrivial background . . . . . . . . . . . . . . . . . . . . . . 57

**Chapter 3. The Kadomtsev–Petviashvili equation** 63
  3.1. $\bar{\partial}$-dressing . . . . . . . . . . . . . . . . . . . . . . . . . . . . . . . . . . . . . . . 63
  3.2. Exact solutions . . . . . . . . . . . . . . . . . . . . . . . . . . . . . . . . . . . . . 71
  3.3. Initial value problem for the KP-I equation . . . . . . . . . . . . . . . . . . 80
  3.4. Initial value problem for the KP-II equation . . . . . . . . . . . . . . . . . 90

**Chapter 4. The modified Kadomtsev–Petviashvili equation** 95
  4.1. $\bar{\partial}$-dressing . . . . . . . . . . . . . . . . . . . . . . . . . . . . . . . . . . . . . . . 96
  4.2. Exact solutions . . . . . . . . . . . . . . . . . . . . . . . . . . . . . . . . . . . . . 102
  4.3. Initial value problem for the mKP-I equation . . . . . . . . . . . . . . . . . 113
  4.4. Initial value problem for the mKP-II equation . . . . . . . . . . . . . . . . 122
  4.5. The Miura map and the 2+1-dimensional Gardner equation . . . . . . . 125

**Chapter 5. The Davey–Stewartson equation** 129
  5.1. $\bar{\partial}$-dressing . . . . . . . . . . . . . . . . . . . . . . . . . . . . . . . . . . . . . . . 129
  5.2. Exact solutions . . . . . . . . . . . . . . . . . . . . . . . . . . . . . . . . . . . . . 139
  5.3. Initial value problem for the DS-I equation . . . . . . . . . . . . . . . . . . 151

5.4. Initial value problem for the DS-II equation . . . . . . . . . . . . . . . . 156

## Chapter 6. The Ishimori equation          164
6.1. $\bar{\partial}$-dressing . . . . . . . . . . . . . . . . . . . . . . . . . . . . . . . . . . . 165
6.2. Exact solutions . . . . . . . . . . . . . . . . . . . . . . . . . . . . . . . . . . 175
6.3. Initial value problem for the Ishimori-I equation . . . . . . . . . . . . . . 189
6.4. Initial value problem for the Ishimori-II equation . . . . . . . . . . . . . 195
6.5. Gauge equivalence of the DS and Ishimori equations . . . . . . . . . . . . 200

## Chapter 7. The 2+1-dimensional integrable sine-Gordon equation          204
7.1. The 2DISG equation and its properties . . . . . . . . . . . . . . . . . . . 204
7.2. $\bar{\partial}$-dressing . . . . . . . . . . . . . . . . . . . . . . . . . . . . . . . . . . . 210
7.3. Exact solutions . . . . . . . . . . . . . . . . . . . . . . . . . . . . . . . . . . 218
7.4. Localized solitons . . . . . . . . . . . . . . . . . . . . . . . . . . . . . . . . 227
7.5. Initial value problem for the 2DISG equation . . . . . . . . . . . . . . . . 234

## Chapter 8. Extensions of the $\bar{\partial}$-dressing method          240
8.1. $\bar{\partial}$-dressing for the DS-equation with constant asymptotics . . . . . . . . . 240
8.2. $\bar{\partial}$-dressing with variable normalization . . . . . . . . . . . . . . . . . . . 244
8.3. Further generalizations . . . . . . . . . . . . . . . . . . . . . . . . . . . . . 252
8.4. Symmetries of the $\bar{\partial}$-problem and $\bar{\partial}$-dressing . . . . . . . . . . . . . . . . 256
8.5. 2D Harry Dym equation and the $\bar{\partial}$-problem with essential singularity . . . 262
8.6. Towards the multidimensional systems . . . . . . . . . . . . . . . . . . . . 268

## References          271

# SOLITONS IN MULTIDIMENSIONS

INVERSE SPECTRAL TRANSFORM METHOD

# Chapter 1

# Introduction

## 1.1. Inverse spectral transform method in 1+1-dimensions

A basis of the mathematical theory of soliton phenomena was established by Gardner, Green, Kruskal and Miura in the pioneer paper in 1967 [1]. They discovered that the solution of an initial value problem for the Korteweg—de Vries (KdV) equation

$$u_t + u_{xxx} + 6uu_x = 0 \qquad (1.1.1)$$

is closely connected with the one-dimensional Schrödinger equation

$$-\frac{\partial^2 \psi}{\partial x^2} - u(x,t)\psi = \lambda^2 \psi. \qquad (1.1.2)$$

They demonstrated that if the potential $u(x,t)$ in (1.1.2) evolves in time $t$ according to the KdV equation then the spectrum of the problem (1.1.2) is time-independent

$$\frac{\partial \lambda^2}{\partial t} = 0$$

and

$$\frac{\partial \psi}{\partial t} + \frac{\partial^3 \psi}{\partial x^3} - 3(\lambda^2 - u)\frac{\partial \psi}{\partial x} - 4i\lambda^3 \psi = 0. \qquad (1.1.3)$$

In other words, the KdV equation (1.1.1) is equivalent to the compatibility condition for the system (1.1.2), (1.1.3) with the time-independent $\lambda$. Equations (1.1.2) and (1.1.3) are basic ones for the whole procedure proposed in [1].

It follows from equation (1.1.3) that the evolution in time $t$ of the scattering data for the Schrödinger equation (1.1.2) is given by linear equations. Indeed, let the potential $u(x,t)$ decreases at $|x| \to \infty$ sufficiently rapidly. Let us consider the solution $\varphi$ of the problem (1.1.2) such that $\varphi \to \exp(-i\lambda x)$ at $x \to -\infty$. At $x \to +\infty$ one has $\varphi \underset{x \to +\infty}{\to} a(\lambda, t)e^{-i\lambda x}$ $+ b(\lambda, t)e^{i\lambda x}$ where $a(\lambda, t)$ and $b(\lambda, t)$ are some complex functions. Considering equation (1.1.3) at $x \to +\infty$ and using this asymptotics, we obtain $\frac{\partial a}{\partial t} = 0$, $\frac{\partial b}{\partial t} = 8i\lambda^3 b$. Hence, the evolution of the reflection coefficient $R(\lambda, t) \doteq \frac{b(\lambda,t)}{a(\lambda,t)}$ is given by the linear equation

$\frac{\partial R}{\partial t} = 8i\lambda^3 R$. This equation is readily integrable: $R(\lambda,t) = e^{8i\lambda^3 t} R(\lambda,0)$. Similarly the evolution of the scattering data which correspond to the discrete spectrum of (1.1.2) is given by the linear equation too.

Just this observation provides an efficiency to the following procedure of solution of the initial value problem for the KdV equation

$$u(x,0) \xrightarrow{I} \{R(\lambda,0),\ldots\} \xrightarrow{II} \{R(\lambda,t),\ldots\} \xrightarrow{III} u(x,t). \qquad (1.1.4)$$

At the first stage, given $u(x,0)$, one must calculate the inverse problem data $\{R(\lambda,0), \lambda_n, b_n(0)\}$ at $t=0$, that is one must solve the forward scattering problem for (1.1.2). The second stage is trivial due to the properties of the scattering data mentioned above. At the third stage, given the scattering data, one must reconstruct the potential $u(x,t)$, i.e. one must solve the inverse scattering problem for the stationary Schrödinger equation (1.1.2). The solution of the inverse scattering problem for equation (1.1.2) is given by the well-known linear integral Gelfand—Levitan—Marchenko equations

$$u(x,t) = 2\frac{d}{dx}K(x,x',t), \qquad (1.1.5a)$$

$$K(x,x',t) + F(x+x',t) + \int_x^\infty ds\, K(x,s)F(s+x',t) = 0, \quad x' > x, \qquad (1.1.5b)$$

$$F(x,t) = \frac{1}{2\pi}\int_{-\infty}^{+\infty} d\lambda\, e^{i\lambda x} R(\lambda,t) + \sum_{n=1}^N b_n(t)\, e^{-\lambda_n x}. \qquad (1.1.5c)$$

Hence, the solution of the initial value problem for the nonlinear KdV equation is reduced to the sequence of linear problems. The third stage in (1.1.4) is technically the most principal and complicated. It gave the initial appellation (the inverse scattering transform (IST) method) for the whole procedure indicated in (1.1.4).

In a general case the explicit solutions of the forward and inverse scattering problems for (1.1.2) cannot be obtained explicitly. Nevertheless, the application of the IST method allows us to investigate the KdV equation with great completeness. In particular, this method provides us the remarkable class of the exact explicit solutions of the KdV equation, the so-called multisoliton solutions. These solutions correspond to the case $R \equiv 0$, i.e. to the function $F(x,t)$ of the form $F(x,t) = \sum_{n=1}^N b_n(0) \exp(8\lambda_n^3 t - \lambda_n x)$. In this case equation (1.1.5b) is the integral equation with the degenerated kernel and it is readily solvable. As a result, one has $u(x,t) = 2\frac{d^2}{dx^2}\ln \det A(x,t)$ where $A(x,t)$ is the $N \times N$ matrix with the elements $A_{nm} = \delta_{nm} + \frac{b_n(0)b_m(0)}{\lambda_n+\lambda_m}\exp(8\lambda_n^3 t - (\lambda_n+\lambda_m)x)$. In the simplest case $N=1$, one has the famous KdV soliton:

$$u(x,t) = 2\lambda_1^2 \cosh^{-2}\{\lambda_1(x - 4\lambda_1^2 t - x_0)\}$$

where $x_0 = \frac{1}{2\lambda_1}\ln\frac{b_1(0)}{2\lambda_1}$. Emphasize that the solitons are essentially nonlinear objects.

An operator formulation of the idea of the paper [1] has been given by Lax [2]. He has shown that the KdV equation (1.1.1) is equivalent to the operator equation

$$\frac{\partial L}{\partial t} = [L, A] \qquad (1.1.6)$$

where $L = -\frac{\partial^2}{\partial x^2} - u(x,t)$ and $A = -4\frac{\partial^3}{\partial x^3} - 3(u\frac{\partial}{\partial x} + \frac{\partial}{\partial x}u) - 4i\lambda^3$. Equation (1.1.6), now known as the Lax equation, manifestly reflects the fact that the evolution of the potential $u(x,t)$ due to the KdV equation (1.1.1) represents the isospectral deformation of the operator $L$. Lax has also shown that, choosing the appropriate operators $A$, one can represent in the form (1.1.6) the other nonlinear differential equations, associated with the spectral problem (1.1.2), the so-called higher KdV equations.

A second important nonlinear differential equation integrable by the IST method has been found by Zakharov and Shabat. In 1974 they showed [3] that the one-dimensional nonlinear Schrödinger (NLS) equation

$$i\frac{\partial q(x,t)}{\partial t} + \frac{\partial^2 q}{\partial x^2} + 2|q|^2 q = 0 \qquad (1.1.7)$$

is representable in the Lax form (1.1.6) too. The operator $L$ in this case is $L = i\begin{pmatrix} -1 & 0 \\ 0 & 1 \end{pmatrix}\frac{\partial}{\partial x} + \begin{pmatrix} 0 & -q \\ \bar{q} & 0 \end{pmatrix}$, i.e. instead of (1.1.2) one has the one-dimensional stationary Dirac equation

$$\frac{\partial \psi}{\partial x} = i\lambda \begin{pmatrix} 1 & 0 \\ 0 & -1 \end{pmatrix}\psi + i\begin{pmatrix} 0 & q(x,t) \\ \bar{q}(x,t) & 0 \end{pmatrix}\psi \qquad (1.1.8)$$

where $\psi = (\psi_1, \psi_2)^T$ and the bar means complex conjugation. Zakharov and Shabat have investigated the forward and inverse problems for the spectral problem (1.1.8). They have calculated the multisoliton solutions of the NLS equation (1.1.7) and have studied their collision. An important feature of the technique used in [3] is that the equations of the inverse scattering problem derived in [3] constitute the system of singular integral equations.

The paper [3] has turned out a great stimulator in a search for the other nonlinear equations integrable by the IST method. Already in 1972 Wadati solved [4] the modified KdV (mKdV) equation

$$\frac{\partial u(x,t)}{\partial t} + \frac{\partial^3 u}{\partial x^3} + 6u^2\frac{\partial u}{\partial x} = 0. \qquad (1.1.9)$$

The corresponding spectral problem is of the form $\frac{\partial \psi}{\partial x} = i\lambda \begin{pmatrix} 1 & 0 \\ 0 & -1 \end{pmatrix}\psi + \begin{pmatrix} 0 & u \\ u & 0 \end{pmatrix}\psi$.

Then Zakharov [5] has demonstrated the applicability of the IST method to the Boussinesq equation

$$u_{tt} - u_{xx} - \frac{1}{4}u_{xxxx} + \frac{3}{4}(u^2)_{xx} = 0. \qquad (1.1.10)$$

At about the same time Zakharov and Manakov have shown [6] that the equations which describe the resonant interaction of the three wave packets, for instance, the system of equations

$$\frac{\partial q_1}{\partial t} + v_1 \frac{\partial q_1}{\partial x} + \gamma_1 \bar{q}_2 \bar{q}_3 = 0,$$

$$\frac{\partial q_2}{\partial t} + v_2 \frac{\partial q_2}{\partial x} + \gamma_2 \bar{q}_1 \bar{q}_3 = 0, \qquad (1.1.11)$$

$$\frac{\partial q_3}{\partial t} + v_3 \frac{\partial q_3}{\partial x} + \gamma_3 \bar{q}_2 \bar{q}_1 = 0$$

is integrable by the IST method too. In this case the spectral problem is the following $3 \times 3$ matrix spectral problem

$$\frac{\partial \psi}{\partial x} = i\lambda \begin{pmatrix} a_1 & 0 & 0 \\ 0 & a_2 & 0 \\ 0 & 0 & a_3 \end{pmatrix} \psi + \begin{pmatrix} 0 & \frac{-q_3}{\beta_{13}\beta_{23}} & \frac{\gamma_1\gamma_2}{\beta_{12}\beta_{23}}\bar{q}_2 \\ \frac{-\gamma_1\gamma_2}{\beta_{13}\beta_{23}}\bar{q}_3 & 0 & \frac{q_1}{\beta_{12}\beta_{13}} \\ -\frac{q_2}{\beta_{12}\beta_{23}} & \frac{\gamma_2\gamma_3}{\beta_{12}\beta_{13}}\bar{q}_1 & 0 \end{pmatrix} \psi$$

where $\beta_{ik} = (a_i - a_k)^{\frac{1}{2}}$ and $a_1 > a_2 > a_3$.

In the same year 1973 Ablowitz, Kaup, Newell and Segur (AKNS) [7] and independently Zakharov, Takhtajan and Faddeev [8] showed that the IST method is applicable to the sine-Gordon equation in the characteristic variables

$$\varphi_{\xi\eta} + \sin\varphi = 0 \qquad (1.1.12)$$

where $\varphi(\xi,\eta)$ is a scalar function. Equation (1.1.12) is equivalent to the compatibility condition for the linear system

$$\frac{\partial \psi}{\partial \xi} = i\lambda \begin{pmatrix} 1 & 0 \\ 0 & -1 \end{pmatrix} \psi + \frac{i}{2} \begin{pmatrix} 0 & \varphi_\xi \\ \varphi_\xi & 0 \end{pmatrix} \psi,$$

$$\frac{\partial \psi}{\partial \eta} = \frac{i}{4\lambda} \begin{pmatrix} \cos\varphi & -i\sin\varphi \\ i\sin\varphi & -\cos\varphi \end{pmatrix} \psi. \qquad (1.1.13)$$

It has since become clear that the IST method is applicable to a wide class of nonlinear differential equations. It was very important that the nonlinear equations of the physical interest with the great universality have been among these integrable equations. For instance, the KdV equation (1.1.1) originally derived for the description of the shallow-water waves without dissipation, has been used for the study of a number of phenomena in which one should take into account simultaneously the nonlinearity and dispersion, e. g. ion acoustic waves in a plasma, waves in anharmonic lattice, hydromagnetic waves in plasma, waves in the mixture of liquid and gas bubbles, longitudinal dispersive waves in elastic rods, self-trapping of the heat pulses in solids, etc. (see review [9]). The NLS equation (1.1.7) has been applied to the study of the self-focusing of optic beams, one-dimensional self-modulation of monochromatic waves, Langmuir waves in plasma and so on [9]. The sine-Gordon equation has appeared in the study of the propagation of flux on a Josephson tunnel junction, plastic deformations in the crystals, propagation of the Block walls in the magnetic crystals, etc. (see [9]).

Since 1974 the number of papers devoted to the study of the structure and properties of the integrable equations has increased sharply.

In Zakharov and Shabat's fundamental paper [10] the first general method for the construction of the integrable equations and simultaneously for the calculation of their solutions has been proposed. The starting point of this method is the problem of factorization of an integral operator on the line into the product of two Volterra type integral operators. Then the Volterra operators are used for the construction of the dressed differential operators $L$ starting from the initial operator $L_0$. The procedure proposed in [10] (dressing method) allows one to construct the pairs of automatically commuting differential operators

$$[L_1, L_2] = 0 \qquad (1.1.14)$$

and calculate the solutions of the system of nonlinear equations for the coefficients of the operators $L_1$ and $L_2$ which is equivalent to the operator equation (1.1.16). The general form of the operators $L_1$ and $L_2$ discussed in the paper [10] is the following

$$L_1 = \alpha \partial_y + \sum_{n=0}^{N} u_n(x,y,t) \partial_x^n,$$
$$L_2 = \beta \partial_t + \sum_{m=0}^{M} v_m(x,y,t) \partial_x^m \qquad (1.1.15)$$

where $\partial_x \doteq \frac{\partial}{\partial x}$, $\partial_y \doteq \frac{\partial}{\partial y}$ and $u_n(x,y,t)$, $v_m(x,y,t)$ are matrix-valued functions of the independent variables $x, y, t$ and $\alpha, \beta$ are constants. The commutativity operator representation (1.1.15) of the integrable equations for such operators $L_1$ and $L_2$ is equivalent to the Lax representation (1.1.6). The 1+1-dimensional integrable equations correspond to the case $\alpha = 0$.

Another approach has been proposed by Ablowitz, Kaup, Newell and Segur. In the paper [11] they have considered the problem of the description of the nonlinear equations which arise as the compatibility condition of the following two linear problems

$$\frac{\partial \psi}{\partial x} = i\lambda \begin{pmatrix} 1 & 0 \\ 0 & -1 \end{pmatrix} \psi + \begin{pmatrix} 0 & q(x,t) \\ r(x,t) & 0 \end{pmatrix} \psi, \qquad (1.1.16a)$$

$$\frac{\partial \psi}{\partial t} = \begin{pmatrix} A(x,t,\lambda) & B(x,t,\lambda) \\ C(x,t,\lambda) & D(x,t,\lambda) \end{pmatrix} \psi \qquad (1.1.16b)$$

where $q(x,t)$, $r(x,t)$ are scalar complex functions and $A, B, C, D$ are the functions on $x$ and $t$ with the polynomial dependence on the spectral parameter $\lambda$. The problem (1.1.16a) is the straightforward generalization of the problem (1.1.8). The equality of the cross derivatives over $x$ and $t$, i. e. the compatibility condition for the system (1.1.16), allows one to express $A, B, C$ and $D$ deductively via $q(x,t)$ and $r(x,t)$ and, as a result, we arrive at the system of two equations for two functions $q$ and $r$ [11]. At the case of functions $A, B, C$ and $D$ quadratic in $\lambda$ one gets, in particular, the system

$$iq_t + q_{xx} + 2q^2 r = 0,$$

$$ir_t - r_{xx} - 2r^2 q = 0. \qquad (1.1.17)$$

At $r = \pm \bar{q}$ this system is reduced to the NLS equation (1.1.7). For functions $A, B, C, D$ cubic in $\lambda$ one obtains the system

$$q_t + q_{xxx} + 6qr q_x = 0,$$

$$r_t + r_{xxx} + 6qr r_x = 0. \qquad (1.1.18)$$

In the particular cases $r = 1$ and $r = q$ the system (1.1.18) is reduced to the KdV and mKdV equations respectively. At the case $r = \alpha + \beta q$ where $\alpha$ and $\beta$ are constants the system (1.1.18) is equivalent to the Gardner equation $q_t + q_{xxx} + 6\alpha q q_x + 6\beta q^2 q_x = 0$, i. e. to the combined KdV and mKdV equation.

One more example of the integrable equation of this type is given by the one-dimensional continuous ferromagnet Heisenberg model equation

$$\frac{\partial \vec{S}}{\partial t} + \vec{S} \times \vec{S}_{xx} = 0 \qquad (1.1.19)$$

where $\vec{S}$ is the unit vector: $\vec{S}\vec{S} = 1$. Equation (1.1.19) is equivalent to the condition of compatibility for the system [12, 13]

$$\frac{\partial \psi}{\partial x} = \frac{\lambda}{2i}(\vec{S}\vec{\sigma})\psi,$$

$$\frac{\partial \psi}{\partial t} = -\frac{i\lambda^2}{2}(\vec{S}\vec{\sigma})\psi - \frac{\lambda}{2}(\vec{S}_x\vec{\sigma})(\vec{S}\vec{\sigma})\psi \qquad (1.1.20)$$

where $\sigma_1$, $\sigma_2$ and $\sigma_3$ are Pauli matrices. Equation (1.1.19) and NLS equation (1.1.7) are equivalent, more exactly, gauge equivalent to each other [14]. The notion of the gauge equivalence is an important ingredient of the IST method.

Next important step in the development of the IST method has been done in the papers [15, 16].

Firstly, Zakharov and Shabat have extended essentially the class of the auxiliary linear problems. They have proposed to consider the linear systems of the form

$$\frac{\partial \psi}{\partial x} = u(x,t;\lambda)\psi, \qquad (1.1.21a)$$

$$\frac{\partial \psi}{\partial t} = v(x,t;\lambda)\psi \qquad (1.1.21b)$$

where $u(x,t;\lambda)$ and $v(x,t;\lambda)$ are $N \times N$ matrix-valued functions on $x$ and $t$ with the rational dependence on the spectral parameter $\lambda$. The compatibility condition for the system (1.1.21) is of the form

$$\frac{\partial u}{\partial t} - \frac{\partial v}{\partial x} + [u,v] = 0. \qquad (1.1.22)$$

Equation (1.1.22) is equivalent to the commutativity condition $[L_1, L_2] = 0$ for the operators $L_1 = \partial_x - u(x,t;\lambda)$ and $L_2 = \partial_t - v(x,t)$. Equation (1.1.21) is equivalent to the system of nonlinear differential equations. A basic feature of equations (1.1.22) and the auxiliary problems (1.1.21) is that they contain the independent variables $x$ and $t$ principally on the equal footing.

The relativistically-invariant two-dimensional principal chiral fields model equation

$$(g^{-1}g_t)_x + (g^{-1}g_x)_t = 0 \qquad (1.1.23)$$

where $g(x,t)$ is a nondegenerate $N \times N$ matrix is the simplest, but important, example of such type of equations. Equation (1.1.23) is equivalent to the compatibility condition of the linear system [16]

$$\frac{\partial \psi}{\partial x} = -\frac{g^{-1}g_x}{\lambda+1}\psi,$$

$$\frac{\partial \psi}{\partial t} = \frac{g^{-1}g_t}{\lambda-1}\psi. \qquad (1.1.24)$$

The most fundamental idea of the paper [15] was the introduction of the new mathematical construction, the Riemann—Hilbert conjugation problem into the IST method. This idea was of great importance for the subsequent progress of the IST method.

In addition to the integrable $1+1$-dimensional equations mentioned above we will present two more important examples.

The one-dimensional Landau—Lifshitz equation

$$\frac{\partial \vec{S}}{\partial t} + \vec{S} \times \vec{S}_{xx} + \vec{S} \times J\vec{S} = 0 \qquad (1.1.25)$$

where $\vec{S}\vec{S} = 1$ and $J$ is a diagonal constant $3 \times 3$ matrix, which describes the anisotropic continuous ferromagnet, is integrable by the IST method too [17, 18]. In this case the $2 \times 2$ matrices $u$ and $v$ in the auxiliary problems (1.1.21) are not rational, but elliptic functions on the spectral parameter $\lambda$.

An essential generalization of the scheme (1.1.21) has been required when the equation

$$(g^{\frac{1}{2}} G_\xi G^{-1})_\eta + (g^{\frac{1}{2}} G_\eta G^{-1})_\xi = 0 \qquad (1.1.26)$$

where $G(\xi, \eta)$ is $2 \times 2$ nondegenerate matrix and $g \doteq \det G$ has been imbedded into the IST method. Equation (1.1.26) is equivalent to the Einstein equation in the case when the metric tensor depends only on two independent variables [19, 20]. Equation (1.1.26) is representable as the commutativity condition $[L_1, L_2] = 0$ of the two operators which contain also the differentiation over the spectral parameter $\lambda$ [19] in addition to the derivatives $\partial_\xi$ and $\partial_\eta$. Equivalently, equation (1.1.26) is representable in the form (1.1.22) with $\lambda$ which is a certain function on $\xi$ and $\eta$ [20]. The development of these ideas has led to the generalization of the IST method to the case of variable spectral parameter [21].

The theory of the nonlinear integrable equations with the two independent variables is now a broad variety of different methods and approaches. Mathematical theory of solitons in $1 + 1$ dimensions is a very important part of modern mathematical physics.

Dozens of surveys and monographs are devoted to the nonlinear integrable equations in $1 + 1$ dimensions, to the methods of their solution, to the structure and properties of these equations and also to the detailed history of the problem [9, 22–79]. We recommend these reviews and texts to readers who are interesting in the theory of one-dimensional solitons and would like to find more bibliography on these problems.

Here, we shall confine ourselves only to the brief description of the Riemann—Hilbert problem. It is the basic tool for solving the $1 + 1$-dimensional integrable equations.

The matrix local Riemann—Hilbert problem is formulated as follows [80, 81, 15]. Let a simple closed contour $\Gamma$ be given on the plane of the complex variable $\lambda$. Let $\Gamma$ divides the entire plane into two parts $D^+$ and $D^-$ ($0 \in D^+, \infty \in D^-$) and let the $N \times N$ matrix function $G(\lambda)$ without singularities be given on this contour $\Gamma$. The problem is to find an analytic everywhere outside $\Gamma$ function $\chi(\lambda)$ such that the equality

$$\chi^+(\lambda) = \chi^-(\lambda) + \chi^-(\lambda) G(\lambda), \quad \lambda \in \Gamma \qquad (1.1.27)$$

holds on the contour $\Gamma$ where $\chi^+$ and $\chi^-$ are the boundary values of the function $\chi(\lambda)$ on $\Gamma$ for the regions $D^+$ and $D^-$ respectively. In such a formulation the solution of the Riemann—Hilbert problem is obviously nonunique since together with the solution $\chi(\lambda)$ the function $g(\lambda)\chi(\lambda)$, where $g(\lambda)$ is an arbitrary nondegenerate $N \times N$ matrix, is the solution too. In order to remove this nonuniqueness it is necessary to normalize the Riemann—Hilbert problem by the fixing the value of the function $\chi(\lambda)$ at one point. The canonical (and usually used) normalization is $\chi(\lambda = \infty) = 1$. Note also that the problem (1.1.27) is also the factorization problem, since $1 + G(\lambda) = (\chi^-)^{-1}(\lambda)\chi^+(\lambda)$.

The Riemann—Hilbert problem if referred as the regular one if $\det \chi(\lambda) \neq 0$ every where in $D^+ \cup D^-$. If $\det \chi(\lambda) = 0$ in the finite number of points in $D^+ \cup D^-$ then the problem (1.1.27) is referred as the singular Riemann—Hilbert problem or the Riemann—Hilbert problem with zeros.

The solution of the Riemann—Hilbert problem is reduced to the solution of the certain system of linear singular integral equations on the contour $\Gamma$ [80, 81]. Here for simplicity we restricted ourselves by the regular Riemann—Hilbert problem with the canonical normalization. It follows from the well-known theorems of the complex analysis that outside the contour $\Gamma$ the function $\chi(\lambda)$ is representable in the form

$$\chi(\lambda) = 1 + \frac{1}{2\pi i} \int_\Gamma d\lambda' \frac{K(\lambda')}{\lambda' - \lambda} \tag{1.1.28}$$

where $K(\lambda) = \chi^+(\lambda) - \chi^-(\lambda)$, $\lambda \in \Gamma$. On the contour $\Gamma$ the use of the Sokhotsky—Plemelj formulae gives

$$\chi^\pm(\lambda) = 1 + \frac{1}{2\pi i} \int_\Gamma d\lambda' \frac{K(\lambda')}{\lambda' - \lambda \mp i0}, \quad \lambda \in \Gamma. \tag{1.1.29}$$

Substituting (1.1.29) into (1.1.27), one obtains

$$K(\lambda) = \left\{ 1 + \frac{1}{2\pi i} \int_\Gamma d\lambda' \frac{K(\lambda')}{\lambda' - \lambda + i0} \right\} \Gamma(\lambda), \quad \lambda \in \Gamma. \tag{1.1.30}$$

The integral equation (1.1.29) and formula (1.1.28) (or the integral equation (1.1.30) and formula (1.1.28)) give the solution of the regular Riemann—Hilbert problem. The solvability conditions of the regular matrix Riemann—Hilbert problem are complicated enough (see e. g. [80, 81]). But if the Riemann—Hilbert problem is solvable then it is solvable (in the fixed normalization) uniquely. It implies that if $\varphi$ solves the same Riemann—Hilbert problem as $\chi$ then $\varphi = v\chi$ where $v(\lambda)$ is some matrix.

Note that equation (1.1.29) is equivalent to the one-dimensional singular integral equation

$$\chi^-(\lambda) = 1 + \frac{1}{2\pi i} \int_\Gamma d\lambda' \frac{\chi^-(\lambda')G(\lambda')}{\lambda' - \lambda + i0}, \quad \lambda \in \Gamma. \tag{1.1.31}$$

For the Riemann—Hilbert problem with zeros one must define a set of discrete variables in addition. The solution of this problem is given by the system of linear singular integral equations too [80, 81, 15].

To construct nonlinear integrable equations associated with the Riemann—Hilbert problem (1.1.27) one have to introduce the independent variables $x$ and $t$ into the problem

(1.1.27). Let the matrix $G$ evolves in $x$ and $t$ via the linear equations

$$\frac{\partial G(\lambda; x, t)}{\partial x} = [A(\lambda), G(\lambda; x, t)], \qquad (1.1.32)$$

$$\frac{\partial G(\lambda; x, t)}{\partial t} = [B(\lambda), G(\lambda; x, t)] \qquad (1.1.33)$$

where $A$ and $B$ are commuting $N \times N$ matrices with rational dependence on $\lambda$. We assume also that $A$ and $B$ have no jump across the contour $\Gamma$.

Differentiating equation (1.1.27) with respect to $x$ and $t$ and using (1.1.32) and (1.1.33), one obtains

$$\left(\frac{\partial \chi}{\partial x} + \chi A\right)^+ = \left(\frac{\partial \chi}{\partial x} + \chi A\right)^- (1 + G), \qquad (1.1.34)$$

$$\left(\frac{\partial \chi}{\partial t} + \chi B\right)^+ = \left(\frac{\partial \chi}{\partial t} + \chi B\right)^- (1 + G). \qquad (1.1.35)$$

So, $\frac{\partial \chi}{\partial x} + \chi A$ and $\frac{\partial \chi}{\partial t} + \chi B$ solve the same Riemann—Hilbert problem as $\chi$. Hence, due to the uniqueness of the solution of (1.1.27) one has

$$\frac{\partial \chi}{\partial x} + \chi A = u(x, t; \lambda)\chi, \qquad (1.1.36)$$

$$\frac{\partial \chi}{\partial t} + \chi B = v(x, t; \lambda)\chi \qquad (1.1.37)$$

where $u(x, t; \lambda)$ and $v(x, t; \lambda)$ are some $N \times N$ matrices. It is readily follows from equations (1.1.36) and (1.1.37) that $u$ has the same singularities in $\lambda$ as $A(\lambda)$ and $v$ has the same singularities in $\lambda$ as $B(\lambda)$.

In terms of the function $\psi(x, t; \lambda) \doteq \chi(x, t; \lambda) e^{Ax + Bt}$ equations (1.1.36) are of the form

$$\frac{\partial \psi}{\partial x} = u(x, t; \lambda)\psi,$$

$$\frac{\partial \psi}{\partial t} = v(x, t; \lambda)\psi. \qquad (1.1.38)$$

The system (1.1.38) is nothing but the $u - v$ system (1.1.21), the compatibility condition of which gives rise to the nonlinear integrable equation in $1 + 1$ dimensions.

The procedure described above is, in fact, the version of the dressing method associated with the local Riemann—Hilbert problem [15].

Emphasize that the above dressing procedure is a purely local with respect to the variables $x$ and $t$ and does not impose any restriction on the behaviour of the functions $u$ and $v$ at the infinity $|x| \to \infty$. Choosing the contour $\Gamma$ and the function $G$ one can construct a wide class of exact solutions, for example, the solutions $u, v$ which exponentially increase as $|x| \to \infty$.

In order to have the complete description of the class of solutions which tend asymptotically to $A$ and $B$ as $|x| \to \infty$ one should consider the Riemann—Hilbert problems with zeros. The most interesting case corresponds to $G \equiv 0$. In this case the solution of the regular Riemann—Hilbert problem is trivial, the functions $\chi^+$ and $\chi^-$ are rational functions and the whole dressing procedure become purely algebraic. Solutions of the integrable equations constructed with the use of such special Riemann—Hilbert problem are usually referred to as the soliton solutions. The application of the Riemann—Hilbert problem to the calculation of the exact solutions of the integrable equations is considered in detail in the monograph [38].

At the present time the method based on the use of the Riemann—Hilbert problem is the very effective and general method of solution of the nonlinear differential equations in $1+1$-dimensions [15, 16, 38].

## 1.2. Basic integrable equations in 2+1-dimensions

The first generalization of the IST method to the nonlinear differential equations with three independent variables has been given by Zakharov and Shabat in their already mentioned paper [10]. They considered nonlinear equations which are equivalent to the commutativity condition $[L_1, L_2] = 0$ of the operators of the form

$$L_1 = \sigma \partial_y + \sum_{n=0}^{N} u_n(x,y,t) \partial_x^n,$$

$$L_2 = \partial_t + \sum_{n=0}^{M} v_n(x,y,t) \partial_x^n \tag{1.2.1}$$

where $u_n$, $v_n$ are matrix valued functions and $\sigma^2 = \pm 1$.

The Kadomtsev—Petviashvili (KP) equation

$$(u_t + u_{xxx} + 6uu_x)_x + 3\sigma^2 u_{yy} = 0 \tag{1.2.2}$$

where $u(x,y,t)$ is a scalar real function, has been the first from such equations. This equation describes the two-dimensional waves propagating in the $x$-direction with the slow variation in the $y$-direction [82]. The KP equation is the two-dimensional generalization of the KdV equation (1.1.1). The operators $L_1$ and $L_2$ for the KP equation are [10, 83]

$$L_1 = \sigma \partial_y + \partial_x^2 + u(x,y,t),$$

$$L_2 = \partial_t + 4\partial_x^3 + 6u\partial_x + 3u_x - 3\sigma \partial_x^{-1} u_y. \tag{1.2.3}$$

In equation (1.2.2) and in (1.2.3) $\sigma^2 = \pm 1$. The properties of the KP equation depend crucially on the sign of $\sigma^2$. For the KP-I equation ($\sigma = i$) the problem $L_1\Psi = 0$ is the nonstationary one-dimensional Schrödinger equation and for the KP-II equation ($\sigma = 1$) it is the heat equation. The KP equation is the most studied integrable equation in $2+1$-dimensions. This equation provided a good laboratory for the elaboration of the two-dimensional version of the IST method.

A number of the $2+1$-dimensional integrable equations are connected with the scalar differential operators $L_1$ and $L_2$. In particular, the modified KP equation [84, 85].

$$u_t + u_{xxx} - 3\sigma^2 \left(\frac{1}{2}u^2 u_x - \partial_x^{-1} u_{yy} + u_x \partial_x^{-1} u_y\right) = 0. \tag{1.2.4}$$

The two-dimensional integrable analog of the system (1.1.17) is given by the system [86, 87]

$$iq_t + \frac{1}{2}(\sigma^2 q_{yy} + q_{xx}) + q^2 r + q\varphi = 0,$$

$$ir_t - \frac{1}{2}(\sigma^2 r_{yy} + r_{xx}) - r^2 q - r\varphi = 0, \tag{1.2.5}$$

$$\sigma^2 \varphi_{yy} - \varphi_{xx} - 2(qr)_{xx} = 0$$

where $\sigma^2 = \pm 1$.

The system (1.2.5) is equivalent to the commutativity condition of the operators [86, 87]

$$L_1 = \sigma \partial_y + \sigma_3 \partial_x + \begin{pmatrix} 0 & q \\ r & 0 \end{pmatrix},$$

$$L_2 = \partial_t - i\sigma_3 \partial_x^2 - i\sigma_3 \begin{pmatrix} 0 & q \\ r & 0 \end{pmatrix}\partial_x + \begin{pmatrix} Q_1 & Q_2 \\ Q_3 & Q_4 \end{pmatrix} \tag{1.2.6}$$

where $Q_1, Q_2, Q_3, Q_4$ are the certain functions on $q$ and $r$. Under the reduction $r = \kappa \bar{q}$ where $\kappa = \pm 1$ the system (1.2.5) is reduced to the equation

$$iq_t + \frac{1}{2}(\sigma^2 q_{yy} + q_{xx}) + \kappa |q|^2 q + q\varphi = 0,$$

$$\sigma^2 \varphi_{yy} - \varphi_{xx} - 2\kappa |q|^2_{xx} = 0. \tag{1.2.7}$$

At $\sigma = i$ equation (1.2.7) is the Davey—Stewartson (DS-1) equation which describes the two-dimensional long surface waves on the water of finite depth [88]. Equation (1.2.7) is the two-dimensional integrable generalization of the NLS equation (1.1.7).

The IST method is also applicable to the equations which describe the resonant three-waves interaction in the multidimensional space [89–91]

$$\frac{\partial q_1}{\partial t} + \vec{v}_1 \vec{\nabla} q_1 + \gamma_1 \bar{q}_2 \bar{q}_3 = 0,$$

Chapter 1                                                                                                      13

$$\frac{\partial q_2}{\partial t} + \vec{v}_2\vec{\nabla}q_2 + \gamma_2\bar{q}_1\bar{q}_3 = 0, \qquad (1.2.8)$$

$$\frac{\partial q_3}{\partial t} + \vec{v}_3\vec{\nabla}q_3 + \gamma_3\bar{q}_1\bar{q}_2 = 0$$

where $\vec{\nabla}$ is the gradient operator and $\vec{v}_1, \vec{v}_2, \vec{v}_3$ are arbitrary multidimensional vectors. In fact the multidimensionality of the system (1.2.8) is illusory. The system (1.2.8), really belongs to the class of the systems with the three independent variables. In terms of the characteristic coordinates $\chi_i$ ($i = 1, 2, 3$) defined, as usually, by the formulae $\partial_i \equiv \frac{\partial}{\partial \chi_i} \doteq \frac{\partial}{\partial t} + \vec{v}_i\vec{\nabla}$, the system (1.2.8) is of the form

$$\partial_1 q_1 + \gamma_1\bar{q}_2\bar{q}_3 = 0,$$

$$\partial_2 q_2 + \gamma_2\bar{q}_1\bar{q}_3 = 0, \qquad (1.2.9)$$

$$\partial_3 q_3 + \gamma_3\bar{q}_1\bar{q}_2 = 0$$

and the corresponding auxiliary linear system is [90, 91]

$$\partial_k \psi_i + \gamma_k \bar{q}_j \psi_k = 0,$$

$$\partial_i \psi_k + \gamma_i q_j \psi_i = 0 \qquad (1.2.10)$$

where indices $i$, $j$, $k$ are cyclic and run through the values 1, 2, 3.

Among other 2+1-dimensional integrable systems one should note the two-dimensional integrable generalization of the isotropic Heisenberg model (1.1.20). This extension is described by the equation [92]

$$\vec{S}_t + \vec{S} \times (\vec{S}_{xx} + \sigma^2 \vec{S}_{yy}) + \Phi_y \vec{S}_x + \Phi_x \vec{S}_y = 0,$$

$$\Phi_{xx} - \sigma^2 \Phi_{yy} + 2\sigma^2 \vec{S} \cdot (\vec{S}_x \times \vec{S}_y) = 0 \qquad (1.2.11)$$

where $\vec{S}\vec{S} = 1$ and $\sigma^2 = \pm 1$. The corresponding operators $L_1$ and $L_2$ are of the form [92]

$$L_1 = \sigma \partial_y + P\partial_x,$$

$$L_2 = \partial_t - 2iP\partial_x^2 - (iP_x + i\sigma P_y P + \sigma^{-1}P\Phi_x - \Phi_y)\partial_x \qquad (1.2.12)$$

where $P(x, y, t) = \vec{\sigma}\vec{S}(x, y, t)$. The system (1.2.11) describes the classical spin system on the plane. It is of interest also at least for two reasons. Firstly, equation (1.2.11) possesses the topological invariant $Q = \frac{1}{4\pi} \iint dx dy \vec{S} \cdot (\vec{S}_x \times \vec{S}_y)$ and its solutions are classified according to the values of $Q$ $(0, \pm 1, \pm 2, \ldots)$. Secondly, equation (1.2.11) is gauge equivalent to the DS system (1.2.7).

Nonlinear equations considered above represent one class of the 2 + 1-dimensional integrable equations. For such equations both operators $L_1$ and $L_2$ are partial differential operators.

There is also another possibility. It consists of the introduction of a new variable only into the second operator $L_2$ while leaving the operator $L_1$ as the ordinary differential operator. For the first time such a method has been proposed by Calogero [93]. Then it has been generalized in [94]. Nontrivial nonlinear equations in 2 + 1-dimensions arise if one chooses $L_1 = L(\partial_x) - \lambda$, $L_2 = \partial_t + f(\lambda)\partial_y + \sum_{m=0}^{M} v_m(x,y,t)\partial_x^m$ or equivalently $L_2 = \partial_t + f(L)\partial_y + \sum_{m=0}^{M} v_m(x,y,t)\partial_x^m$ where $f(L)$ is some function and $L(\partial_x)$ is the ordinary differential operator.

The two-dimensional generalization of the KdV equation of the form [93, 95]

$$u_t + \alpha u_{xxx} + \beta u_{xxy} + 6\alpha u u_x + 4\beta u u_y + 4\beta u_x \partial_x^{-1} u_y = 0 \quad (1.2.13)$$

is the example of such type of integrable equations. For equation (1.2.13)

$$L_1 = \partial_x^2 + u(x,y,t) - \lambda,$$

$$L_2 = \partial_t + 4\alpha \partial_x^3 + 4\beta(\partial_x^2 + u)\partial_y + f\partial_x + g \quad (1.2.14)$$

where $f = 6\alpha u + 4\beta \partial_x^{-1} u_y$, $g = 3\alpha u_x + 3\beta u_y$ and $\alpha$ and $\beta$ are arbitrary constants.

Equation (1.2.13) describes the interaction of the Riemann wave propagating along axis $y$ and long wave propagating along $x$-axis. In the particular case $\alpha = 0$ it has been studied in detail in [96, 97] (see also [71]).

Equation (1.2.13) and other integrable equations of such type have one interesting feature. If one allows the $t$ and $y$ dependence of $\lambda$ then it obeys a nonlinear equation. For equation (1.2.14) it is of the form [44]

$$\lambda_t = \lambda \lambda_y.$$

In this case equation (1.2.14) possesses overlapping solitons [96, 97].

This approach admits an essentially more general formulation [75]. Namely, one can consider the pair of the commuting operators of the form

$$L_1 = \sum_{i=1}^{3} a_i(\lambda)\partial_{x_i} + u(x,\lambda),$$

$$L_2 = \sum_{k=1}^{3} b_k(\lambda)\partial_{x_k} + v(x,\lambda) \quad (1.2.15)$$

where $u(x,\lambda)$ and $v(x,\lambda)$ are matrix-valued functions and $a_i(\lambda)$, $b_k(\lambda)$ are commuting matrices with the polynomial dependence on the spectral parameter $\lambda$.

A simple and beautiful 2 + 1-dimensional integrable equation with the operators of the form (1.2.15) is given by [98, 99]

$$(g^{-1}g_t)_t - (g^{-1}g_\xi)_\eta = 0 \qquad (1.2.16)$$

where $g(x,y,t)$ is a nondegenerate $N \times N$ matrix, $\xi = \frac{1}{2}(x + \sigma y)$, $\eta = \frac{1}{2}(x - \sigma y)$ and $\sigma^2 = \pm 1$. The corresponding operators $L_1$ and $L_2$ are [98, 99]

$$L_1 = \lambda^2 \partial_\eta - \partial_\xi + \lambda g^{-1}g_t + g^{-1}g_\xi,$$

$$L_2 = \lambda \partial_\eta + \partial_t + g^{-1}g_t. \qquad (1.2.17)$$

The two methods of the two-dimensionalization of integrable equations considered above (see [87, 95]) represent the ways of introducing a new independent variable which preserve the basic feature of the standard theory of the 1 + 1-dimensional integrable equations. Namely, they preserve the representability of the integrable equation as the compatibility condition for the system of two linear equations $L_1\Psi = 0$, $L_2\Psi = 0$ with the equation $[L_1, L_2] = 0$ as the operator form of this compatibility condition. The commutativity requirement $[L_1, L_2] = 0$ is a very strong condition. It guarantees the existence of the common spectrum for the operators $L_1$ and $L_2$. In the case $L_1 = L - \lambda$, $L_2 = \partial_t + A$ this condition ensures the time-independence of the whole spectrum of the operator $L$.

The characteristic feature of all the partial differential operators $L_1$ and $L_2$ considered heretofore in this section is that only the partial derivative with respect to one of the independent variable is of the order greater than one. This is the penalty for the requirement of the commutativity $[L_1, L_2] = 0$. Indeed, if one tries to extend the class of the operators $L_1$ and $L_2$ by including the operators which contain higher order derivatives in two or three variables, then it is readily seen that the commutativity condition $[L_1, L_2] = 0$ is satisfied only if the coefficients of these operators are constants or if the operators $L_1, L_2$ are of the form (1.2.15). A simple reason for this is that the number of the conditions which arise from the requirement of vanishing the total coefficients in front of the cross derivatives in $[L_1, L_2]$ is greater than the number of the coefficient functions in the operators $L_1$ and $L_2$. So, the condition $[L_1, L_2] = 0$ is too restrictive and cannot produce the nontrivial evolution equations in the case of the operators $L_1$ and $L_2$ which contain the higher order derivatives in two or three independent variables.

There is a way to solve this problem. First of all one must note that the condition $[L_1, L_2] = 0$ is not necessary. In fact, the only requirement is that the equations $L_1\Psi = 0$, $L_2\Psi = 0$ should have a sufficiently broad family of common solutions. Then, for the one-dimensional operators $L_1$ of the form $L_1 = L(\partial_x) - \lambda$ the necessity of the condition

$\frac{\partial \lambda}{\partial t} = 0$ for all $\lambda$ has been dictated by the fact that in this case one must know the scattering data for all $\lambda$ in order to reconstruct the potential. This is not necessary in the two-dimensional case. Indeed, the inverse scattering problem for the multidimensional operator can be solved even if the spectral characteristics are "collected" only from the solutions which correspond to only one value $\lambda_0$ of the spectral parameter $\lambda$ (see e. g. [100]). Hence, in the two-dimensional case it is sufficient that the system $(L - \lambda_0)\Psi = 0$, $L_2\Psi = 0$ should be compatible, at least, for one eigenvalue of the operator $L$. In other words, the operator $L_2$ should leave invariant the subspace of the eigenfunctions $\Psi_0$ of the operator $L$ with the eigenvalue $\lambda_0$, i. e. the condition $L_1 L_2 \Psi_0 = 0$ should satisfied [101]. This condition is fulfilled if

$$[L_1, L_2] = \gamma L_1 \qquad (1.2.18)$$

where $\gamma$ is an appropriate operator. The operator equation (1.2.18) is the generalization of the commutativity condition which guarantees the existence of the sufficiently broad family of the common solutions for the pair of equations $L_1 \Psi = 0$ and $L_2 \Psi = 0$.

The operator equation (1.2.18) is the triad operator representation introduced for the first time by Manakov [101]. In the case $L_2 = \partial_t + A$ equation (1.2.18) is equivalent to the equation $L_t = LA - (A + B)L$ which is apparently the generalization of the Lax equation.

This remarkable Manakov's observation has played an important role in the development of the two-dimensional version of the IST method. Equation (1.2.18) is less restrictive than the commutativity condition and it allows us to construct the nontrivial nonlinear equations in the case when the operator $L_1$ contains the second order derivatives in two independent variables.

The interesting and important equation in $2+1$-dimensions representable in the form (1.2.18) looks like

$$u_t + k_1 u_{\xi\xi\xi} + k_2 u_{\eta\eta\eta} + 3k_1 (u\partial_\xi^{-1} u_\eta)_\eta + 3k_2 (u\partial_\eta^{-1} u_\xi)_\xi = 0 \qquad (1.2.19)$$

where $u(x, y, t)$ is a scalar function, $k_1$, $k_2$ are arbitrary constants, $\partial_\xi = \partial_x - \sigma \partial_y$, $\partial_\eta = \partial_x + \sigma \partial_y$ and $\sigma^2 = \pm 1$. For this equation

$$L_1 = (\partial_x^2 - \sigma^2 \partial_y^2) + u(x, y, t),$$

$$L_2 = \partial_t + k_1 \partial_\xi^3 + k_2 \partial_\eta^3 + 3k_1 \partial_\xi^{-1} u_\eta \partial_\eta + 3k_2 \partial_\eta^{-1} u_\xi \partial_\xi, \qquad (1.2.20)$$

$$B = -3(k_1 \partial_\xi^{-1} u_{\eta\eta} + k_2 \partial_\eta^{-1} u_{\xi\xi}).$$

Equation (1.2.19) has been considered for the first time by Nizhnik [102] in the case $\sigma = 1$ and by Veselov and Novikov [103] in the case $\sigma = i$, $k_1 = k_2 = 1$. The Nizhnik—Veselov—Novikov (NVN) equation is other two-dimensional generalization of the KdV equation.

*Chapter 1* 17

In contrast to the KP equation it contains the spatial variables $x$ and $y$ in a symmetric manner.

Many other 2+1-dimensional and 1+1-dimensional integrable equations have the triad operator representation (see e. g. [76]).

Triad operator representation (1.2.18) admits an obvious extension. Indeed, the compatibility for the linear system

$$L_1 \Psi = 0,$$

$$L_2 \Psi = 0 \qquad (1.2.21)$$

means only that the equation

$$[L_1, L_2]\Psi = 0 \qquad (1.2.22)$$

holds simultaneously with equations (1.2.21). If the operators $L_1$ and $L_2$ form a basic in the space of annihilators of $\Psi$ then equation (1.2.22) is equivalent to the operator equation

$$[L_1, L_2] = \gamma_1 L_1 + \gamma_2 L_2 \qquad (1.2.23)$$

where $\gamma_1$ and $\gamma_2$ are some operators.

Several 2+1-dimensional integrable equations have the quartet operator representation (1.2.23), [76, 104]. Such an operator equation appears as the operator representation for the compatibility condition for the system (1.2.21) with the operators $L_1$ and $L_2$ of the form $L_i = A_i(\partial_{x_3})\partial_{x_i} + B_i(\partial_{x_3})$ ($i = 1, 2$) where $A_i(\partial_{x_3})$ and $B_i(\partial_{x_3})$ are matrix-valued ordinary differential operators. In this case $\gamma_1 = D_1(\partial_{x_3})\partial_{x_2} + C_1(\partial_{x_3})$ and $\gamma_2 = D_2(\partial_{x_3})\partial_{x_1} + C_2(\partial_{x_3})$ where $D_i$, $C_i$ ($i = 1, 2$) are certain operators. Such operators $L_1$ and $L_2$ arise as the two-dimensionalization ($\lambda \to \partial_{x_3}$) of the problems (1.1.21) with the generic rational dependence on $\lambda$ [95].

A simple and interesting example of the nonlinear system with the quartet operator representation is given by the system [105]

$$(e^{i\theta}(\theta + \tilde{\theta})_{tx})_x - \sigma^2(e^{i\theta}(\theta + \tilde{\theta})_{ty})_y = 0,$$

$$(e^{-i\theta}(\theta - \tilde{\theta})_{tx})_x - \sigma^2(e^{-i\theta}(\theta - \tilde{\theta})_{ty})_y = 0 \qquad (1.2.24)$$

where $\theta$, $\tilde{\theta}$ are scalar functions and $\sigma^2 = \pm 1$. For the system (1.2.24)

$$L_1 = \sigma \partial_y + \sigma_3 \partial_x - \frac{i}{2}\begin{pmatrix} 0 & \sigma\theta_y + \theta_x \\ \sigma\theta_y - \theta_x & 0 \end{pmatrix},$$

$$L_2 = (\sigma\partial_y - \sigma_3\partial_x)\partial_t - \frac{i}{2}\begin{pmatrix} 0 & \sigma\theta_y - \theta_x \\ \sigma\theta_y + \theta_x & 0 \end{pmatrix}\partial_t + \frac{1}{2}\begin{pmatrix} \sigma\tilde{\theta}_{ty} - \tilde{\theta}_{tx} & 0 \\ 0 & \sigma\tilde{\theta}_{ty} + \tilde{\theta}_{tx} \end{pmatrix},$$

$$\gamma_1 = \frac{i}{2}\begin{pmatrix} 0 & 2\theta_y\partial_t + \sigma\theta_{ty} + \theta_{tx} \\ 2\theta_y\partial_t + \sigma\theta_{ty} - \theta_{tx} & 0 \end{pmatrix}, \quad \gamma_2 = -i\theta_y \begin{pmatrix} 0 & 1 \\ 1 & 0 \end{pmatrix}. \quad (1.2.25)$$

In the one-dimensional limit ($\theta_y = \tilde{\theta}_y = 0$) the integration of equations (1.2.24) with respect to $x$ and summation gives rise to the equation

$$\theta_{tx} = n_1 e^{i\theta} + n_2 e^{-i\theta}$$

where $n_1$ and $n_2$ are arbitrary constants. At $n_1 = -n_2 = -\frac{1}{2i}$ this equation is nothing but the sine-Gordon equation (1.1.13). So, the system (1.2.24) is the 2+1-dimensional integrable generalization of the sine-Gordon equation. The system (1.2.24) has a number of interesting properties. For instance, it is invariant under rotations in $(x,y)$, $(x,t)$ and $(y,t)$ planes.

The system (1.2.24) has been derived in [105] within the wide class of the 2+1-dimensional integrable equations of the Loewner type. It is interesting to note that C. Loewner discussed in [106] the idea of representability of nonlinear partial differential equations as the compatibility condition for certain system of linear partial differential equations as far as 25 years before the discovery of the IST method.

The system (1.2.25) is equivalent to another 2+1-dimensional system which under certain reduction converts into the single equation [105] ($\sigma = 1$)

$$\left(\frac{\theta_{tx}}{\sin\theta}\right)_x - \left(\frac{\theta_{ty}}{\sin\theta}\right)_y + \frac{\theta_x\theta_{ty} - \theta_y\theta_{tx}}{\sin^2\theta} = 0. \quad (1.2.26)$$

This equation is the simplest 2+1-dimensional integrable generalization of the sine-Gordon equation.

All the integrable equations considered in this section up to now have one common and principal feature. Their integrability is a hereditary 1+1-dimensional one. These equations are integrable via the two auxiliary linear problems $L_1\Psi = 0$, $L_2\Psi = 0$ as in the 1+1-dimensional case. These two-dimensional linear problems can be obtained from certain one-dimensional spectral problems with the special matrix structure in the infinite matrix order limit. For the KP equation this has been proved in the paper [107]. This can be done in all the known cases too. As a consequence, the linear problems $L_1\Psi = 0$, $L_2\Psi = 0$ and the whole integration procedure of the corresponding integrable equations in 2+1-dimensions as well inherit the fundamental properties of the 1+1-dimensional integrable systems.

A quite new situation arises when one considers the truly three-dimensional problem which contains all three independent variables on the equal footing and more or less symmetrically.

It occurs that in this case one should consider not two but three compatible auxiliary linear problems. The first example of such system has been found by Zakharov and Manakov [108, 109]. The original version of Zakharov—Manakov system is [108, 109]

$$D_i(\lambda_l)D_k(\lambda_l)\chi_l - D_k(\lambda_i)\chi_i \cdot \chi_i^{-1}D_i(\lambda_l)\chi_l - D_i(\lambda_k)\chi_k \cdot \chi_k^{-1}D_k(\lambda_l)\chi_l = 0$$

$$i,k,l = 1,2,3 \quad i \neq k, \quad k \neq l, \quad l \neq i \tag{1.2.27}$$

where $\chi_l(x_1, x_2, x_3)$ ($l = 1, 2, 3$) are matrix-valued functions of $x_1$, $x_2$, $x_3$,

$$D_i(\lambda)\chi \doteq \left(\partial_{x_i}\chi + \chi\frac{A_i}{\lambda - \lambda_i}\right)$$

and $A_i$ ($i = 1, 2, 3$) are commuting matrices.

In terms of new variables $g_i$ ($i = 1, 2, 3$) defined by

$$g_1 = \chi_1 \exp\left(\frac{A_2 x_2}{\lambda_1 - \lambda_2} + \frac{A_3 x_3}{\lambda_1 - \lambda_3}\right),$$

$$g_2 = \chi_2 \exp\left(\frac{A_1 x_1}{\lambda_2 - \lambda_1} + \frac{A_3 x_3}{\lambda_2 - \lambda_3}\right), \tag{1.2.28}$$

$$g_3 = \chi_3 \exp\left(\frac{A_1 x_1}{\lambda_3 - \lambda_1} + \frac{A_2 x_2}{\lambda_3 - \lambda_2}\right),$$

the system (1.2.27) looks like

$$\frac{\partial^2 g_l}{\partial x_i \partial x_k} - \frac{\partial g_i}{\partial x_k} g_i^{-1} \frac{\partial g_l}{\partial x_i} - \frac{\partial g_k}{\partial x_i} g_k^{-1} \frac{\partial g_l}{\partial x_k} = 0 \tag{1.2.29}$$

$$i,k,l = 1,2,3; \quad i \neq k, \quad k \neq l, \quad l \neq i.$$

This system is equivalent to the compatibility condition for the system of three linear problems

$$L_i \psi = 0 \tag{1.2.30}$$

where

$$L_1 = \partial_{x_2}\partial_{x_3} - g_{2x_3}g_2^{-1}\partial_{x_2} - g_{3x_2}g_3^{-1}\partial_{x_3},$$

$$L_2 = \partial_{x_3}\partial_{x_1} - g_{3x_1}g_3^{-1}\partial_{x_3} - g_{1x_3}g_1^{-1}\partial_{x_1}, \tag{1.2.31}$$

$$L_3 = \partial_{x_1}\partial_{x_2} - g_{1x_2}g_1^{-1}\partial_{x_1} - g_{2x_1}g_2^{-1}\partial_{x_2}.$$

Operator form of the compatibility condition is given by the following system of operator equations [110]

$$[L_i, L_k] = \sum_{l=1}^{3} \gamma_{ikl} L_l, \quad i, k = 1, 2, 3 \tag{1.2.32}$$

where $\gamma_{ikl}$ are matrix-valued functions. In the scalar case they are

$$\gamma_{121} = \left(\ln \frac{g_2}{g_3}\right)_{x_1 x_3}, \quad \gamma_{122} = \left(\ln \frac{g_3}{g_1}\right)_{x_2 x_3}, \quad \gamma_{123} = \left(\ln \frac{g_2}{g_3}\right)_{x_3 x_3} \quad \text{etc.} \quad (1.2.33)$$

Matrix Zakharov—Manakov system is, in fact, a three-dimensional integrable generalization of the principal chiral fields model equation (1.1.23). Geometrical meaning of the scalar system (1.2.29) has been discussed in [111]. It is associated with the description of the triple orthogonal systems of surfaces.

Special cases (reductions) of the system (1.2.29) are closely connected with the 2+1-dimensional integrable sine-Gordon equations mentioned above. In particular, with the choice

$$g_1 = \cos \frac{\theta}{2}, \quad g_2 = \sin \frac{\theta}{2}, \quad g_3 = \frac{\theta_t}{2}$$

and redefinition

$$x_1 = \frac{1}{2}(x+y), \quad x_2 = \frac{1}{2}(x-y), \quad x_3 = t$$

the system (1.2.29) is equivalent to equation (1.2.26) [112]. Under the reduction

$$g_1 = \cos \frac{\theta}{2}, \quad g_2 = \sin \frac{\theta}{2}, \quad g_3 = \frac{\varphi}{2}$$

the system (1.2.29) is converted into the system which is gauge equivalent to (1.2.24).

The system (1.2.32) seems to be a typical representative of the class of the genuine three-dimensional integrable systems. In the general case the nonlinear integrable systems with three independent variables are equivalent to the compatibility condition of the three linear auxiliary problems $L_i(\partial_{x_1}, \partial_{x_2}, \partial_{x_3})\Psi = 0$ ($i = 1, 2, 3$) and the operator form of these compatibility condition is (1.2.32). The main feature of the integration procedure for the genuine three-dimensional integrable equations is that in this case one should solve simultaneously (and compatibly) all three linear problems $L_i\Psi = 0$ ($i = 1, 2, 3$). The properties of different operator representations of integrable equations are discussed in [113].

The operator representation (1.2.32) of the integrable equations is the logical completion of the original idea of the Lax and commutativity operator representations. The generic three-dimensional case is important also because it may contain the integrable systems which are not straightforward generalization of the corresponding 1+1-dimensional equations. The structure and properties of the generic three-dimensional integrable equations are of great interest.

Here we have presented a few principal 2+1-dimensional nonlinear systems integrable by the IST method. Many other examples can be found in the monographs [75, 76].

## 1.3. Solitons and initial value problem in 2+1-dimensions

The properties of the 2+1-dimensional nonlinear integrable equations on the one hand are, in many respects, similar to those of the integrable equations in 1+1-dimensions. On the other hand, these equations exhibit much richer class of exact solutions in comparison with the 1+1-dimensional case. A study of the 2+1-dimensional equations has led to the discovery of new and very general versions of the IST method. Here we will give a brief review of these results.

The 2+1-dimensional integrable equations possess soliton type solutions which are plain generalization of the corresponding 1+1-dimensional counterparts. The simplest such solution for the KP-II equation (KP equation with $\sigma = 1$) is of the form

$$u(x,y,t) = \frac{(\lambda - \mu)^2}{2} \cosh^{-2}\{(\lambda - \mu)x - (\lambda^2 - \mu^2)y - 4(\lambda^3 - \mu^3)t + \alpha\} \qquad (1.3.1)$$

where $\lambda$, $\mu$ and $\alpha$ are arbitrary real constants. The solution (1.3.1) is apparently the KdV soliton moving along certain direction on $(x,y)$ plane.

General multisoliton solution describes the elastic scattering of an arbitrary number of simple solitons. Phase shift is calculated similar to the 1+1-dimensional case. This general $n$-soliton solution does not decrease along $n$ lines on $(x,y)$ plane. Solutions of such type of the 2+1-dimensional equations are referred usually as line solitons.

Novel classes of exact solutions of the integrable equations in 2+1 dimensions include lumps, exponentially localized solitons and solutions with functional parameters. Lumps, typically, are the nonsingular moving solutions with rational dependence on the coordinates $x$ and $y$. For the KP-I equation (KP equation with $\sigma^2 = -1$) the simplest one lump solution looks like [114]

$$u(x,y,t) = 4\frac{-(\tilde{x} + a\tilde{y})^2 + b^2\tilde{y}^2 + b^{-2}}{((\tilde{x} + a\tilde{y})^2 + b^2\tilde{y}^2 + b^{-2})^2} \qquad (1.3.2)$$

where $\tilde{x} = x - 3(a^2 + b^2)t - x_0$, $\tilde{y} = y + 6at - y_0$ and $a$, $b$, $x_0$, $y_0$ are arbitrary real constants. As follows from (1.3.2) the lump has no singularities for all $x$, $y$ and $t$, decreases in all directions on the plane as $O(\frac{1}{x^2}, \frac{1}{y^2})$ as $|x|$, $|y| \to \infty$ and moves with velocity $\vec{v} = (v_x, v_y)$ where $v_x = 3(a^2 + b^2)$ and $v_y = -6a$. General multilump solution describes the scattering of the lumps of the form (1.3.2). The collision of the lumps is completely trivial. Even the phase shift is absent [114].

Lumps are the new phenomenon which appears in 2+1-dimensional case. The lumps are the reflectionless potentials. However they are the potentials transparent at the fixed energy while the one-dimensional solitons are the potentials transparent at all values of the energy.

Solutions of the integrable equations which depend on the functional parameters constitute another new and interesting class of the solutions in 2+1-dimensions. The solutions of the integrable equations which contain the arbitrary functions of single variable have been found for the first time by the dressing method [10]. The simplest solution of such type for the KP-I equation is of the form (see e. g. [38])

$$u(x,y,t) = 2\partial_x^2 \ln(1 + \int_{-\infty}^{x} dx' |\xi(x',y,t)|^2) \qquad (1.3.3)$$

where

$$\xi(x,y,t) = \iint_C d\lambda\, f(\lambda) \exp(i\lambda x - i\lambda^2 y + 4i\lambda^3 t) \qquad (1.3.4)$$

and $f(\lambda)$ is an arbitrary function. In the particular case $f(\lambda) = f_0 \delta(\lambda - i\lambda_0)$ where $\delta(\lambda - \lambda_0)$ is the Dirac-delta function, the solution (1.3.3) represents the line soliton of the KP-I equation.

Solutions with functional parameters correspond to the degenerate kernels of the corresponding linear integral equations. They exhibit the most wide class of exact explicit solutions of the 2+1-dimensional integrable equations. Line solitons and lumps represent the very special subclasses of such solutions.

Solitons of the 2+1-dimensional integrable equations which decrease exponentially in all directions on $(x,y)$ plane are a very recent discovery. For the first time they have been found for the DS-I equation (DS equation with $\sigma^2 = -1$) in the paper [115]. The simplest such solution looks like [115]

$$q(x,y,t) = \frac{4\rho(\alpha\beta)^{\frac{1}{2}} \exp(-(\alpha\eta + \beta\xi) - i[(\gamma\eta + \delta\xi) - (\alpha^2 + \gamma^2 + \beta^2 + \delta^2)t + \gamma])}{(1 + e^{-2\alpha\eta})(1 + e^{-2\beta\xi}) + \rho^2} \qquad (1.3.5)$$

where $\alpha$, $\beta$, $\gamma$, $\delta$, $\rho$ are arbitrary real constants and $\xi = x + y + 2\delta t$, $\eta = x - y + 2\gamma t$.

The localized soliton (1.3.5) corresponds to the nontrivial boundary values of the function $\Phi$ at $x+y \to -\infty$, $x-y \to -\infty$ in equation (1.2.7). These boundary values are, in fact, the line solitons of the KP-I equation [116].

Multisoliton boundaries give rise to the multisoliton exponentially localized solutions of the DS-I equation. Properties of such solitons are generated and determined, in essence, by the corresponding boundaries. In particular, one can drive the localized solitons by driving the boundaries [117]. Entire existence of the exponentially localized solitons of the type (1.3.5) is due to the nontrivial boundary conditions. So, the nature of the localized solitons in 2+1 dimensions is essentially different from that of the 1+1-dimensional solitons. Remind that the origin of the existence of last one is a proper balance between dispersion and nonlinearity.

The construction of the explicit exact solutions is one of the most important problems of the theory of the integrable equations. Solutions which are free from singularities are naturally of special interest in connection with the possible applications in physics.

## Chapter 1

Solutions of the 2+1-dimensional integrable equations with functional parameters, line solitons and lumps have been constructed for the first time by the original Zakharov—Shabat dressing method [10]. Construction of the exponentially localized solitons has demanded new and more sophisticated approaches. The description of these modern methods of construction of the exact solutions will be the topics of the next chapters.

Another important problem of the theory of integrable equations is the solution (more precisely, linearization) of the initial value problem. The study of this problem for the 2+1-dimensional integrable equations has originated the very important new ideas in the IST method.

In general the two-dimensional scattering problems are essentially different from the one-dimensional problems within the framework of the IST method. Firstly, the eigenvalues of the partial differential operators $L_1$ do not play the same role as that in the one-dimensional case. Indeed, for the operators $L_1$ of the form $L_1 = \sigma \partial_y + u(\partial_x)$ the problem $L_1 \psi_\varepsilon = \varepsilon \psi_\varepsilon$ may be converted into the problem $L_1 \psi = 0$ by the change $\psi_\varepsilon = \exp(\frac{\varepsilon y}{\sigma}) \psi$. So the parameter $\varepsilon$ is eliminated completely. On the other hand, for the integrable equations which possess the triad representation (1.2.18), for instance, for the NVN equation (1.2.19) it is necessary to consider the problem $L_1 \psi_\varepsilon = \varepsilon \psi_\varepsilon$ at the fixed value of $\varepsilon$, as it was mentioned in the previous section.

In order to formulate and effectively solve the inverse problems for the two-dimensional scattering problems it is necessary first of all to introduce an auxiliary spectral variable in one or another way. This variable should play a similar role as that of the spectral parameter plays in the one-dimensional case. The problem is to introduce such variable in a way which is adequate to the problem under consideration. It is achieved not for all ways of introduction of this auxiliary variable.

A standard way to do this is to consider a special class of solutions of the linear problem. We will briefly describe this approach for the KP equation. The basic point is to consider the class of solutions of the linear system $L_1 \psi = 0$, $L_2 \psi = 0$ of the type [118]

$$\psi = \mu(x, y, t, \lambda) \exp(i\lambda x + \frac{\lambda^2}{\sigma} y) \tag{1.3.6}$$

where $\lambda$ is a complex parameter and the function $\mu$ has a property: $\mu \to 1$ as $\lambda \to \infty$. In terms of $\mu$ one has the following linear problems

$$\tilde{L}_1 \mu = (\sigma \partial_y + \partial_x^2 + 2i\lambda \partial_x + u)\mu = 0, \tag{1.3.7a}$$

$$\tilde{L}_2 \mu = (\partial_t + 4\partial_x^3 + 12i\lambda \partial_x^2 - 12\lambda^2 \partial_x + 6u\partial_x + 3i\lambda u + 3iu\partial_x + 3u_x - 3\sigma(\partial_x^{-1} u_y))\mu = 0. \tag{1.3.7b}$$

The compatibility condition for the system (1.3.7) ($[\tilde{L}_1, \tilde{L}_2] = 0$) is apparently equivalent to the KP equation. So, we have now the auxiliary linear problem for the KP equation which contains a parameter $\lambda$. This turns out to be the adequate way to introduce spectral parameter into the problem. It allows to construct the solutions of the linear problem (1.3.7) with the desired analytic properties in $\lambda$ [118].

Let us consider first the case $\sigma = i$. We start with the solutions $\mu^+$ and $\mu^-$ of equation (1.3.7a) which are bounded for all complex $\lambda$, tend to unity as $\lambda \to \infty$ and obey the integral equations

$$\mu^\pm(x,y,\lambda) = 1 - \iint dx'dy' G^\pm(x-x', y-y', \lambda) u(x',y') \mu^\pm(x',y',\lambda) \quad (1.3.8)$$

where $G^\pm$ are the Green functions of the operator $L_0 = i\partial_y + \partial_x^2 + 2i\lambda\partial_x$. The Green functions $G^+$ and $G^-$ can be constructed which are analytic at $\operatorname{Im}\lambda > 0$ and $\operatorname{Im}\lambda < 0$ respectively [118]. As a result, the corresponding solutions $\mu^+$ and $\mu^-$ of equations (1.3.8) are meromorphic in the upper and lower half planes $\operatorname{Im}\lambda > 0$ and $\operatorname{Im}\lambda < 0$, respectively. The existence of the poles for the functions $\mu^+$, $\mu^-$ is connected with the existence of the nontrivial solutions $\mu_k$ for the homogeneous integral equations (1.3.8).

It follows from equations (1.3.8) that the boundary values of the functions $\mu^+$ and $\mu^-$ on the real axis are connected by the relation [118]

$$\mu^+(\lambda) = \mu^-(\lambda) + \int_{-\infty}^{+\infty} d\lambda' \, \mu^-(\lambda') \mathcal{F}(\lambda', \lambda) \exp\left(i(\lambda'-\lambda)x - i(\lambda'^2 - \lambda^2)y\right) \quad (\operatorname{Im}\lambda = 0)$$
$$(1.3.9)$$

where $\mathcal{F}(\lambda', \lambda)$ is the appropriate function. Similar formula holds for the functions $\psi^\pm = \mu^\pm \exp(i\lambda x - i\lambda^2 y)$ too. So, we arrive at the conjugation problem, namely, to the Riemann—Hilbert problem of construction of the function $\mu(\lambda)$ which is analytic at $\operatorname{Im}\lambda \neq 0$ and the boundary values of which on the real axis $\operatorname{Im}\lambda = 0$ are connected by the relation (1.3.9). In contrast to the local problem (1.1.27), the relation (1.3.9) is the nonlocal one. So it is quite natural to refer to the corresponding conjugation problem as the nonlocal Riemann—Hilbert problem [118].

The solution of the nonlocal Riemann—Hilbert problem is given by the singular linear integral equation (1.1.29). The function $\mathcal{F}(\lambda', \lambda)$ is the inverse problem data and $u$ is given by

$$u(x,y,t) = \frac{1}{\pi}\frac{\partial}{\partial x} \int_{-\infty}^{+\infty} \int_{-\infty}^{+\infty} d\lambda' d\lambda \, \mathcal{F}(\lambda', \lambda) \mu^-(x,y,t,\lambda)$$

$$\times \exp(i(\lambda'-\lambda)x - i(\lambda'^2 - \lambda^2)y) + 2\frac{\partial}{\partial x}\sum_{k=1}^n (\mu_k^+ + \mu_k^-).$$

The poles of the functions $\mu^\pm(\psi^\pm)$ correspond to the lumps of the KP-I equation. Then, equation (1.3.7b) determines the evolution of the inverse problem data $\mathcal{F}(\lambda', \lambda, t)$ in time: $\mathcal{F}(\lambda', \lambda, t) = \mathcal{F}(\lambda', \lambda, 0) \exp 4i(\lambda'^3 - \lambda^3)t$. Hence, one can solve, in principle, the initial value problem for the KP-I equation by the standard for the IST method procedure

$$u(x,y,0) \xrightarrow{I} \{\mathcal{F}(\lambda',\lambda,0),\ldots\} \xrightarrow{II} \{\mathcal{F}(\lambda',\lambda,t),\ldots\} \xrightarrow{III} u(x,y,t).$$

The Manakov's discovery of the fact, that in the two-dimensional case the nonlocal Riemann—Hilbert problem arises rather than a local one, has played a significant role in the development of the two-dimensional IST method.

The nonlocal Riemann—Hilbert problem arises and effectively works for the wide class of the 2+1-dimensional integrable equations.

However, the possibility to formulate the inverse problem as the Riemann—Hilbert problem is based on the existence of the sectionally-meromorphic solutions of the given spectral problem with the jumps on some contour $\Gamma$. For the one-dimensional spectral problems, at least, in all known cases such solutions exist.

A quite different situation arises in the two-dimensional case. The KP-II equation (1.2.2) ($\sigma = 1$) is the simplest example. For the KP-II equation we again have the equations (1.3.7) but with the change $i\partial_y \to \partial_y$. This crucially changes the properties of the solutions of equations (1.3.7). Namely, the Green function of the operator $L_0 = \partial_y + \partial_x^2 + 2i\lambda\partial_x$ and, consequently, the solutions $\mu$ of the corresponding integral equation (1.3.7) are analytic nowhere with respect to $\lambda$ and have no jump across the real axis. So, the possibility to formulate any conjugation problem is absent.

The solution of this problem for the KP-II equation has been given by Ablowitz, Bar Yaacov and Fokas in their remarkable paper [119]. It turns out that there is no need in the consideration of the solutions with the good analytic properties at all. It is sufficient to consider the solution $\mu(x,y,t,\lambda)$ of equation (1.3.7a) ($\sigma = 1$) which is bounded for all $\lambda$ and then calculate the derivative $\partial\mu/\partial\bar\lambda$. The use of the integral equation of the type (1.3.8), then of the properties of the corresponding bounded Green function and the symmetry property of the solution $\mu$ allows one to perform such calculation. For the KP-II equation one gets [119]

$$\frac{\partial \mu(x,y,\lambda)}{\partial \bar\lambda} = \mathcal{F}(\lambda,\lambda')\mu(x,y,-\bar\lambda)e^{-(2\lambda_R x + 4\lambda_R \lambda_I y)} \qquad (1.3.10)$$

where $\mathcal{F}(\lambda,\bar\lambda) = \tilde{\mathcal{F}}(\lambda_R,\lambda_I)$ is the certain integral expression over $u(x,y,t)$ and $\mu(x,y,t)$ and $\lambda_R = \operatorname{Re}\lambda$, $\lambda_I = \operatorname{Im}\lambda$.

The possibility to derive the equations of the inverse problem is connected now with the existence of the following generalization of the integral Cauchy formula (see e. g. [120])

$$\mu(\lambda) = \frac{1}{2\pi i}\iint_\Omega d\lambda \wedge d\bar\lambda' \frac{\partial \mu/\partial \bar\lambda'}{\lambda' - \lambda} + \frac{1}{2\pi i}\int_{\partial\Omega} d\lambda' \frac{\mu(\lambda')}{\lambda' - \lambda} \qquad (1.3.11)$$

where $\Omega$ is a domain on the complex plane, $\partial\Omega$ is its boundary and $d\lambda \wedge d\bar\lambda = -2id\lambda_R d\lambda_I$. Using the formula (1.3.11) in the concrete case (1.3.10), with $\mu \to 1$ as $\lambda \to \infty$ and the entire complex plane $C$ as $\Omega$, one obtains

$$\mu(x,y,\lambda) = 1 + \frac{1}{2\pi i} \iint_C d\lambda' \wedge d\bar\lambda' \frac{\mathcal{F}(\lambda',\bar\lambda')\mu(x,y,-\bar\lambda')e^{-i(2\lambda'_R x + 4\lambda'_R \lambda'_I y)}}{\lambda' - \lambda}. \tag{1.3.12}$$

In addition one has [119]

$$u(x,y,t) = \frac{1}{\pi}\frac{\partial}{\partial x} \iint_C d\lambda \wedge d\bar\lambda\, \mathcal{F}(\lambda,\bar\lambda)\mu(x,y,-\bar\lambda)e^{-i(2\lambda_R x + 4\lambda_R \lambda_I y)}. \tag{1.3.13}$$

Equation (1.3.12) and formula (1.3.13) provide the solution of the inverse problem for the KP-II equation. The function $\mathcal{F}(\lambda,\bar\lambda)$ is the inverse problem data. The evolution of the inverse problem data $\mathcal{F}(\lambda,\bar\lambda,t)$ in the time is determined by equation (1.3.7a) with $\sigma = 1$. It is of the form $\mathcal{F}(\lambda,\bar\lambda,t) = \mathcal{F}(\lambda,\bar\lambda,0)\exp(-4i(\lambda^3 + \bar\lambda^3)t)$. This allows us to solve the initial value problem for the KP-II equation in a standard manner.

The so-called $\bar\partial$-equation (1.3.10) plays a pivotal role in the approach described above and it gives the name ($\bar\partial$-method) for the whole method proposed in [119]. The $\bar\partial$-equation provides the equations of the inverse problem for the KP-II equation. It is important that the r.h.s. of the $\bar\partial$-equation (1.3.10) is linear with respect to $\mu$. Only in this case the equations of the inverse problem which follows from (1.3.10) are the linear singular integral equations.

The connection of the $\bar\partial$-equation with the IST method has been discovered by Beals and Coifman [122]. Considering the one-dimensional matrix spectral problems they have noted that one can treat the corresponding Riemann—Hilbert problem as a particular case of the $\bar\partial$-problem.

The discovery of Ablowitz, Fokas and Bar Yaakov [119] of the role of the $\bar\partial$-equation and the demonstration of the necessity and efficiency of the $\bar\partial$-approach for the two-dimensional problems was the principal event in the development of the two-dimensional version of the IST method.

A number of the 2+1-dimensional nonlinear equations (the KP-II, the DS-II, the NVN-II equations etc.) have been integrated with help of the $\bar\partial$-method. The main stages within the framework of this method are the same as for KP-II equation [127].

The fact that the nonlocal Riemann—Hilbert problem and quasi-local $\bar\partial$-problem are associated with the different versions $\sigma^2 = -1$ and $\sigma^2 = 1$ of the same nonlinear equation clearly indicated the existence of the more general structure. It naturally has led to the consideration of the nonlocal $\bar\partial$-problem [109].

The general linear matrix $\bar{\partial}$-problem is formulated as the following nonlocal $\bar{\partial}$-problem [109]

$$\frac{\partial \chi(\lambda, \bar{\lambda})}{\partial \bar{\lambda}} = (\chi * R)(\lambda, \bar{\lambda}) = \iint_C d\lambda' \wedge d\bar{\lambda}' \chi(\lambda', \bar{\lambda}') R(\lambda', \bar{\lambda}'; \lambda, \bar{\lambda}) + \eta(\lambda, \bar{\lambda}) \qquad (1.3.14)$$

where $R(\lambda', \bar{\lambda}', \lambda, \bar{\lambda})$ is a matrix-valued function. For the function $\chi$ with the normalization $\chi_{\lambda \to \lambda_0} n(\lambda)$ the $\bar{\partial}$-equation (1.3.14) is equivalent, due to (1.3.10), to the integral equation

$$\chi(\lambda, \bar{\lambda}) = n(\lambda) + \frac{1}{2\pi i} \iint_C d\lambda' \wedge d\bar{\lambda}' \frac{(\chi * R)(\lambda', \bar{\lambda}')}{\lambda' - \lambda}. \qquad (1.3.15)$$

Equation (1.3.15) gives the solution of the general nonlocal $\bar{\partial}$-problem (1.3.14).

Equation (1.3.15) is the two-dimensional singular equation in the independent variables $\lambda_R$ and $\lambda_I$. Let us compare this equation with the equation (1.1.31) which solves the local Riemann—Hilbert problem. It is easy to see that their forms are very similar. Their main common feature is that they both are singular linear integral equations. The formulation of the inverse problem equations in the form (1.3.15) is the logical development of the treatment of the inverse problem equations as the singular integral equations. Such treatment has been proposed for the first time in [3] in the one-dimensional case for the NLS equation. The basic difference between equation (1.3.15) and equation (1.1.31) is that equation (1.1.31) which solves the local Riemann—Hilbert problem is the one-dimensional equation ($\lambda \in \Gamma$) while equation (1.3.15) which solves the nonlocal $\bar{\partial}$-problem (1.3.14) is the two-dimensional integral equation. Integral equation which solves the nonlocal Riemann—Hilbert problem occupies the intermediate position between them.

The nonlocal $\bar{\partial}$-problem (1.3.14) includes the problems discussed above as the particular cases. The simplest example corresponds to the function $R = \delta(\lambda' - \lambda)\tilde{R}(\lambda, \bar{\lambda})$. In this case the problem (1.3.14) is reduced to the local $\bar{\partial}$-problem

$$\frac{\partial \chi(\lambda, \bar{\lambda})}{\partial \bar{\lambda}} = \chi(\lambda, \bar{\lambda}) R(\lambda, \bar{\lambda}). \qquad (1.3.16)$$

In the case $R = \delta(\lambda' - \xi(\lambda, \bar{\lambda}))\tilde{R}(\lambda, \bar{\lambda})$ where $\xi(\lambda, \bar{\lambda})$ is some function one gets the quasi-local $\bar{\partial}$-problem or the $\bar{\partial}$-problem with shift

$$\frac{\partial \chi(\lambda, \bar{\lambda})}{\partial \bar{\lambda}} = \chi(\xi(\lambda, \bar{\lambda}), \bar{\xi}(\lambda, \bar{\lambda}))\tilde{R}(\lambda, \bar{\lambda}). \qquad (1.3.17)$$

For instance, the $\bar{\partial}$-equation (1.3.14) corresponds to the case $\xi = -\bar{\lambda}$. Finally, if $R(\lambda', \bar{\lambda}'; \lambda, \bar{\lambda}) = \delta_\Gamma(\lambda') R(\lambda', \lambda) \delta_\Gamma(\lambda)$ where $\delta_\Gamma(\lambda)$ is the Dirac delta-function concentrated on the contour $\Gamma$, then $\partial \chi / \partial \bar{\lambda} = \chi^+ - \chi^-$ on the contour $\Gamma$ and the nonlocal $\bar{\partial}$-problem (1.3.14) is reduced to the nonlocal Riemann—Hilbert problem (1.3.9). The general integral equation (1.3.15) is reduced under these constrains in a corresponding ways too.

The different particular cases of the general $\bar{\partial}$-problem mentioned above arise in the study of the different 2+1-dimensional integrable equations.

The $\bar{\partial}$-approach essentially changes the analytic ideology of the IST method. The $\bar{\partial}$-method is the useful tool for the $1 + 1$-dimensional integrable equations too. The reformulation of the one-dimensional inverse problems in the form of the $\bar{\partial}$-problem allows one not only extend the class of the explicit solutions but also to understand more deeply the structure of the integrable equations [109, 121, 123–135].

The $\bar{\partial}$-approach also sheds light on the more deep interrelation between the IST method and the theory of complex variables. The IST method in the $\bar{\partial}$-formulation turns out, as has been pointed out in [136, 137], to be deeply connected with the theory of generalized (or pseudo) analytic functions introduced by Bers [138] and Vekua [139]. The generalized analytic functions are the functions which obey the generalized Cauchy—Riemann equation

$$\frac{\partial \chi(\lambda)}{\partial \bar{\lambda}} = A(\lambda)\chi(\lambda) + B(\lambda)\bar{\chi}(\lambda) \qquad (1.3.18)$$

where $A(\lambda)$ and $B(\lambda)$ are scalar functions.

There exists established theory of generalized analytic functions which on the one hand is the essential extension of the classical theory of analytic functions and on the other hand preserves many of its important features [140–143].

Comparision of (1.3.18) and (1.3.16) shows that the solutions of the local $\bar{\partial}$-problem and also of the $\bar{\partial}$-problem which is reducible to the form (1.3.18) (e. g. for the KP-II) are nothing but the generalized analytic functions. This fact together with properties of the generalized analytic functions allows us to strengthen considerably certain results concerning the inverse problems [137].

General nonlocal $\bar{\partial}$-problem (1.3.14) can also be set as the basic equation of the general version of the dressing method [109]. This $\bar{\partial}$-dressing method is the strongest modern tool for the construction of the 2+1-dimensional integrable equations and their exact solutions [109, 144–148]. Explicit solutions with functional parameters, lumps, line and localized solitons are obtained in a unified and rather simple manner within the framework of this method.

The description of the $\bar{\partial}$-dressing method and corresponding results for several 2+1-dimensional integrable equations is one of the two main goals of this book.

# Chapter 2

# Dressing method

The original dressing method was proposed by Zakharov and Shabat in 1974 [10]. It was based on the one-dimensional Volterra type integral equation associated with the factorization problem for the integral operators on a line. Next version of the dressing method was connected with the local Riemann—Hilbert problem [15]. The most advanced and powerful version of the dressing method is based on the nonlocal $\bar{\partial}$-problem. Here we will consider directly this $\bar{\partial}$-dressing method.

## 2.1. $\bar{\partial}$-equation

For convenience we will expose some elementary results concerning the $\bar{\partial}$-equation in this section. We will deal with the functions $\chi(\lambda, \bar{\lambda})$ defined on the plane (Riemann sphere) of complex variable $\lambda$. As usual $\lambda = \lambda_R + i\lambda_I$, where $\lambda_R$ and $\lambda_I$ are the real and imaginary parts of $\lambda$, respectively, and $i^2 = -1$. Together with $\lambda_R$ and $\lambda_I$ one can treat $\lambda$ and $\bar{\lambda}$ (bar means complex conjugation) as the independent variables on the plane. The corresponding differential operators are

$$\frac{\partial}{\partial \lambda} = \frac{1}{2}\left(\frac{\partial}{\partial \lambda_R} - i\frac{\partial}{\partial \lambda_I}\right),$$

$$\frac{\partial}{\partial \bar{\lambda}} = \frac{1}{2}\left(\frac{\partial}{\partial \lambda_R} + i\frac{\partial}{\partial \lambda_I}\right). \tag{2.1.1}$$

In general, functions on the plane depend on both $\lambda_R$ and $\lambda_I$ or, equivalently, on $\lambda$ and $\bar{\lambda}$. Theory of analytic functions is the special but very important case of the theory of functions on the plane. Analytic functions are specified by the celebrated Cauchy—Riemann condition

$$\frac{\partial}{\partial \bar{\lambda}}\chi = 0. \tag{2.1.2}$$

In fact, the whole richness of the theory of analytic functions is the consequence of equation (2.1.2).

The departure from analyticity of the function $\chi$ is apparently described by the nonzero contributions to the r.h.s. of (2.1.2). The corresponding extension of the Cauchy—Riemann equation (2.1.2), namely, the equation

$$\frac{\partial \chi(\lambda,\bar{\lambda})}{\partial \bar{\lambda}} = f(\lambda,\bar{\lambda}) \tag{2.1.3}$$

is referred, usually, as the $\bar{\partial}$ (DBAR) equation. The properties of the function $\chi$ are completely determined by the function $f$.

Several generalizations of the well-known formulae of the theory of analytic functions hold for the functions defined by the $\bar{\partial}$-equation (2.1.3). Extension of the Cauchy formulae is of special interest to us.

We will denote a domain (usually simply connected one) on the plane by $D$ and its piecewise differentiable and counterclockwise oriented boundary by $\partial D$. The entire complex plane will be denoted by $C$. For the integration over domain we will use the standard Lebesgue measure

$$d\lambda \wedge d\bar{\lambda} = -2i d\lambda_R \wedge d\lambda_I, \tag{2.1.4}$$

where $\wedge$ means the edge product. So (2.1.4) is skewsymmetric.

In such a notation the well-known Gauss—Green formula looks like (see e.g.[120])

$$\oint_{\partial D} d\lambda \, \chi(\lambda) = -\iint_D d\lambda \wedge d\bar{\lambda} \frac{\partial \chi(\lambda)}{\partial \bar{\lambda}}. \tag{2.1.5}$$

Formula (2.1.5) is apparently the extension of the Cauchy formula $\oint_{\partial D} d\lambda \, \chi(\lambda) = 0$ for the function analytic in $D$.

To derive another Cauchy formula one needs to introduce the Dirac delta-function on the complex plane. The standard definition is (see e.g. [149])

$$-\frac{1}{2i} \iint_D d\lambda' \wedge d\bar{\lambda}' \delta(\lambda' - \lambda) \chi(\lambda',\bar{\lambda}') = \chi(\lambda,\bar{\lambda}), \quad \lambda \in D. \tag{2.1.6}$$

Comparing (2.1.6) with the obvious formula

$$\iint_D d\lambda'_R \wedge d\lambda'_I \, \delta(\lambda'_R - \lambda_R) \delta(\lambda'_I - \lambda_I) \chi(\lambda'_R, \lambda'_I) = \chi(\lambda_R, \lambda_I), \tag{2.1.7}$$

one concludes, that effectively

$$\delta(\lambda' - \lambda) = \delta(\lambda'_R - \lambda_R) \delta(\lambda'_I - \lambda_I). \tag{2.1.8}$$

So, for instance

$$\overline{\delta(\lambda)} = \delta(\lambda). \tag{2.1.9}$$

The derivative delta-functions $\delta^{(n,m)}(\lambda' - \lambda)$ are defined as follows [149]

$$-\frac{1}{2i} \iint_D d\lambda' \wedge d\bar{\lambda}' \, \delta^{(n,m)}(\lambda' - \lambda) \chi(\lambda', \bar{\lambda}') = (-1)^{n+m} \frac{\partial^{n+m} \chi(\lambda, \bar{\lambda})}{\partial \lambda^n \partial \bar{\lambda}^m}, \qquad (2.1.10)$$

i. e.

$$\delta^{(n,m)}(\lambda, \bar{\lambda}) = \frac{\partial^{n+m} \delta(\lambda, \bar{\lambda})}{\partial \lambda^n \partial \bar{\lambda}^m}. \qquad (2.1.11)$$

Solution of the $\bar{\partial}$-equation (2.1.3) with delta-function r.h.s. plays a fundamental role in the further constructions. Considering the Gauss—Green formula (2.1.5) for the function $\chi = \frac{\tilde{\chi}(\lambda)}{\lambda - \lambda_0}$ where $\tilde{\chi}(\lambda)$ is analytic in $D$ function and using the Cauchy formula

$$\frac{1}{2\pi i} \int_{\partial D} d\lambda \frac{\tilde{\chi}(\lambda)}{\lambda - \lambda_0} = \tilde{\chi}(\lambda_0),$$

one can prove that

$$\frac{\partial}{\partial \bar{\lambda}} \left( \frac{1}{\pi(\lambda - \lambda_0)} \right) = \delta(\lambda - \lambda_0). \qquad (2.1.12)$$

Analogously, one can get the formula

$$\frac{\partial}{\partial \bar{\lambda}} \left( \frac{(-1)^n n!}{\pi} \frac{1}{(\lambda - \lambda_0)^{n+1}} \right) = \delta^{(n,0)}(\lambda - \lambda_0) \qquad (2.1.13)$$

and similar formula for $\delta^{(0,n)}(\lambda)$.

So $\frac{1}{\pi \lambda}$ is the Green function of the $\bar{\partial}$-equation (2.1.3) for the class of functions $\chi$ which vanish at infinity.

Now let us consider the Gauss—Green formula (2.1.5) for the function $\frac{\chi(\lambda, \bar{\lambda})}{\lambda - \lambda_0}$ instead of $\chi$. It is not assumed here that $\chi$ is analytic in $D$. Using the formulae (2.1.12) and (2.1.7), one readily obtains

$$\chi(\lambda, \bar{\lambda}) = \frac{1}{2\pi i} \oint_{\partial D} d\lambda' \frac{\chi(\lambda', \bar{\lambda}')}{\lambda' - \lambda} + \frac{1}{2\pi i} \iint_D \frac{d\lambda' \wedge d\bar{\lambda}'}{\lambda' - \lambda} \frac{\partial \chi(\lambda', \bar{\lambda}')}{\partial \bar{\lambda}'}, \quad \lambda \in D. \qquad (2.1.14)$$

This is the desired extension of the Cauchy formula for nonanalytic functions [120].

The generalized Cauchy formula provides the solution of the $\bar{\partial}$- equation (2.1.3). Given $f(\lambda, \bar{\lambda})$ it reduces equation (2.1.14) to the solution of the integral equation

$$\chi(\lambda, \bar{\lambda}) = \frac{1}{2\pi i} \int_{\partial D} d\lambda' \frac{\chi(\lambda', \bar{\lambda}')}{\lambda' - \lambda} + \frac{1}{2\pi i} \iint_D \frac{d\lambda' \wedge d\bar{\lambda}'}{\lambda' - \lambda} f(\lambda', \bar{\lambda}'). \qquad (2.1.15)$$

As far as the $\bar{\partial}$-equation (2.1.3) together with the boundary condition are concerned

$$\chi(\lambda, \bar{\lambda}) = \chi_0(\lambda, \bar{\lambda}), \quad \lambda \in \partial D \qquad (2.1.16)$$

it is quite clear that this problem has no solution for the generic boundary function $\chi_0(\lambda, \bar{\lambda})$. It should obey the linear integral equation which follows from (2.1.5) as $\lambda \to \partial D$.

The special, very important case corresponds to $D = C$ and $\chi_0 = 1$. In this case the solution of the $\bar{\partial}$-equation is given by

$$\chi(\lambda, \bar{\lambda}) = 1 + \frac{1}{2\pi i} \iint_D \frac{d\lambda' \wedge d\bar{\lambda}'}{\lambda' - \lambda} f(\lambda', \bar{\lambda}'). \qquad (2.1.17)$$

In particular for $f = \delta(\lambda - \lambda_0)$, one has

$$\chi(\lambda, \bar{\lambda}) = 1 + \frac{1}{\pi(\lambda - \lambda_0)}. \qquad (2.1.18)$$

Note in conclusion the useful formula

$$\frac{\partial}{\partial \bar{\lambda}} \left( \varphi(\lambda) + \frac{1}{2\pi i} \iint_D \frac{d\lambda' \wedge d\bar{\lambda}'}{\lambda' - \lambda} \chi(\lambda', \bar{\lambda}') \right) = \chi(\lambda, \bar{\lambda}) \qquad (2.1.19)$$

where $\varphi(\lambda)$ is an arbitrary function analytic in $D$.

All formulae presented in this section obviously hold for the matrix-valued functions.

## 2.2. Nonlocal $\bar{\partial}$-problem

General matrix $\bar{\partial}$-problem is the $\bar{\partial}$-equation (2.1.3) with the general linear in $\chi$ r. h. s. It is of the form [145]

$$\frac{\partial \chi(\lambda, \bar{\lambda})}{\partial \bar{\lambda}} = \iint_D d\lambda' \wedge d\bar{\lambda}' \chi(\lambda', \bar{\lambda}') R(\lambda', \bar{\lambda}'; \lambda, \bar{\lambda}) + \chi_0(\lambda, \bar{\lambda}), \qquad (2.2.1)$$

where $\chi, R, \chi_0$ are matrix-valued functions. The inhomogeneous term $\chi_0$ and the kernel $R$ are given functions. It is convenient to represent $\chi_0$ as $\chi_0 = \frac{\partial \eta(\lambda, \bar{\lambda})}{\partial \bar{\lambda}}$. In virtue of the generalized Cauchy formula (2.1.14) the $\bar{\partial}$-problem (2.2.1) is equivalent to the following integral equation

$$\chi(\lambda, \bar{\lambda}) = \eta(\lambda, \bar{\lambda}) + a(\lambda)$$

$$+ \frac{1}{2\pi i} \iint_D \frac{d\lambda' \wedge d\bar{\lambda}'}{\lambda' - \lambda} \iint_D d\lambda'' \wedge d\bar{\lambda}'' \chi(\lambda'', \bar{\lambda}'') R(\lambda'', \bar{\lambda}''; \lambda', \bar{\lambda}') \qquad (2.2.2)$$

where $a(\lambda)$ is an arbitrary function analytic in $D$.

We will assume that equation (2.2.2) is solvable. It is so, at least, for $R$ small in norm. The solution of the $\bar{\partial}$-problem (2.2.1) is not unique due to the presence of an arbitrary analytic function $a(\lambda)$ in (2.2.2). One can achieve the uniqueness by demanding that the function $\chi$ solves the integral equation (2.2.2) with the fixed inhomogeneous term $n(\lambda, \bar{\lambda})$, i. e. the equation

$$\chi(\lambda, \bar{\lambda}) = n(\lambda, \bar{\lambda}) + \frac{1}{2\pi i} \iint_D \frac{d\lambda' \wedge d\bar{\lambda}'}{\lambda' - \lambda} \iint_D d\lambda'' \wedge d\bar{\lambda}'' \chi(\lambda'', \bar{\lambda}'') R(\lambda'', \bar{\lambda}''; \lambda', \bar{\lambda}'). \qquad (2.2.3)$$

Note that since $\chi_0 = \frac{\partial n}{\partial \bar{\lambda}}$ only a singular part of $n$ contributes to the inhomogeneous term in $\bar{\partial}$-problem (2.2.1). In our further constructions we will assume that the kernel $R(\lambda'', \bar{\lambda}''; \lambda', \bar{\lambda}')$ is such that equation (2.2.3) is a Fredholm integral equation of the second kind. So equation (2.2.3) is uniquely solvable within a certain class of functions $\chi$ at least for $R$ small in norm (see. e.g. [150, 151]).

In virtue of the Fredholm alternative the uniqueness of solution of equation (2.2.3) with the fixed $n$ implies that the homogeneous equation (2.2.3) has no nontrivial solutions.

It is easy to see from (2.2.3) that the inhomogeneous term $n$ coincides with the asymptotics of $\chi$ at the infinity:

$$\chi(\lambda, \bar{\lambda}) \underset{\lambda \to \infty}{\longrightarrow} n(\lambda, \bar{\lambda}). \tag{2.2.4}$$

On the other hand, if $n$ has a singularity at the point $\lambda_0$ (for instance, $n = n_0 = \frac{\alpha}{\lambda - \lambda_0}$) then

$$\chi(\lambda, \bar{\lambda}) \underset{\lambda \to \lambda_0}{\longrightarrow} n_0(\lambda). \tag{2.2.5}$$

So, the inhomogeneous term in equation (2.2.3) and consequently the solution of the $\bar{\partial}$-problem (2.2.1) are completely fixed by fixing the normalization of the solution $\chi$ at some point. The unique solvability of the $\bar{\partial}$-problem implies that if $\tilde{\chi}$ is the solution of the $\bar{\partial}$-problem such that

$$\tilde{\chi} \underset{\lambda \to \lambda_0}{\longrightarrow} 0 \tag{2.2.6}$$

then $\tilde{\chi} = 0$.

The normalization

$$\tilde{\chi} \underset{\lambda \to \infty}{\longrightarrow} 1 \tag{2.2.7}$$

is referred usually as the canonical normalization. In this case, of course, $D = C$. Another very useful normalization is given by (2.2.5) where $n_0(\lambda)$ is some rational function. For the canonical normalization one has the homogeneous $\bar{\partial}$-problem (2.2.1) ($n = 1$) while for the rational normalization one should consider the inhomogeneous $\bar{\partial}$-problem (2.2.1) [145].

Note, that the situation with normalizations for the $\bar{\partial}$-problem is essentially different from that for the local Riemann—Hilbert problem (1.1.27). In the last case the solutions of (1.1.27) with the two different normalizations $\chi_i(\lambda) \underset{\lambda \to \lambda_0}{\longrightarrow} n_i(\lambda)$ ($i = 1, 2$) are connected by the local gauge transformation $\chi_1(\lambda) \to \chi_2(\lambda) = \frac{n_2(\lambda)}{n_1(\lambda)} \chi_1(\lambda)$, i. e. they are gauge equivalent to one another. For the $\bar{\partial}$-problem (2.2.1) the two solutions with the different normalizations are not connected by the gauge transformation (at least, local one). So they are essentially different.

Solutions of the $\bar{\partial}$-problems with different normalizations arise is various problems. Moreover, the use of the solutions of the same $\bar{\partial}$-problem with variable normalizations essentially simplifies the analysis of the generic case.

The integral equation (2.2.3) with $D = C$ will be our basic linear integral equation in what follows.

Equation (2.2.3) implies the following equation

$$K(\lambda, \lambda') = \iint_D d\mu \wedge d\bar{\mu}\, n(\mu, \bar{\mu}) R(\mu, \bar{\mu}; \lambda, \bar{\lambda})$$

$$+ \iint_D \iint_D \frac{d\mu \wedge d\bar{\mu}\, d\lambda' \wedge d\bar{\lambda}'}{\lambda' - \mu} K(\lambda', \bar{\lambda}') R(\mu, \bar{\mu}; \lambda, \bar{\lambda}) = 0 \qquad (2.2.8)$$

for

$$K(\lambda, \bar{\lambda}) \doteq \iint_D d\mu \wedge d\bar{\mu}\, \chi(\mu, \bar{\mu}) R(\mu, \bar{\mu}; \lambda, \bar{\lambda}). \qquad (2.2.9)$$

The solution of equation (2.2.8) gives the solution of the $\bar{\partial}$-problem by the formula

$$\chi(\lambda, \bar{\lambda}) = n(\lambda, \bar{\lambda}) + \frac{1}{2\pi i} \iint_D \frac{d\mu \wedge d\bar{\mu}}{\mu - \lambda} K(\mu, \bar{\mu}). \qquad (2.2.10)$$

This formula clearly indicates that only certain integral characteristic of $R(\mu, \bar{\mu}; \lambda, \bar{\lambda})$ contributes to the solution of the $\bar{\partial}$-problem. For instance, in the case $n \equiv 1$ and $D = C$ the first nontrivial term $\chi_{-1}$ in the asymptotic expansion $\chi \to 1 + \frac{1}{\lambda}\chi_{-1} + \ldots$ as $\lambda \to \infty$ for small $R$ is given by

$$\chi_{-1} = -\iint_C d\lambda \wedge d\bar{\lambda} \iint_C d\mu \wedge d\bar{\mu}\, R(\mu, \bar{\mu}; \lambda, \bar{\lambda}). \qquad (2.2.11)$$

So the different $R$ and $\tilde{R}$ such that

$$\iint_C \iint_C d\lambda \wedge d\bar{\lambda}\, d\mu \wedge d\bar{\mu}\, (R - \tilde{R})(\mu, \bar{\mu}; \lambda, \bar{\lambda}) = 0 \qquad (2.2.12)$$

give rise to the same $\chi_{-1}$.

From the general point of view the $\bar{\partial}$-problem (2.2.1) is the system of the two-dimensional differential-integral equations. The theory of integral equations tells us that together with (2.2.1) it is important to consider also the adjoint equation

$$\frac{\partial \tilde{\chi}^*(\lambda, \bar{\lambda})}{\partial \bar{\lambda}} = -\iint_D d\lambda' \wedge d\bar{\lambda}'\, R(\lambda, \bar{\lambda}; \lambda', \bar{\lambda}') \chi^*(\lambda', \bar{\lambda}') - \frac{\partial n(\lambda)}{\partial \bar{\lambda}}. \qquad (2.2.13)$$

In the case of the canonical normalization $n = 1$ equations (2.2.1) and (2.2.13) imply the important bilinear identity for the function $\chi$ and adjoint function $\chi^*$. Indeed, multiplying equation (2.2.1) by $\chi^*(\lambda, \bar{\lambda})$ from the right and equation (2.2.13) by $\chi(\lambda, \bar{\lambda})$ from the left, summing the resulting equations and integrating with the measure $d\lambda \wedge d\bar{\lambda}$, one gets

$$\iint_D d\lambda \wedge d\bar{\lambda}\, \frac{\partial}{\partial \bar{\lambda}} (\chi(\lambda, \bar{\lambda}) \chi^*(\lambda, \bar{\lambda})) = 0. \qquad (2.2.14)$$

Using the Gauss—Green formula (2.1.5), one finally obtains

$$\int_{\partial D} d\lambda\, \chi(\lambda, \bar{\lambda}) \chi^*(\lambda, \bar{\lambda}) = 0. \qquad (2.2.15)$$

The bilinear identity (2.2.15) plays an important role in the theory of the 2+1-dimensional integrable equations.

Now let us proceed to the construction of explicit solutions of the $\bar{\partial}$-problem (2.2.1). As typical for integral equations equation (2.2.3) can be solved explicitly in the case of the degenerate kernel $R(\mu, \bar{\mu}; \lambda, \bar{\lambda})$. So let

$$R(\mu, \bar{\mu}; \lambda, \bar{\lambda}) = \sum_{k=1}^{N} f_k(\mu, \bar{\mu}) g_k(\lambda, \bar{\lambda}), \qquad (2.2.16)$$

where $f_k$ and $g_k$ are arbitrary functions and $N$ is an arbitrary integer. Substituting (2.2.16) into (2.2.3), one gets

$$\chi(\lambda, \bar{\lambda}) = n(\lambda, \bar{\lambda}) + \frac{1}{2\pi i} \iint_D \frac{d\lambda' \wedge d\bar{\lambda}'}{\lambda' - \lambda} \sum_{l=1}^{N} h_l g_l(\lambda', \bar{\lambda}') \qquad (2.2.17)$$

where

$$h_l = \iint_D d\mu \wedge d\bar{\mu}\, \chi(\mu, \bar{\mu}) f_k(\mu, \bar{\mu}). \qquad (2.2.18)$$

So to find $\chi$ one should calculate all $h_k$. The system of equations for $h_k$ follows from equation (2.2.17). Indeed, multiplying (2.2.7) by $f_k(\lambda, \bar{\lambda})$ from the left and integrating over $\lambda$, one gets

$$h_k + \sum_{l=1}^{N} h_l A_{lk} = \xi_k \quad (k = 1, \ldots, N) \qquad (2.2.19)$$

where

$$\xi_k = \iint_D d\lambda \wedge d\bar{\lambda}\, n(\lambda, \bar{\lambda}) f_k(\lambda, \bar{\lambda}) \qquad (2.2.20)$$

and

$$A_{lk} = \frac{1}{2\pi i} \iint_D d\lambda' \wedge d\bar{\lambda}' \iint_D d\lambda \wedge d\bar{\lambda} \frac{1}{\lambda - \lambda'} g_l(\lambda', \bar{\lambda}') f_k(\lambda, \bar{\lambda}) \quad (l, k = 1, \ldots, N). \qquad (2.2.21)$$

Thus, the linear integral equation (2.2.3) with the degenerate kernel (2.2.16) is reduced to the linear algebraic system. Solving the system (2.2.14), one finally obtains

$$\chi(\lambda, \bar{\lambda}) = n(\lambda, \bar{\lambda}) + \frac{1}{2\pi i} \iint_D \frac{d\lambda' \wedge d\bar{\lambda}'}{\lambda' - \lambda} \sum_{l,k=1}^{N} \xi_k (1+A)^{-1}_{kl} g_l(\lambda', \bar{\lambda}') \qquad (2.2.22)$$

where $A$ is the matrix with the matrix elements $A_{lk}$ given by (2.2.21).

The formula (2.2.22) gives the explicit solution of the $\bar{\partial}$-problem (2.2.1) which is parametrized by $2N$ arbitrary matrix-valued functions $f_k(\lambda, \bar{\lambda})$ and $g_k(\lambda, \bar{\lambda})$. These solutions represent the most wide class of the explicit exact solutions of the $\bar{\partial}$-problem.

Choosing particular functions $f_k(\lambda, \bar{\lambda})$ and $g_k(\lambda, \bar{\lambda})$, one gets the special classes of explicit solutions of the $\bar{\partial}$-problem. An interesting class of solutions corresponds to

$$f_k(\mu, \bar{\mu}) = f_k \delta(\mu - \mu_k), \quad g_k(\lambda, \bar{\lambda}) = g_k \delta(\lambda - \lambda_k)$$

or

$$R(\mu, \bar{\mu}; \lambda, \bar{\lambda}) = \sum_{k=1}^{N} f_k g_k \delta(\mu - \mu_k) \delta(\lambda - \lambda_k) \quad (2.2.23)$$

where $\delta(\mu - \mu_k)$, $\delta(\lambda - \lambda_k)$ are Dirac delta-functions and $f_k$, $g_k$, $\mu_k$, $\lambda_k$ are arbitrary complex constants. It is assumed that $\mu_i$ are distinct from the positions of poles of the normalization $n(\lambda, \bar{\lambda})$. In this case the solution of the $\bar{\partial}$-problem is of the form

$$\chi(\lambda, \bar{\lambda}) = n(\lambda, \bar{\lambda}) - \frac{1}{\pi} \sum_{l,k=1}^{N} \frac{\xi_k (1+A)^{-1}_{kl} g_l}{\lambda_l - \lambda} \quad (2.2.24)$$

where

$$A_{lk} = \frac{2}{\pi i} \frac{g_l f_k}{\mu_l - \lambda_k} \quad (l, k = 1, \ldots, N) \quad (2.2.25)$$

and

$$\xi_k = -2in(\mu_k, \bar{\mu}_k) f_k. \quad (2.2.26)$$

This solution is parametrized by $2p^2 + 2N$ arbitrary constants $(g_k)_{\alpha\beta}$, $(f_k)_{\alpha\beta}$, $\mu_l$, $\lambda_k$ where $\alpha, \beta = 1, \ldots, p$. The characteristic feature of the solutions (2.2.24) is that the function $\chi$ has the simple poles in the points $\lambda_l$ ($l = 1, \ldots, N$). In the similar manner one can find explicit solutions which correspond to more general than (2.2.24) kernel

$$R(\mu, \bar{\mu}; \lambda, \bar{\lambda}) = \sum_{i,k=1}^{N} C_{ik} \delta(\mu - \mu_i) \delta(\lambda - \lambda_k). \quad (2.2.27)$$

One can construct also the explicit solutions of the $\bar{\partial}$-problem with the multiple poles. Indeed, choosing

$$g_l(\lambda, \bar{\lambda}) = g_l \delta^{(n,0)}(\lambda - \lambda_k) \quad (2.2.28)$$

where $\delta^{(n,0)}(\lambda - \lambda_k)$ is the derivative delta-function (2.1.11), one gets

$$\chi(\lambda, \bar{\lambda}) = n(\lambda, \bar{\lambda}) - \frac{(-1)^n}{\pi} \sum_{l,k=1}^{N} \frac{\xi_k (1+A)^{-1}_{kl} g_l}{(\lambda_l - \lambda)^n} \quad (2.2.29)$$

where

$$A_{lk} = \frac{(-1)^n}{\pi} \iint_D d\lambda \wedge d\bar{\lambda} \frac{g_l(\lambda, \bar{\lambda}) f_k}{(\lambda_k - \lambda)^n}. \quad (2.2.30)$$

The solutions (2.2.29) are well defined if $\lambda_k \neq \mu_l$ ($k, l = 1, \ldots, N$).

The case when some of $\lambda_k$ coincide with some $\mu_l$ requires a special consideration. Here we confine ourselves by the simplest case

$$R(\mu, \bar{\mu}; \lambda, \bar{\lambda}) = \sum_{k=1}^{N} f_k(\mu) g_k(\lambda) \delta(\mu - \lambda_k) \delta(\lambda - \lambda_k) \qquad (2.2.31)$$

where $f_k(\mu)$, $g_k(\lambda)$ are smooth functions and $\lambda_i$ are isolated point, different from the positions of poles for the normalization function $n(\lambda, \bar{\lambda})$. Substitution of (2.2.31) into (2.2.3) gives

$$\chi(\lambda, \bar{\lambda}) = n(\lambda, \bar{\lambda}) + \frac{2i}{\pi} \sum_{k=1}^{N} \frac{\chi_k f_k g_k}{\lambda_k - \lambda} \qquad (2.2.32)$$

where $\chi_k = \chi(\lambda_k, \bar{\lambda}_k)$, $f_k \equiv f_k(\lambda_k)$, $g_k \equiv g_k(\lambda_k)$. In virtue of possible singularities a slightly more sophisticated analysis of equation (2.2.3) is now needed in order to derive the system of equations for the quantities $\chi(\lambda_k)$ ($k = 1, ..., N$). Proceeding in equation (2.2.3) with the kernel $R$ of the form (2.2.31) to the limit $\lambda \to \lambda_i$, one gets

$$\chi(\lambda_i, \bar{\lambda}_i) = n(\lambda_i; \bar{\lambda}_i)$$

$$+ \frac{1}{\pi} \sum_{k=1}^{N} \iint_D \frac{d\lambda' \wedge d\bar{\lambda}'}{\lambda' - \lambda_i} \chi(\lambda_k, \bar{\lambda}_k) f_k(\lambda_k) g_k(\lambda') \delta(\lambda' - \lambda_k), \quad i = 1, ..., N. \qquad (2.2.33)$$

In virtue of the well-known formula of complex analysis the term in (2.2.33) with $k = i$ is equal to

$$-\frac{2i}{\pi} \operatorname{Res} \frac{\chi(\lambda_i, \bar{\lambda}_i) f_i(\lambda_i) g_i(\lambda_i)}{(\lambda - \lambda_i)^2} \bigg|_{\lambda = \lambda_i} = \frac{2i}{\pi} \chi(\lambda_i, \bar{\lambda}_i) f_i(\lambda_i) g_i'(\lambda_i) \qquad (2.2.34)$$

where

$$g_i'(\lambda_i) \doteq \frac{\partial g_i(\lambda)}{\partial \lambda}\bigg|_{\lambda = \lambda_i}.$$

Consequently, equations (2.2.33) with different $i$ give rise to the system

$$\chi_i = n_i + \frac{2i}{\pi} \sum_{k \neq i}^{N} \frac{\chi_k f_k g_k}{\lambda_k - \lambda_i} + \frac{2i}{\pi} \chi_i f_i g_i' \quad (i = 1, \ldots, N) \qquad (2.2.35)$$

where $n_i \doteq n(\lambda_i, \bar{\lambda}_i)$. Solving this system for given $f_k$ and $g_k$, one gets the solution (2.2.32) of the $\bar{\partial}$-problem (2.2.1).

One can also construct more complicated explicit solutions of the $\bar{\partial}$-problem in the matrix case. They correspond to the delta-function type kernel $R$ of the form

$$R_{\alpha\beta}(\mu, \bar{\mu}; \lambda, \bar{\lambda}) = \sum_{k=1}^{N} (f_k)_{\alpha\gamma} \delta(\mu - \mu_{\alpha\gamma}^{(k)})$$

$$\times (g_k)_{\gamma\beta}\delta(\lambda - \lambda_{\gamma\beta}^{(k)}) \quad (\alpha,\beta = 1,\ldots,n). \tag{2.2.36}$$

where $n$ is the order of the matrices. In this case the function $\chi$ looks like

$$\chi_{\alpha\beta}(\lambda,\bar{\lambda}) = n_{\alpha\beta}(\lambda,\bar{\lambda}) + \frac{2}{\pi i}\sum_{k=1}^{N}\sum_{\gamma=1}^{N}\frac{\chi_{\alpha\delta}(\mu_{\delta\gamma}^{(k)})(f_k)_{\delta\gamma}(g_k)_{\alpha\beta}}{\lambda - \lambda_{\gamma\beta}^{(k)}} \quad (\alpha,\beta = 1,\ldots,n). \tag{2.2.37}$$

So each matrix element of $\chi$ has its own set of poles.

In the particular case $\mu_{\alpha\beta}^{(k)} = \lambda_{\alpha\beta}^{(k)}$ one obtains the generalization of the solutions (2.2.35). The quantities $\chi_{\alpha\beta}(\lambda_{\delta\gamma}^{(k)})$ obey the following system of equations [152]

$$\chi_{\alpha\beta}(\lambda_{\beta\sigma}^{(i)}) = n_{\alpha\beta}(\lambda_{\beta\sigma}^{(i)}) + \frac{2i}{\pi}\sum_{k=1}^{N}\chi_{\alpha\gamma}(\lambda_{\gamma\delta}^{(k)})\frac{f_{(k)\delta\gamma}(\lambda_{\gamma\delta}^{(k)})g_{(k)\gamma\beta}(\lambda_{\rho\beta}^{(k)})}{\lambda_{\beta\sigma}^{(i)} - \lambda_{\rho\beta}^{(k)}},$$

$$\alpha,\beta,\sigma = 1,\ldots,n, \quad \sigma \neq \beta, \tag{2.2.38}$$

$$\chi_{\alpha\beta}(\lambda_{\beta\beta}^{(i)}) = n_{\alpha\beta}(\lambda_{\beta\beta}^{(i)}) + \frac{2i}{\pi}\chi_{\alpha\gamma}(\lambda_{\gamma\delta}^{(i)})f_{(i)\delta\beta}(\lambda_{\gamma\delta}^{(i)})g'_{(i)\delta\beta}(\lambda_{\beta\beta}^{(i)})$$

$$+\sum_{k\neq i}\sum_{i,\rho\neq\beta}\frac{\chi_{\alpha\delta}(\lambda_{\gamma\delta}^{(k)})f_{(k)\delta\gamma}(\lambda_{\gamma\delta}^{(k)})g_{(k)\gamma\rho}(\lambda_{\rho\beta}^{(k)})}{\lambda_{\beta\beta}^{(i)} - \lambda_{\rho\beta}^{(i)}} \quad (\alpha,\beta = 1,\ldots,n). \tag{2.2.39}$$

The derivation of the system (2.2.38), (2.2.39) is similar to that of the system (2.2.35).

All the classes of explicit solutions of the $\bar{\partial}$-problem considered above will be used in construction of the explicit exact solutions of the 2+1-dimensional integrable equations.

## 2.3. $\bar{\partial}$-dressing method

The nonlocal $\bar{\partial}$-problem is a basis of the $\bar{\partial}$-dressing method. Here we will confine ourselves mainly by the $\bar{\partial}$-problem with the canonical normalization

$$\frac{\partial \chi}{\partial \bar{\lambda}} = \hat{R}\chi \equiv \chi * R, \tag{2.3.1}$$

where $\hat{R}$ is the linear integral operator acting as

$$(\hat{R}\chi)(\lambda,\bar{\lambda}) = (\chi * R)(\lambda,\bar{\lambda}) = \iint_C d\lambda' \wedge d\bar{\lambda}'\, \chi(\lambda',\bar{\lambda}')R(\lambda',\bar{\lambda}';\lambda,\bar{\lambda}) \tag{2.3.2}$$

and $\chi \to 1$ at $\lambda \to \infty$. We assume that the $\bar{\partial}$-problem (2.3.1) is uniquely solvable.

The aim of the $\bar{\partial}$-dressing method [108, 144–148] is to construct the compatible system of linear equations for $\chi$ and, consequently the nonlinear differential equations associated with the $\bar{\partial}$-problem (2.3.1). Such nonlinear equations are formulated in the "physical" space. Let $x_1$, $x_2$ and $x_3$ are the local coordinates in this space.

*Chapter 2* 39

So, one introduces the dependence on the new variables $x_1$, $x_2$, $x_3$ in (2.3.1) and should start with an appropriate dependence of $R(\lambda', \bar{\lambda}'; \lambda, \bar{\lambda})$ on $x_1$, $x_2$, $x_3$. According to the main idea of the IST method the evolution of the operator $\hat{R}$ or, equivalently, of the kernel $R(\lambda', \bar{\lambda}'; \lambda, \bar{\lambda}; x_1, x_2, x_3)$ in the variables $x_1$, $x_2$, $x_3$ should be covered by the linear and solvable equations.

So we assume that
$$\frac{\partial \hat{R}}{\partial x_i} + [\hat{B}_i^*, \hat{R}] = 0 \quad (i = 1, 2, 3) \tag{2.3.3}$$
where $\hat{B}_i^*$ ($i = 1, 2, 3$) are the operators adjoint to some operators $\hat{B}_i$ with respect to the standard bilinear form $\langle f, g \rangle \doteq \int d\xi \operatorname{tr}(f(\xi)g(\xi))$. The operators $\hat{B}_i^*$ are chosen in (2.3.3) instead of $\hat{B}_i$ by the reason of convenience.

In terms of the operators
$$D_i \doteq \partial_{x_i} + \hat{B}_i^*, \quad (i = 1, 2, 3)$$
equations (2.3.3) look like
$$[D_i, \hat{R}] = 0 \quad (i = 1, 2, 3). \tag{2.3.4}$$
For the kernel $R$ of the operator $\hat{R}$ equation (2.3.3) is of the form
$$\frac{\partial R(\lambda', \bar{\lambda}'; \lambda, \bar{\lambda}; x)}{\partial x_i} = \iint_C d\mu \wedge d\bar{\mu} \left\{ B_i(\lambda', \mu) R(\mu, \bar{\mu}; \lambda, \bar{\lambda}; x) - R(\lambda', \bar{\lambda}'; \mu, \bar{\mu}; x) B_i(\mu, \lambda) \right\}$$
$$i = 1, 2, 3 \tag{2.3.5}$$
where the kernels $B_i(\lambda, \mu)$ of the operators $\hat{B}_i$ are defined by
$$(\hat{B}_i f)(\lambda, \bar{\lambda}) = \iint_C d\mu \wedge d\bar{\mu} \, B_i(\lambda, \mu) f(\mu, \bar{\mu})$$
and respectively
$$(\hat{B}_i^* f)(\lambda, \bar{\lambda}) = \iint_C d\mu \wedge d\bar{\mu} \, f(\mu, \bar{\mu}) B(\mu, \lambda).$$

For differential operators $\hat{B}_i$ one has
$$\frac{\partial R(\lambda', \bar{\lambda}'; \lambda, \bar{\lambda}; x)}{\partial x_i} = \hat{B}_i(\lambda') R(\lambda', \bar{\lambda}'; \lambda, \bar{\lambda}; x) - \hat{B}_i^*(\lambda) R(\lambda', \bar{\lambda}'; \lambda, \bar{\lambda}; x) \quad (i = 1, 2, 3). \tag{2.3.6}$$

The compatibility condition for the system (2.3.3) implies
$$[D_i, D_k] = 0, \quad (i, k = 1, 2, 3), \tag{2.3.7}$$
i. e.
$$\frac{\partial \hat{B}_k^*}{\partial x_i} - \frac{\partial \hat{B}_i^*}{\partial x_k} + [\hat{B}_i^*, \hat{B}_k^*] = 0 \quad (i, k = 1, 2, 3). \tag{2.3.8}$$

So, we postulate the linear commutative evolution of the kernel $R$. The problem is to convert the linear system (2.3.3) for $\hat{R}$ into the linear compatible system of equations with variable coefficient for $\chi$. This can be done in the following manner [144–148].

First, we note that, in virtue of the condition (2.3.3) or (2.3.4), one has

$$D_i \frac{\partial \chi}{\partial \bar{\lambda}} = D_i \chi * R \quad (i = 1, 2, 3). \tag{2.3.9}$$

Then, taking into account the condition (2.3.7) and the fact that the $\bar{\partial}$-problem admits the multiplication from the left by an arbitrary matrix-valued function which depends only on $x_1$, $x_2$, $x_3$, one also concludes that

$$M \frac{\partial \chi}{\partial \bar{\lambda}} = M \chi * R \tag{2.3.10}$$

where $M$ is an arbitrary differential operator of the form

$$M = \sum_{n_1, n_2, n_3} u_{n_1 n_2 n_3}(x_1, x_2, x_3) D_1^{n_1} D_2^{n_2} D_3^{n_3} \tag{2.3.11}$$

and $u_{n_1 n_2 n_3}(x_1, x_2, x_3)$ are arbitrary matrix-valued functions.

In virtue of (2.3.10), the function $M\chi$ is not a solution of the $\bar{\partial}$-problem for generic operator $M$. This property will take place only for particular operator $\tilde{M}$ which obeys the condition

$$\left[ \frac{\partial}{\partial \bar{\lambda}}, \tilde{M} \right] \chi = 0. \tag{2.3.12}$$

For such operators $\tilde{M}$ one has

$$\frac{\partial \tilde{M} \chi}{\partial \bar{\lambda}} = \tilde{M} \chi * R, \tag{2.3.13}$$

i. e. the function $\tilde{M}\chi$ solves the $\bar{\partial}$-problem (2.3.1) together with $\chi$. Note that the condition (2.3.12) is equivalent to

$$\frac{\partial \tilde{M}}{\partial \bar{\lambda}} \chi(\lambda) = 0. \tag{2.3.14}$$

The conditions (2.3.12) or (2.3.14) mean that the corresponding function $\tilde{M}\chi$ has no singularities at those points where the operators $D_1$, $D_2$, $D_3$ have singularities. So such a function $\tilde{M}\chi$ has singularities at the same points as $\chi$. The condition (2.3.14) clearly shows that the coefficients $u_{n_1, n_2, n_3}$ of the operators $\tilde{M}$ of the form (2.3.11) which obey (2.3.12) are expressed via the function $\chi$ evaluated in the singularity points of $\tilde{M}$.

One particular type of operators $\tilde{M}$ is of special interest. They are the operators $L$ which obey the condition (2.3.12), i. e.

$$\left[ \frac{\partial}{\partial \bar{\lambda}}, L \right] \chi = 0 \tag{2.3.15}$$

and, in addition, the condition

$$(L\chi)(\lambda) \underset{\lambda \to \infty}{\longrightarrow} 0. \tag{2.3.16}$$

The unique solvability of the $\bar{\partial}$-problem (2.3.1) and the condition (2.3.16) imply that for such operators $L$ one has

$$L\chi = 0. \tag{2.3.17}$$

Equation (2.3.17) demonstrates the importance of the operators $L$ which obey the conditions (2.3.15) and (2.3.16). Such operators provide the differential equations for the function $\chi$.

The operators $L$ form the left ideal of the ring of operators $M$. The construction of the basis for this ideal is the problem of the principal importance. If operators $L_i$ ($i = 1, \ldots, k$) form a basis for the ideal of the operators $L$ which obey the conditions (2.3.15) and (2.3.16) then the corresponding linear equations

$$L_i\chi = 0 \quad (i = 1, \ldots, k) \tag{2.3.18}$$

form the complete set of the independent linear equations for $\chi$ associated with the $\bar{\partial}$-problem (2.3.1) and equation (2.3.3).

The system (2.3.18) is just the desired linear system for $\chi$ [108, 144–148]. The system (2.3.18) is deliberately compatible. Indeed this system by construction has the common solution $\chi$ which is the solution of the original $\bar{\partial}$-problem (2.3.1) and the compatible system (2.3.3). The system (2.3.18) is the system of linear equations with the variable coefficients $u_{n_1 n_2 n_3}(x_1, x_2, x_3)$. So, the compatibility of the system (2.3.18) is expressed by the certain set of the nonlinear differential relations, i. e. nonlinear differential equations for $u_{n_1 n_2 n_3}(x_1, x_2, x_3)$. This system of nonlinear differential equations can be obtained directly from (2.3.18) by evaluating its l.h.s. at the same points $\lambda_i$ at which $\chi(\lambda_i)$ are connected with $u_{n_1 n_2 n_3}$.

Thus, the system (2.3.18) provides the system of nonlinear differential equations associated with the $\bar{\partial}$-problem (2.3.1) and the linear evolution (2.3.3). Different $\hat{B}_i$ in (2.3.3) give rise to the different nonlinear equations associated with the $\bar{\partial}$-problem (2.3.1). The main technical problem here is to construct for given $\hat{B}_i$ ($i = 1, 2, 3$) the basis $L_i$ ($i = 1, \ldots, k$) for the ideal of operators $L$ which obey the conditions (2.3.15) and (2.3.16).

The system of nonlinear equations which is associated with the linear system (2.3.18) is solvable. Indeed, the linear system (2.3.3) (or (2.3.6)) defines, at least, in principal, the explicit dependence of the kernel $R$ on the variables $x_1$, $x_2$, $x_3$. Then considering the degenerate kernels $R$, one can applies to our problem the results of the previous section. The formulae (2.2.22), (2.2.24) and (2.2.32), (2.2.35) with the functions $f_k$, $g_k$ depending also on $x_1$, $x_2$, $x_3$ allow us to construct the function $\chi(\lambda, \bar{\lambda}; x_1, x_2, x_3)$ explicitly. Then, since the functions $u_{n_1 n_2 n_3}(x_1, x_2, x_3)$ are expressed via the function $\chi$ evaluated at certain points, one gets $u_{n_1 n_2 n_3}(x_1, x_2, x_3)$. So, finally, we obtain the exact explicit solutions of the system of nonlinear equations associated with (2.3.18). The solutions constructed in such a way form the broadest class of exact explicit solutions of nonlinear equations with the three independent variables $x_1$, $x_2$, $x_3$ which can be found by the IST method.

Thus, the method described above allow us both to construct the nonlinear equations and their exact solutions. This is, in essence, the dressing method the original version of which has been proposed near 20 years by Zakharov and Shabat [10]. The version presented here is referred, usually, as the $\bar{\partial}$-dressing method. It has been formulated mainly in the papers [104, 144–148].

Emphasize also that the $\bar{\partial}$-dressing method is, primarily, the method for solution of the linear differential equations (2.3.17) with variable coefficients. So this method is very important and useful even without any connection with nonlinear partial differential equations.

In this section we will consider the simplest case when $\hat{B}_i$ are the operators of multiplication by the matrices $I_i$ which are independent of $x_1$, $x_2$, $x_3$. In this case equation (2.3.7) implies that $I_i(\lambda)$ $(i = 1, 2, 3)$ are commuting matrices and operators $D_{x_i}$ act as follows

$$D_i f = \partial_{x_i} f + f I_i(\lambda). \tag{2.3.19}$$

The evolution of the kernel $R$ in $x_1$, $x_2$, $x_3$ is given consequently by the linear equations with constant coefficients

$$\frac{\partial R(\lambda', \bar{\lambda}'; \lambda, \bar{\lambda}; x_1, x_2, x_3)}{\partial x_i} = I_i(\lambda') R - R I_i(\lambda) \quad (i = 1, 2, 3). \tag{2.3.20}$$

Equations (2.3.20) imply that

$$R(\lambda', \bar{\lambda}'; \lambda, \bar{\lambda}; x_1, x_2, x_3) = e^{F(\lambda', x)} R_o(\lambda', \bar{\lambda}'; \lambda, \bar{\lambda}) e^{-F(\lambda, x)} \tag{2.3.21}$$

where

$$F(\lambda, x) \doteq \sum_{i=1}^{3} I_i(\lambda) x_i \tag{2.3.22}$$

and $R_0$ is an arbitrary matrix-valued function.

In this case the singularities of the generic operators $M$ (2.3.11) coincide with the singularities of $I_i(\lambda)$ and their products. Here we will consider the solutions $\chi$ of the $\bar{\partial}$-problem (2.3.1) bounded at all $\lambda$. So, the problem of construction of the operators $L$ which obey the conditions (2.3.15) and (2.3.16) is reduced at the first step to construction of such operators $\tilde{L}_i$ that $\tilde{L}_i \chi$ are bounded.

After derivation of the basic operators $L_i$ one arrives at the linear system (2.3.18) where the action of the operators $D_i$ is given by (2.3.19). The $\lambda$-dependence can be

removed from the system (2.3.18) in the considered case. To do this let us introduce the function
$$\psi(x, \lambda; \bar{\lambda}) = \chi(\lambda, \bar{\lambda}; x) e^{F(\lambda, x)} \qquad (2.3.23)$$
where $F(\lambda, x)$ is given by (2.3.22). Since
$$D_i \chi = \partial_{x_i} \psi \cdot e^{-F} \qquad (2.3.24)$$
the transition from the function $\chi$ to the function $\psi$ effectively convents the operators $D_i$ into the usual derivatives $\partial_{x_i}$. Consequently, the system (2.3.18) with the operators $L_i$ of the form (2.3.11) is equivalent to the following system
$$L_i(\partial_x)\psi = \sum_{n_1, n_2, n_3} u^{(i)}_{n_1, n_2, n_3}(x) \partial_{x_1}^{n_1} \partial_{x_2}^{n_2} \partial_{x_3}^{n_3} \psi = 0 \quad (i = 1, \dots, k). \qquad (2.3.25)$$
So, one gets the system of the linear partial differential equations which does not contain the parameter $\lambda$ at all. Similar to (2.3.18) the system (2.3.25) is automatically compatible and it gives rise to the same system of nonlinear differential equations for the coefficients $u^{(i)}_{n_1 n_2 n_3}(x)$ as that provided by (2.3.18). The system (2.3.25) represents itself the usual form of the auxiliary linear systems for the integrable nonlinear equations.

In the terms of function $\psi$ the $\bar{\partial}$-problem (2.3.1) looks like
$$\frac{\partial \psi(\lambda, \bar{\lambda}; x)}{\partial \bar{\lambda}} = \psi * R_0 \equiv \iint_C d\lambda' \wedge d\bar{\lambda}' \, \psi(\lambda', \bar{\lambda}'; x) R_o(\lambda', \bar{\lambda}'; \lambda, \bar{\lambda}). \qquad (2.3.26)$$
So, in such a formulation all the dynamics on the variables $x_1$, $x_2$, $x_3$ is contained in the function $\psi(\lambda, x)$ only. The $\bar{\partial}$-problem of the type (2.3.26) is the basic equation for the calculation of the so-called asymptotic modules [153–159]. The method of the asymptotic modules has been proposed in [153–159] for the construction of the 2+1-dimensional and multidimensional integrable equations. It can be treated as the version of the $\bar{\partial}$-dressing method which uses only the algebraic structures arising in the whole $\bar{\partial}$-dressing method.

The formulation of the nonlocal $\bar{\partial}$-problem in the form (2.3.26) leads also to another interesting interrelation. Let us consider together with (2.3.26) the formally adjoint $\bar{\partial}$-problem
$$\frac{\partial \psi^*(\lambda, \bar{\lambda}; x')}{\partial \bar{\lambda}} = - \iint_C d\lambda' \wedge d\bar{\lambda}' R_0(\lambda, \bar{\lambda}; \lambda', \bar{\lambda}')) \psi^*(\lambda', \bar{\lambda}'; x'). \qquad (2.3.27)$$
Multiplying equation (2.3.26) by $\psi^*(\lambda, \bar{\lambda}; x')$ from the right, equation (2.3.27) by $\psi(\lambda, \bar{\lambda}; x)$ from the left, summing the obtained equations and integrating, one gets
$$\iint_C d\lambda \wedge d\bar{\lambda} \frac{\partial}{\partial \bar{\lambda}} (\psi(\lambda, \bar{\lambda}; x) \psi^*(\lambda, \bar{\lambda}; x')) = 0. \qquad (2.3.28)$$
Using the formula (2.1.5), one finally obtains
$$\int_{\partial C} d\lambda \psi(\lambda, \bar{\lambda}; x) \psi^*(\lambda, \bar{\lambda}; x') = 0 \qquad (2.3.29)$$

where coordinates $x = (x_1, x_2, x_3, \ldots)$ and $x' = (x'_1, x'_2, x'_3, \ldots)$ are arbitrary. Note that this relation holds for any $R_0(\lambda', \bar{\lambda}'; \lambda, \bar{\lambda})$.

The bilinear and bilocal identity (2.3.29) provides the bridge between the $\bar{\partial}$-dressing method and the other beautiful method in the theory of the integrable equations, the so-called, $\tau$-function method [160, 161]. The bilinear identity (2.3.29) plays the fundamental role in that method.

Before proceeding to concrete examples let us discuss some general properties of the integrable equations considered above.

Exact explicit solutions of such integrable equations obtainable by the $\bar{\partial}$-dressing method correspond to the degenerate kernels $R(\lambda', \bar{\lambda}'; \lambda, \bar{\lambda}, x)$. In the case (2.3.21) one chooses

$$R_0(\lambda', \bar{\lambda}'; \lambda, \bar{\lambda}) = \sum_{k=1}^{N} f_{0k}(\lambda', \bar{\lambda}') g_{0k}(\lambda, \bar{\lambda}). \tag{2.3.30}$$

Hence, the functions $f_k$ and $g_k$ defined in (2.2.16) are of the form

$$f_k(\mu, \bar{\mu}) = e^{F(\mu, x)} f_{0k}(\mu, \bar{\mu}),$$

$$g_k(\lambda, \bar{\lambda}) = g_{0k}(\lambda, \bar{\lambda}) e^{-F(\lambda, x)}. \tag{2.3.31}$$

As a result the formula (2.2.22) where $f_k$ and $g_k$ are given by (2.3.31) gives us the explicit solution of the $\bar{\partial}$-problem. Then, since the functions $u_{n_1 n_2 n_3}(x_1, x_2, x_3)$ are expressed via the function $\chi$ evaluated in the certain points $\lambda_i$, the formulae (2.2.22) and (2.3.31) provide us the explicit expressions for $u_{n_1 n_2 n_3}$. Thus, we get the explicit exact solutions with the functional parameters $f_{0k}(\mu, \bar{\mu})$, $g_{0k}(\lambda, \bar{\lambda})$ of the nonlinear 2+1-dimensional equations which correspond to (2.3.21).

For the particular case (2.2.23) one has

$$f_k = e^{F(\mu_k, x)} f_{0k}, \quad g_k = g_{0k} e^{-F(\lambda_k, x)} \tag{2.3.32}$$

where $f_{0k}$ and $g_{0k}$ are constant matrices. In this case the function $\chi$ and, consequently, the solutions of the integrable equations are the combination of the exponents (2.3.32). At last, at $\mu_k = \lambda_k$ the explicit solutions are given by (2.2.32) and (2.2.35) where

$$g'_i(\lambda_i) = -g_{0k} e^{-F(\lambda_k, x)} F' \tag{2.3.33}$$

and

$$F' = \sum_{i=1}^{3} x_i \frac{\partial I_i(\lambda)}{\partial \lambda}\bigg|_{\lambda = \lambda_i}. \tag{2.3.34}$$

In virtue of (2.3.33) and (2.3.34), the corresponding explicit solutions in general contain both exponent and rational dependence on $x_1, x_2, x_3$. In the scalar case $f_k g_k = f_{0k} g_{0k}$

and, consequently, the function $\chi$ and the solutions of the integrable equations are pure rational functions on $x_1$, $x_2$, $x_3$. To construct the pure rational solutions in the matrix case one has to consider a special kernels $R$.

The solutions described above form the classes of exact solutions with functional parameters, line solitons and lumps for the 2+1-dimensional integrable equations.

These solutions are parametrized at most by several arbitrary functions of two variables. In the generic case the kernels $R$ or $R_0$ depend on four independent variables. But the particular value of $\chi$ are determined by the certain integral characteristics of $R$. For instance, if $\chi \to 1 + \frac{1}{\lambda}\chi_{-1} + \ldots$ as $\lambda \to \infty$ then in the case of small $R$ the quantity $\chi_{-1}(x)$ is defined by the expression (2.2.11). In virtue of (2.2.12) the quantity $\chi_{-1}$ is invariant under the transformations $R_0 \to R_0' = R_0 + \tilde{R}_0$ such that [147]

$$\iint_C d\mu \wedge d\bar{\mu} \iint d\lambda \wedge d\bar{\lambda}\, e^{F(\mu,x)} \tilde{R}_0(\mu,\bar{\mu};\lambda,\bar{\lambda}) e^{-F(\lambda,x)} = 0. \qquad (2.3.35)$$

So, only certain invariant parts of $R_0(\mu,\bar{\mu};\lambda,\bar{\lambda})$ contribute to the particular values of $\chi$ and, consequently, to the functions $u_{n_1 n_2 n_3}(x_1,x_2,x_3)$.

Emphasize now one important point. In the whole $\bar{\partial}$-dressing method constructions there was no any assumption about the behaviour of the functions $u_{n_1 n_2 n_3}(x_1,x_2,x_3)$ at infinity. All the constructed solutions are essentially local in the space $(x_1,x_2,x_3)$. They can behave arbitrarily (e. g. grow) as $|x| \to \infty$. Their properties depend on the choice of the kernel $R$. For applications in physics the nonsingular solutions, of course, are of the special interest.

The class of nonsingular solutions can be characterized in terms of the kernel $R_0(\mu,\bar{\mu};\lambda,\bar{\lambda})$ at least for small $R_0$ [147]. Indeed, for small $R_0$ the solution of the $\bar{\partial}$-problem is given by the first iteration of the integral equation (2.2.3), i. e.

$$\chi(\lambda,\bar{\lambda}) = n(\lambda,\bar{\lambda}) + \frac{1}{2\pi i}\iint_C \iint_C \frac{d\lambda' \wedge d\bar{\lambda}' \cdot d\mu \wedge d\bar{\mu}}{\lambda' - \lambda}$$

$$\times n(\mu,\bar{\mu})e^{F(\mu,x)} R_0(\mu,\bar{\mu};\lambda,\bar{\lambda}) e^{-F(\lambda',x)}. \qquad (2.3.36)$$

Assuming that $I_i(\lambda)$ are the diagonal matrices $(I_i)_{\alpha\beta} = I_i^{(\alpha)}\delta_{\alpha\beta}$, one has the following component form of (2.3.36)

$$\chi_{\alpha\beta}(\lambda,\bar{\lambda}) = n_{\alpha\beta}(\lambda,\bar{\lambda}) + \frac{1}{2\pi i}\iint_C \iint_C \frac{d\lambda' \wedge d\bar{\lambda} \cdot d\mu \wedge d\bar{\mu}}{\lambda' - \lambda}$$

$$\times \sum_{\gamma} n_{\alpha\gamma}(\mu,\bar{\mu}) R_{0\gamma\beta}(\mu,\bar{\mu};\lambda,\bar{\lambda}) \exp\left(\sum_{i=1}^{3}(I_i^{(\gamma)}(\mu) - I_i^{(\beta)}(\lambda'))x_i\right). \qquad (2.3.37)$$

Now let us choose one variable say $x_3$ as the time $t$ and treat the variables $x_1$, $x_2$ as the spatial variables. In other words we will consider the corresponding integrable equation as the evolution equation with respect to $x_3 = t$.

The particular values of the function $\chi$, for instance, the coefficients of the asymptotic expansion as $\lambda \to \infty$ will be regular if the exponents in (2.3.37) with $x_1$ and $x_2$ are pure oscillating, i. e.

$$\text{Re}(I_1^{(\gamma)}(\mu) - I_1^{(\beta)}(\lambda')) = 0,$$

$$\text{Re}(I_2^{(\gamma)}(\mu) - I_2^{(\beta)}(\lambda')) = 0 \qquad (2.3.38)$$

for all $\gamma$ and $\beta$. The conditions (2.3.38) define the two-dimensional surface $s$ in the four dimensional space $\mu, \bar{\mu}, \lambda', \bar{\lambda}'$. For boundness of $\chi$ the functions $R_{0\gamma\beta}(\mu, \bar{\mu}; \lambda', \bar{\lambda}')$ should be concentrated on the manifold (2.3.38), i. e.

$$R_{0\gamma\beta}(\mu, \bar{\mu}; \lambda, \bar{\lambda}) = \delta(\text{Re}(I_1^{(\gamma)}(\mu) - I_1^{(\beta)}(\lambda)))$$

$$\times \delta(\text{Re}(I_2^{(\gamma)}(\mu) - I_2^{(\beta)}(\lambda)))\tilde{R}_{0\gamma\beta}(\mu, \bar{\mu}; \lambda, \bar{\lambda}) \qquad (2.3.39)$$

where $\tilde{R}_{0\gamma\beta}$ are arbitrary functions.

Thus, the kernels of the $\bar{\partial}$-problem of the form (2.3.39) give rise to the bounded in $x_1$, $x_2$ solutions of the corresponding integrable equations at least for small $R_0$ [147].

The surfaces defined by the constraints (2.3.38) are essentially different for different $I_i(\lambda)$. For the kernels $R_0$ of the type (2.3.39) the nonlocal $\bar{\partial}$-problem (2.3.1) is reduced in some cases to the nonlocal Riemann—Hilbert problem and in other cases to the quasi-local $\bar{\partial}$-problem (1.3.17). Concrete examples of the constraints (2.3.38) and the corresponding reductions of the $\bar{\partial}$-problem (2.3.1) will be discussed in the next chapters.

## 2.4. Examples

Now we will consider three examples as the illustrations of the general scheme of the $\bar{\partial}$-dressing method. For the first example we choose

$$I_1 = A\lambda, \quad I_2 = B\lambda^n, \quad I_3 = C\lambda^m \qquad (2.4.1)$$

where $A$, $B$, $C$ are generic diagonal matrices of the arbitrary order and $n$, $m$ are arbitrary integers ($m > n$). In this case

$$D_1 f = \partial_x f + \lambda f A,$$

$$D_2 f = \partial_y f + \lambda^n f B, \qquad (2.4.2)$$

$$D_3 f = \partial_t f + \lambda^m f C,$$

where we denote $x_1 = x$, $x_2 = y$, $x_3 = t$ and

$$F(\lambda, x) = \lambda A x + \lambda^n B y + \lambda^m C t. \tag{2.4.3}$$

We have to construct the operators $L_i$ which obey the conditions (2.3.15) and (2.3.16).

In the case (2.4.2) and in more general case of arbitrary $I_i(\lambda)$ polynomial in $\lambda$ any operator $L$ of the form (2.3.11) formally obeys the condition (2.3.15). So it is necessary, in fact, to meet only the condition (2.3.16). But, in order to treat this case in the same manner as those with rational $I_i(\lambda)$ we will discuss the conditions (2.3.15) and (2.3.16) for it separately. Let us start with the function $D_2\chi$. Since $\chi \to 1$ as $\lambda \to \infty$ the function $D_2\chi$ has the $n$-th order singularity as the infinity. The function $D_1^n \chi$ also has the $n$-th order singularity as $\lambda \to \infty$. Hence, subtracting $u_n D_1^n \chi$ with appropriate $u_n$ from $D_2\chi$, one can annihilate the $n$-order singularity. It is easy to see that the proper choice is $u_n = BA^{-n}$. Thus, the function $D_2\chi - BA^{-n}D_1^n \chi$ has the $n-1$-th and lower order singularities as $\lambda \to \infty$. To annihilate the $n-1$-th order singularity one can subtract from $D_2\chi - u_n D_1^n \chi$ the function $u_{n-1} D_1^{n-1}\chi$. With the proper choice of $u_{n-1}(x,y,t)$ the function $D_2\chi - u_n D_1^n \chi - u_{n-1} D_1^{n-1}\chi$ will only have $n-2$-th and lower order singularities as $\lambda \to \infty$. This condition completely fixes $u_{n-1}(x,y,t)$ in terms of coefficients of the asymptotic expansion of $\chi$ as $\lambda \to \infty$. Continuing this process one can construct the function

$$\tilde{\chi} = D_2\chi - \sum_{k=0}^{n} u_k(x,y,t) D_1^k \chi$$

which has no singularities and such that $\tilde{\chi} \to 0$ as $\lambda \to \infty$. Hence,

$$L_1 \chi = D_2\chi - \sum_{k=0}^{n} u_k(x,y,t) D_1^k \chi = 0. \tag{2.4.4}$$

The matrix-valued coefficients $u_k$ in (2.4.4) are expressed via the coefficients of the asymptotic expansion

$$\chi \Rightarrow 1 + \frac{1}{\lambda}\chi_{-1} + \frac{1}{\lambda^2}\chi_{-2} + \ldots \tag{2.4.5}$$

as $\lambda \to \infty$.

Equation (2.4.4) is the first desired linear equation from the basis set (2.3.18). The second equation can be found in completely similar manner with the use of the operator $D_3$. It is of the form

$$L_2 \chi = D_3\chi - \sum_{k=0}^{m} v_k(x,y,t) D_1^k \chi = 0 \tag{2.4.6}$$

where the coefficients $v_m$ are expressed via the coefficients of the expansion (2.4.5).

It is quite clear that in the generic case the operators $L_1$ and $L_2$ given by (2.4.4) and (2.4.6) are the two lowest order differential operators which obey the conditions (2.3.15) and (2.3.16). All other such operators are of the form $\sum_{\alpha_1\alpha_2} C_{\alpha_1\alpha_2} L_1^{\alpha_1} L_2^{\alpha_2}$ where $C_{\alpha_1\alpha_2}$ are some operators polynomial in $D_i$ $(i = 1, 2, 3)$.

If $n$ and $m$ have a common divisor then one can construct several different operators $L_2$ of the same order using the operator $D_2$ instead of certain powers of the operator $D_1$. Nevertheless, as it is not difficult to show, all these operators $L_2$ are equivalent modulo the operator $L_1$.

So, in any case the operators $L_1$ and $L_2$ given by (2.4.4) and (2.4.6) form the basis for the ideal of operators $\{L\}$ for the case (2.4.10).

In terms of the function $\psi$ defined by

$$\psi = \chi(\lambda, \bar{\lambda}; x, y, t) \exp(\lambda Ax + \lambda^n By + \lambda^m Ct) \tag{2.4.7}$$

one has the following system of linear equations

$$\partial_y \psi - \sum_{k=0}^{n} u_k(x, y, t) \partial_x^k \psi = 0,$$

$$\partial_t \psi - \sum_{k=0}^{m} v_k(x, y, t) \partial_x^k \psi = 0. \tag{2.4.8}$$

This is the standard form of the auxiliary linear system for the choice (2.4.1). Equations (2.4.8) are compatible by construction and the functions $u_k$, $v_k$ obey the system of nonlinear differential equations. Several well-known 2+1-dimensional integrable equations (e. g. KP and DS equations) are associated with the auxiliary linear systems of the type (2.4.8). For such integrable equations the variable $x$ is the marked variable and its role is different from that of the variables $y$ and $t$.

In a similar manner one can construct the linear problems and corresponding nonlinear integrable equations for

$$I_1 = \frac{1}{\lambda} A, \quad I_2 = \frac{1}{\lambda^n} B, \quad I_3 = \frac{1}{\lambda^m} C \tag{2.4.9}$$

where $A$, $B$, $C$ are commuting matrices and $n$, $m > 0$. In this case the powers of the operators $D_1$, $D_2$, $D_3$ have the singularities at the origin but the rest in the construction of the operators $L_1$ and $L_2$ is the same as in the previous example. So, one again arrives at the two linear problems of the form (2.4.4), (2.4.6) or (2.4.8). But in this case the coefficients $u_n$, $v_m$ may be the functions on $x$, $y$, $t$ and $u_{n-1}(x, y, t)$, $v_{m-1}(x, y, t)$ are the nonzero functions.

The mKP equation and the Ishimori equation belong to this type of equations. Note that for both cases (2.4.1) and (2.4.9) the matrices $I_i$ and the operators $D_i$ have the singularities at the single point.

## Chapter 2

Our second example corresponds to the scalar case with the singularities at the two points, say infinity and origin. So let

$$I_1 = \lambda, \quad I_2 = \frac{\varepsilon}{\lambda}, \quad I_3 = -k_1\lambda^3 - k_2\varepsilon^3\lambda^{-3}. \tag{2.4.10}$$

where $\varepsilon$, $k_1$, $k_2$ are arbitrary constants. The operators $D_i$ are

$$D_1 = \partial_\xi + \lambda, \quad D_2 = \partial_\eta + \frac{\varepsilon}{\lambda}, \quad D_3 = \partial_t - k_1\lambda^3 - k_2\varepsilon^3\lambda^{-3} \tag{2.4.11}$$

and

$$F = \lambda\xi + \frac{\varepsilon}{\lambda}\eta - (k_1\lambda^3 + k_2\varepsilon^3\lambda^{-3})t.$$

It is clear that in order to construct the operator $L_1$ one has to combine the operators $D_1$ and $D_2$. Let us consider the function

$$D_1 D_2 \chi = \partial_\xi \partial_\eta \chi + \varepsilon\chi + \lambda\partial_\eta\chi + \frac{\varepsilon}{\lambda}\partial_\xi\chi. \tag{2.4.12}$$

Since $\chi \to 1$ at $\lambda \to \infty$ the function (2.4.12) has no singularity at infinity. The situation at the origin $\lambda = 0$ is different. Since

$$\chi(\xi, \eta, t, \lambda, \bar{\lambda}) = \chi_0(\xi, \eta, t) + \lambda\chi_1 + \lambda^2\chi_2 + \ldots \tag{2.4.13}$$

as $\lambda \to 0$ then $D_1 D_2 \chi$ has a simple pole at the origin with the residue $\varepsilon\partial_\xi\chi_0$. In order to compensate this contribution let us subtract $-v(\varepsilon, \eta, t) D_2 \chi$ from $D_1 D_2 \chi$. The expression

$$D_1 D_2 \chi + v D_2 \chi = \partial_\xi \partial_\eta \chi + \varepsilon\chi + \lambda\partial_\eta\chi + v\partial_\eta\chi + \frac{\varepsilon}{\lambda}(\partial_\xi\chi + v(\xi, \eta, t)\chi) \tag{2.4.14}$$

is free from the singularity at $\lambda = 0$ if

$$v(\xi, \eta, t) = -\partial_\xi \ln \chi_0(\xi, \eta, t). \tag{2.4.15}$$

So, we have found the operator $\tilde{M}$ which obeys the condition (2.3.12).

Then, since as $\lambda \to \infty$ the function $\chi$ has the asymptotic expansion (2.4.5), one has

$$D_2 D_1 \chi + v D_2 \chi \to \varepsilon + \partial_\eta \chi_{-1}(\xi, \eta, t). \tag{2.4.16}$$

Taking into account (2.4.16), one readily concludes that the operator

$$L_1 = D_2 D_1 + v D_2 + u - \varepsilon \tag{2.4.17}$$

where

$$u(\xi, \eta, t) = -\partial_\eta \chi_{-1}(\xi, \eta, t) \tag{2.4.18}$$

obeys the conditions (2.3.15) and (2.3.16). Therefore

$$L_1 \chi = D_2 D_1 \chi + v D_2 \chi + u\chi - \varepsilon\chi = 0 \tag{2.4.19}$$

where the coefficients $v$ and $u$ are given by (2.4.15) and (2.4.18).

The second desired auxiliary linear problem can be constructed using the operator $D_3 = \partial_t + k_1\lambda^3 + k_2\varepsilon^3\lambda^{-3}$. Indeed the expression

$$D_3\chi + k_1 D_1^3\chi + k_2 D_2^3\chi = \partial_t\chi + k_1\partial_\xi^3\chi + 3k_1\lambda\partial_\xi^2\chi + 3k_1\lambda^2\partial_\xi\chi$$

$$+ k_2\partial_\eta^3\chi + 3k_2\varepsilon\lambda^{-1}\partial_\eta^2\chi + 3k_2\varepsilon^2\lambda^{-2}\partial_\eta\chi \qquad (2.4.20)$$

has no third order poles at the infinity and origin. It is easy to see that it also has no second order singularity as $\lambda \to \infty$. To compensate the second order pole at $\lambda = 0$ and the first order singularities at $\lambda = \infty$ and $\lambda = 0$ one should subtract from (2.4.20) the terms $v_1(\xi,\eta,t)D_2^2$, $v_2(\xi,\eta,t)D_1$, $v_3(\xi,\eta,t)D_2$ with the appropriate functions $v_1$, $v_2$, $v_3$. Finally, one can construct the operator

$$L_2 = D_3 + k_1 D_1^3 + k_2 D_2^3 - v_1 D_2^2 - v_2 D_1 - v_3 D_2 - v_4 \qquad (2.4.21)$$

such that $L_2\chi$ obeys the conditions (2.3.15) and (2.3.16). As a result, one gets the linear problem

$$L_2\chi = (D_t + k_1 D_1^3 + k_2 D_2^3 - v_1 D_2^2 - v_2 D_1 - v_3 D_2 - v_4)\chi = 0. \qquad (2.4.22)$$

The operators $L_1$ (2.4.17) and $L_2$ (2.4.21) form a desired basis and equations (2.4.19) and (2.4.22) are the linear problems associated with the $\bar{\partial}$-problem for the choice (2.4.10).

Transiting to the function

$$\psi = \chi \exp\left(\lambda\xi + \frac{\varepsilon}{\lambda}\eta - (k_1\lambda^3 + k_2\varepsilon^3\lambda^{-3})t\right), \qquad (2.4.23)$$

one gets the linear problems

$$(\partial_\xi\partial_\eta + v\partial_\eta + u)\psi = \varepsilon\psi,$$

$$(\partial_t + k_1\partial_\xi^3 + k_2\partial_\eta^3 - v_1\partial_\eta^2 - v_2\partial_\xi - v_3\partial_\eta - v_4)\psi = 0. \qquad (2.4.24)$$

The compatibility condition for the system (2.4.19), (2.4.22) or the system (2.4.24) gives the 2+1-dimensional nonlinear system of equations for $u$ and $v$.

The particular case of this system with $v \equiv 0$ is of special interest. In virtue of (2.4.15) the condition $v \equiv 0$ is satisfied if $\chi_0 = $ const say $\chi_0 = 1$. It is not difficult to show that in this case $v_1 = 0$, $v_4 = $const and

$$v_2 = -3k_1\partial_\eta^{-1}u_\xi, \quad v_3 = -3k_2\partial_\xi^{-1}u_\eta. \qquad (2.4.25)$$

So under the reduction $v \equiv 0$ one gets the linear system

$$L_1\psi = (\partial_\xi\partial_\eta + u - \varepsilon)\psi = 0, \qquad (2.4.26a)$$

Chapter 2  51

$$L_2\psi = (\partial_t + k_1\partial_\xi^3 + k_2\partial_\eta^3 + 3k_1\partial_\eta^{-1}u_\xi\partial_\xi + 3k_2\partial_\xi^{-1}u_\eta\partial_\eta + \text{const})\psi = 0. \quad (2.4.26b)$$

The compatibility condition for the system (2.4.26) is nothing but the NVN equation (1.2.19) with the substitution $u \to u - \varepsilon$. This integrable equation can be obtained directly from equation (2.4.22) with $v_1$, $v_2$, $v_3$, $v_4$ given by (2.4.25) by the asymptotic expansion of the l.h.s. near the infinity and keeping the terms of the order $\lambda^{-1}$.

The exact solutions of this generalized NVN equation can be constructed with the use of the scalar $\bar{\partial}$-problem (2.3.1) with the canonical normalization under the additional constraint $\chi(\lambda = 0) = 1$ and the formula (2.4.18).

In a similar manner one can treat within the framework of the $\bar{\partial}$-dressing method the other integrable equations with $\hat{B}_1$, $\hat{B}_2$, $\hat{B}_3$ having the singularities at two points, for instance, the 2DISG equation (1.2.24).

Our last example here is essentially different from the previous two. Let [108, 109]

$$I_i(\lambda) = \frac{A_i}{\lambda - \lambda_i} \quad (i = 1, 2, 3) \tag{2.4.27}$$

where $A_i$ are commuting matrices $[A_i, A_k] = 0$ ($i, k = 1, 2, 3$) and all the complex numbers $\lambda_i$ are distinct. So $F(\lambda, x) = \sum_{i=1}^{3} \frac{x_i A_i}{\lambda - \lambda_i}$.

In this case all the independent variables $x_1$, $x_2$, $x_3$ are contained in the problem on the equal footing. The quantities

$$D_i(\lambda)\chi = \partial_{x_i}\chi + \chi\frac{A_i}{\lambda - \lambda_i}, \tag{2.4.28}$$

have the poles at the points $\lambda_i$. It is rather obvious that it is impossible to construct the combination without singularities using only the first order operators $D_i$. One should proceed to the second order operators. The expression $D_i D_k \chi$ with $i \neq k$ has the simple poles at the points $\lambda_i$ and $\lambda_k$. One can remove these singularities by adding the terms $u_{ik}^i(x)D_i\chi$ and $u_{ik}^k(x)D_k\chi$ to $D_i D_k\chi$. The requirement of vanishing the residues at the poles $\lambda_i$ and $\lambda_k$ in the quantities

$$D_i D_k \chi + u_{ik}^i D_i \chi + u_{ik}^k D_k \chi \quad (i \neq k) \quad (i, k = 1, 2, 3) \tag{2.4.29}$$

gives

$$u_{ik}^i(x) = -\left(\partial_{x_k}\chi_i + \frac{\chi_i A_k}{\lambda_i - \lambda_k}\right)\chi_i^{-1},$$

$$u_{ik}^k(x) = -\left(\partial_{x_i}\chi_k + \frac{\chi_k A_i}{\lambda_k - \lambda_i}\right)\chi_k^{-1}, \tag{2.4.30}$$

where $\chi_i(x) \doteq \chi(\lambda = \lambda_i)$. Note that there is no summation over repeated indices in these and next formulae.

So the expressions (2.4.29) define the operators of type $\tilde{M}$. Moreover, it gives us, in fact, the operators $L$. Indeed, using (2.4.28) one can readily show that the quantities (2.4.29) have zero asymptotics at $\lambda \to 0$.

Thus, the desired linear problems are of the form [108, 109]

$$L_{ik}\chi \equiv (D_i D_k + u^i_{ik} D_i + u^k_{ik} D_k)\chi = 0 \quad (i,k = 1,2,3; \quad i \neq k) \tag{2.4.31}$$

where $u^i_{ik}$ and $u^k_{ik}$ are given by (2.4.30). So we have three linear problems. All of them are of the second order and contain the variables $x_1$, $x_2$, $x_3$ symmetrically.

In order to obtain the nonlinear system of differential equations associated with (2.4.31) it is sufficient to evaluate the l.h.s. of (2.4.31) at the three points $\lambda_l$ ($l = 1,2,3$) such that $l \neq i$, $l \neq k$ for given $i$ and $k$ ($i \neq k$). Using (2.4.28) and (2.4.30), one gets [108, 109]

$$\partial_{x_i}\partial_{x_k}\chi_l + \partial_{x_i}\chi_l \frac{A_k}{\lambda_l - \lambda_k} + \partial_{x_k}\chi_l \frac{A_i}{\lambda_l - \lambda_i} + \chi_l \frac{A_i A_k}{(\lambda_l - \lambda_i)(\lambda_l - \lambda_k)}$$

$$- \left(\partial_{x_i}\chi_k + \frac{\chi_k A_i}{\lambda_k - \lambda_i}\right)\chi_k^{-1}\left(\partial_{x_k}\chi_l + \chi_l \frac{A_k}{\lambda_l - \lambda_k}\right)$$

$$- \left(\partial_{x_k}\chi_i + \frac{\chi_i A_k}{\lambda_i - \lambda_k}\right)\chi_i^{-1}\left(\partial_{x_i}\chi_l + \chi_l \frac{A_i}{\lambda_l - \lambda_i}\right) = 0 \tag{2.4.32}$$

$$(i,k,l = 1,2,3; \quad i \neq k, \quad k \neq l, \quad l \neq i).$$

The system of three equations (2.4.32) represents the nonlinear system integrable by the $\bar{\partial}$-problem (2.3.1) with $I_i(\lambda)$ given by (2.4.27). For the first time the system (2.4.32) has been derived by Zakharov and Manakov in [108, 109] within the version of the dressing method based on the nonlocal Riemann—Hilbert problem.

The system (2.4.32) is the representative of the genuine three-dimensional integrable nonlinear systems. It is associated with the three auxiliary linear problems and contains the independent variables $x_1$, $x_2$, $x_3$ symmetrically.

Transiting to the function

$$\psi \doteq \chi \exp\left(\sum_{i=1}^{3} \frac{x_i A_i}{\lambda - \lambda_i}\right), \tag{2.4.33}$$

one gets the auxiliary linear system in the form (2.4.25). Namely, one has

$$\tilde{L}_{ik}\psi \equiv (\partial_{x_i}\partial_{x_k} + u^i_{ik}\partial_{x_i} + u^k_{ik}\partial_{x_k})\psi = 0 \quad (i,k = 1,2,3; i \neq k). \tag{2.4.34}$$

This system is compatible and the corresponding nonlinear system is

$$\frac{\partial u^i_{ik}}{\partial x_l} - u^i_{ik}u^i_{il} + u^l_{lk}u^i_{li} + u^k_{lk}u^i_{ik} = 0 \quad (i,k,l = 1,2,3; i \neq k, k \neq l, l \neq i). \tag{2.4.35}$$

Using (2.4.30), it is not difficult to show that the system (2.4.35) is equivalent to the system (2.4.32).

The operator form of the compatibility condition for the system (2.4.34) is of the form [110]

$$[\tilde{L}_{ik}, \tilde{L}_{nk}] = \alpha_{ink}\tilde{L}_{in} + \beta_{ink}\tilde{L}_{ik} + \delta_{ink}\tilde{L}_{nk} \quad (i,k,n,=1,2,3;\, i\neq k, k\neq n, n\neq i) \quad (2.4.36)$$

where

$$\alpha_{ink} = \frac{\partial u_{nk}^n}{\partial x_k} - \frac{\partial u_{ik}^i}{\partial x_i} + [u_{ik}^i, u_{nk}^n],$$

$$\beta_{ink} = \frac{\partial u_{nk}^k}{\partial x_i} - \frac{\partial u_{ik}^i}{\partial x_n} + [u_{ik}^i, u_{nk}^k], \quad (2.4.37)$$

$$\delta_{ink} = \frac{\partial u_{nk}^n}{\partial x_i} - \frac{\partial u_{ik}^k}{\partial x_k} + [u_{ik}^k, u_{nk}^n],$$

where $i,k,n = 1,2,3$.

Redefining $\tilde{L}_{12} \to L_3$, $\tilde{L}_{31} \to L_2$, $\tilde{L}_{23} \to L_1$, one arrives at the following operator system

$$[L_i, L_k] = \sum_{l=1}^{3} \gamma_{ikl} L_l \quad (2.4.38)$$

where the "structure constants" $\gamma_{ikl}$ are simply expressed via $\alpha_{ink}$, $\beta_{ink}$ and $\delta_{ink}$. For instance,

$$\gamma_{121} = \beta_{213}, \quad \gamma_{122} = \delta_{213}, \quad \gamma_{123} = \alpha_{213}.$$

The Lie algebra type operator system (2.4.38) is the generic operator form of the three-dimensional integrable equations [113, 96].

Zakharov—Manakov system (2.4.32) contains three arbitrary parameters $\lambda_1$, $\lambda_2$, $\lambda_3$ and three arbitrary diagonal matrices $A_1$, $A_2$, $A_3$. But, in fact, all these free parameters can be eliminated. Indeed, introducing new dependent variables $g_i$ ($i = 1, 2, 3$) via

$$g_1 = \chi_1 e^{\frac{A_2 x_2}{\lambda_1 - \lambda_2} + \frac{A_3 x_3}{\lambda_1 - \lambda_3}},$$

$$g_2 = \chi_2 e^{\frac{A_1 x_1}{\lambda_2 - \lambda_1} + \frac{A_3 x_3}{\lambda_2 - \lambda_3}}, \quad (2.4.39)$$

$$g_3 = \chi_3 e^{\frac{A_1 x_1}{\lambda_3 - \lambda_1} + \frac{A_2 x_2}{\lambda_3 - \lambda_2}},$$

one rewrites the system (2.4.32) in the following form

$$\partial_{x_i}\partial_{x_k} g_l - \partial_{x_k} g_i \cdot g_i^{-1} \cdot \partial_{x_i} g_l - \partial_{x_i} g_k \cdot g_k^{-1} \partial_{x_k} g_l = 0 \quad (2.4.40)$$

$$i,k,l = 1,2,3 \quad i \neq k,\, k \neq l,\, l \neq i.$$

In terms of $g_i$ the operators $L_i$ look like

$$L_1 = \partial_{x_2}\partial_{x_3} - g_{2x_3}g_2^{-1}\partial_{x_2} - g_{3x_2}g_3^{-1}\partial_{x_3},$$
$$L_2 = \partial_{x_3}\partial_{x_1} - g_{3x_1}g_3^{-1}\partial_{x_3} - g_{1x_3}g_1^{-1}\partial_{x_1}, \qquad (2.4.41)$$
$$L_3 = \partial_{x_1}\partial_{x_2} - g_{1x_2}g_1^{-1}\partial_{x_1} - g_{2x_1}g_2^{-1}\partial_{x_2}.$$

So the system (2.4.40) is the basic nonlinear integrable system associated with (2.4.27). It is independent of the choice of $A_i$ and $\lambda_i$.

It is easy to see that the system (2.4.40) is nothing but the particular case of the original Zakharov—Manakov system (2.4.32) with $A_i \equiv 0$ ($i = 1, 2, 3$). Unfortunately, the choice $A_i = 0$ ($i = 1, 2, 3$) causes the problem since in this case the kernel $R$ of the $\bar{\partial}$-problem is independent of $x_1$, $x_2$, $x_3$ due to (2.4.21). So the system (2.4.40) requires a special treatment.

The same is true for the scalar version of the system which looks like

$$g_{1_{x_2x_3}} = \frac{g_{1_{x_2}}g_{2_{x_3}}}{g_2} + \frac{g_{1_{x_3}}g_{3_{x_2}}}{g_3},$$
$$g_{2_{x_3x_1}} = \frac{g_{2_{x_3}}g_{3_{x_1}}}{g_3} + \frac{g_{2_{x_1}}g_{1_{x_3}}}{g_1}, \qquad (2.4.42)$$
$$g_{3_{x_1x_2}} = \frac{g_{3_{x_1}}g_{1_{x_2}}}{g_1} + \frac{g_{3_{x_2}}g_{2_{x_1}}}{g_2}.$$

The system (2.4.42) was discussed recently by V. Dryuma [111]. He realised that this system is a part of a wider system of six nonlinear differential equations which describes the triple orthogonal families of surfaces in the three-dimensional Euclidean space associated with the metric

$$ds^2 = \sum_{i=1}^{3} g_i^2(x_1, x_2, x_3)dx_i^2. \qquad (2.4.43)$$

Moreover he demonstrated that the Einstein equations for the special space-time metric of the type

$$ds^2 = \psi^2(x_1, x_2, x_3)dt - \sum_{i=1}^{3} g_i^2(x_1, x_2, x_3)dx_i^2 \qquad (2.4.44)$$

contain the system (2.4.42) and the corresponding linear system $L_i\psi = 0$ ($i = 1, 2, 3$) as a part.

Thus, the three-dimensional nonlinear system (2.4.42) is of great interest both for physics and mathematics. Some particular exact solutions of this system were constructed in [111].

The change of variables (2.4.39) indicates a way for construction of wide classes of exact solutions of the system (2.4.40), including the scalar system (2.4.42). Indeed, let us consider the solutions of the system (2.4.40) with the asymptotics

$$g_1 \to e^{\frac{A_2 x_2}{\lambda_1 - \lambda_2}} + e^{\frac{A_3 x_3}{\lambda_1 - \lambda_3}},$$
$$g_2 \to e^{\frac{A_1 x_1}{\lambda_2 - \lambda_1}} + e^{\frac{A_3 x_3}{\lambda_2 - \lambda_3}}, \qquad (2.4.45)$$
$$g_3 \to e^{\frac{A_1 x_1}{\lambda_3 - \lambda_1}} + e^{\frac{A_2 x_2}{\lambda_3 - \lambda_2}}$$

as $x_1^2 + x_2^2 + x_3^2 \to \infty$, where $\lambda_1$, $\lambda_2$, $\lambda_3$ are some distinct constants and $A_1$, $A_2$, $A_3$ are diagonal matrices. In general all $\lambda_i$, $A_i$ are arbitrary. Then introducing the variables $\chi_i$ by (2.4.39) with $\chi_i \to 1$ as $x_1^2 + x_2^2 + x_3^2 \to \infty$, one obtains the generic Zakharov—Manakov system (2.4.32) for $\chi_i$ with some $\lambda_i$ and $A_i$.

The generic system (2.4.32) as well as the equations considered in the previous examples can be effectively solved by the $\bar{\partial}$-dressing method. Using the formulae of the sections (2.3) and (2.4) and the concrete form of $F(\lambda, x)$, one can construct the classes of the explicit exact solutions with functional parameters, line solitons and lumps of the system (2.4.32) and, consequently, of the system (2.4.40).

Having in mind the applications of the system (2.4.42), we present here exact solutions of the scalar system (2.4.40). So, we will look for the solutions of the system (2.4.42) with the asymptotics (2.4.45). Then we apply the general formulae of sections (2.3) and (2.4) and obtain the exact solutions for $\chi_i$ ($i = 1, 2, 3$).

Solutions with functional parameters correspond to the kernel

$$R_0(\mu, \bar{\mu}; \lambda, \bar{\lambda}) = i \sum_{\alpha=1}^{N} f_\alpha(\mu, \bar{\mu}) \phi_\alpha(\lambda, \bar{\lambda}) \qquad (2.4.46)$$

where $f_\alpha$ and $\phi_\alpha$ are arbitrary functions. The corresponding solutions are of the form

$$\chi_i(x) = 1 - \frac{1}{2\pi A_i} \sum_{\alpha,\beta=1}^{N} \frac{\partial \eta_\alpha}{\partial x_i} (1+B)^{-1}_{\alpha\beta} \xi_\beta \quad (i=1,2,3) \qquad (2.4.47)$$

where

$$\xi_\alpha(x) \doteq \iint_C d\lambda \wedge d\bar{\lambda} f_\alpha(\lambda, \bar{\lambda}) e^{\sum_{i=1}^{3} \frac{A_i x_i}{\lambda - \lambda_i}},$$

$$\eta_\alpha(x) \doteq \iint_C d\mu \wedge d\bar{\mu} \phi_\alpha(\mu, \bar{\mu}) e^{-\sum_{i=1}^{3} \frac{A_i x_i}{\mu - \lambda_i}},$$

and

$$B_{\alpha\beta} = \frac{1}{2\pi} \iint_C \iint_C \frac{d\lambda \wedge d\bar{\lambda} \cdot d\mu \wedge d\bar{\mu}}{\mu - \lambda} f_\alpha(\lambda, \bar{\lambda}) \phi_\beta(\mu, \bar{\mu})$$

$$\times \exp\left\{\sum_{i=1}^{3} A_i x_i \left(\frac{1}{\lambda - \lambda_i} - \frac{1}{\mu - \lambda_i}\right)\right\}. \qquad (2.4.48)$$

The matrix $B_{\alpha\beta}$ (2.4.48) can be represented in a more compact form. Indeed, it follows from (2.4.48) that

$$\frac{\partial B_{\alpha\beta}}{\partial x_i} = -\frac{1}{2\pi A_i} \frac{\partial \xi_\alpha(x)}{\partial x_i} \cdot \frac{\partial \eta_\beta(x)}{\partial x_i}$$

for any $i$ ($i = 1, 2, 3$). Note that there is no summation over $i$ in this and the next formulae. Hence,

$$B_{\alpha\beta} = -\frac{1}{2\pi A_i} \int^{x_i} dx'_i \frac{\partial \xi_\alpha(x')}{\partial x'_i} \cdot \frac{\partial \eta_\beta(x')}{\partial x'_i}.$$

The solution of the scalar system (2.4.42) is given by (2.4.39). The solutions (2.4.39), (2.4.47) are real-valued for real $\lambda_i$, $A_i$ ($i = 1, 2, 3$) and $f_\alpha$, $\phi_\alpha$ ($\alpha = 1, \ldots, N$).

Note also that the functions $\xi_\alpha$ and $\eta_\alpha$ are the generic solutions of the linear systems

$$(\lambda_i - \lambda_k)\frac{\partial^2 \xi}{\partial x_i \partial x_k} - A_k \frac{\partial \xi}{\partial x_i} + A_i \frac{\partial \xi}{\partial x_k} = 0,$$

$$(\lambda_i - \lambda_k)\frac{\partial^2 \eta}{\partial x_i \partial x_k} + A_k \frac{\partial \eta}{\partial x_i} - A_i \frac{\partial \eta}{\partial x_k} = 0$$

$$i, k = 1, 2, 3; \quad i \neq k.$$

Line solitons correspond to the choice

$$f_\alpha(\lambda) = C_\alpha \delta(\lambda - \nu_\alpha), \quad \phi_\alpha(\lambda) = \tilde{C}_\alpha \delta(\lambda - \mu_\alpha)$$

where $C_\alpha$, $\tilde{C}_\alpha$, $\nu_\alpha$, $\mu_\alpha$ are arbitrary constants. The line-soliton solutions of the system (2.4.42) is given by (2.4.39) where

$$\chi_i(x) = 1 + \frac{2}{\pi} \sum_{\alpha,\beta=1}^{N} \frac{\tilde{C}_\alpha C_\beta}{\mu_\alpha - \lambda_i} (1 + B)^{-1}_{\alpha\beta} \exp\left(\sum_{k=1}^{3} A_k x_k \left(\frac{1}{\nu_\beta - \lambda_k} - \frac{1}{\mu_\beta - \lambda_k}\right)\right) \qquad (2.4.49)$$

and

$$B_{\alpha\beta} = \frac{2}{\pi} \frac{C_\alpha \tilde{C}_\beta}{\mu_\beta - \nu_\alpha} \exp\left(\sum_{k=1}^{3} A_k x_k \left(\frac{1}{\nu_\alpha - \lambda_k} - \frac{1}{\mu_\beta - \lambda_k}\right)\right).$$

For real $C_\alpha$, $\tilde{C}_\alpha$, $\lambda_\alpha$, $\mu_\alpha$ the solution (2.4.49) is also real.

One line-soliton solution ($N = 1$) is of the form

$$\chi_i(x) = 1 + \frac{\mu_1 - \nu_1}{\mu_1 - \lambda_i}\left(1 + \exp\left(-\sum_{k=1}^{3} A_k x_k \left(\frac{1}{\nu_1 - \lambda_k} - \frac{1}{\mu_1 - \mu_k}\right) + \phi_0\right)\right)^{-1}. \qquad (2.4.50)$$

Rational solutions are associated with the kernel

$$R_0 = \frac{\pi}{2i} \sum_{\alpha=1}^{N} R_\alpha \delta(\mu - \nu_\alpha) \delta(\lambda - \nu_\alpha) \qquad (2.4.51)$$

where $\nu_\alpha$ are some constants such that $\nu_\alpha \neq \lambda_i$, $\nu_\alpha \neq \nu_\beta$.

The corresponding solution of the system (2.4.42) is given by

$$\chi_i(x) = 1 + \sum_{\alpha=1}^{N} \frac{R_\alpha h_\alpha}{\nu_\alpha - \lambda_i} \qquad (2.4.52)$$

where the quantities $h_\alpha$ are calculated from the system

$$h_\alpha \left(1 + \gamma_\alpha + R_\alpha \sum_{k=1}^{3} \frac{A_k x_k}{(\nu_\alpha - \lambda_k)^2}\right) + \sum_{\beta \neq \alpha}^{N} \frac{h_\beta R_\beta}{\nu_\beta - \nu_\alpha} = 1 \quad \alpha = 1, \ldots, N \qquad (2.4.53)$$

where $\gamma_\alpha$ are arbitrary constants. The simplest rational solution is of the form

$$\chi_i(x) = 1 + \frac{R_0}{\nu_1 - \lambda_i} \frac{1}{1 + \gamma_0 + R_0 \sum_{k=1}^{3} \frac{A_k x_k}{(\nu_1 - \lambda_k)^2}}. \qquad (2.4.54)$$

Exact solutions constructed in [111] are very particular cases of solutions (2.4.47) and (2.4.52).

Here we have discussed the simple examples which correspond to the simple functions $I_i(\lambda)$. For more complicated $I_i(\lambda)$, for instance, for the generic rational $I_i(\lambda)$, the problem of construction of the basis of the operators $L_i$ and the corresponding integrable equations becomes very complicated. This problem can be effectively solved if instead of the solution $\chi$ with single canonical normalization one will use the solutions of the $\bar{\partial}$-problem with different normalizations [145]. The $\bar{\partial}$-dressing method with the variable normalization will be discussed in chapter VIII.

## 2.5. $\bar{\partial}$-dressing on nontrivial background

In the previous section we considered the $\bar{\partial}$-dressing method for the simplest choice of $\hat{B}_i$ in (2.3.3) as the operators of multiplication by matrices $I_i(\lambda)$ independent of $x_i$. In this case the general dependence of $R$ on $x_i$ can be readily found (2.3.21). With such choice of operators $\hat{B}_i$ one captures a number of the 2+1-dimensional integrable equations. However, it is not wide enough. In particular, for integrable equations considered above it allows us to construct the exact solutions only on the so-called trivial (zero) background. An analysis of opportunities for the construction of the 2+1-dimensional integrable systems of new type is of great interest. So the extension of the $\bar{\partial}$-dressing method to more general operators $\hat{B}_i$ in (2.3.3) is an important problem [147].

In dealing with such extensions one should obey the two principal demands. First, the operators $\hat{B}_i$ obviously should be such that it would be possible effectively construct the operators $L_i$ which satisfy the conditions (2.3.15) and (2.3.16). Second, one should be able to construct the explicit exact solutions of the corresponding integrable equation using the $\bar{\partial}$-problem. For this purpose one, at least, should be able to find the explicit exact solutions of equation (2.3.6).

The explicit form of equation (2.3.6) and the commutativity condition (2.3.8) point out a way on which one can find a broad class of nontrivial operators $\hat{B}_i$ which obey the second demand. Indeed, the form of equation (2.3.6) indicates that it can be solved by separation of the variables $\lambda$ and $\lambda'$

$$R(\mu, \bar{\mu}; \lambda, \bar{\lambda}, x) = f(\mu, \bar{\mu}; x) g^*(\lambda, \bar{\lambda}; x). \tag{2.5.1}$$

The functions $f$ and $g^*$ obey the equations

$$\frac{\partial f}{\partial x_i} - \hat{B}_i f = 0, \tag{2.5.2}$$

$$\frac{\partial g^*}{\partial x_i} + \hat{B}_i^* g^* = 0, \quad (i = 1, 2, 3). \tag{2.5.3}$$

The equation adjoint to (2.5.3) is

$$\frac{\partial g}{\partial x_i} - \hat{B}_i g = 0, \quad (i = 1, 2, 3), \tag{2.5.4}$$

i. e. the same as for $f$. So, solving the system (2.5.2), one will be able to find exact solutions of equation (2.3.6).

Further, the commutativity condition (2.3.8) is equivalent to the following

$$\frac{\partial \hat{B}_k}{\partial x_i} - \frac{\partial \hat{B}_i}{\partial x_k} + [\hat{B}_i, \hat{B}_k] = 0 \quad (i, k = 1, 2, 3). \tag{2.5.5}$$

However, equation (2.5.5) is nothing but the compatibility condition for the system (2.5.2). Recalling now, for instance (1.1.21), (1.1.22), one concludes that equations (2.5.5) and (2.5.2) represent, in fact, the integrable nonlinear system and the corresponding linear system.

Thus, the use of the known results for the 2+1-dimensional integrable systems provides a broad class of operators $\hat{B}_i$ for which one has wide classes of exact solutions of equation (2.5.2) and, consequently, equation (2.3.6).

The operators $\hat{B}_i$ act on the variables $\lambda, \bar{\lambda}$ while the variables $x_1$, $x_2$, $x_3$ play a role of parameters for them. On the other hand, equations (2.5.2) and (2.5.4) for $f$ and $g$

are differential equations with respect to $x_1$, $x_2$, $x_3$. This indicates that the operators $\hat{B}_i$ should have rather special structure to provide the effective solution of equations (2.5.2).

Here following the paper [148] we will briefly consider the particular realization of this scheme in which the operators $\hat{B}_i$ are integral operators on $\lambda$ of convolutive type. The choice of $\hat{B}_i$ in (2.3.3) as the convolutive integral operators is a felicitous choice. The use of the basic property of the convolutive operators, namely, that the Fourier transform of convolutive operator is the operator of point multiplication, and the fact that under the Fourier transform $\lambda$ is converted into the operator of differentiation, will allow us to solve effectively equation (2.5.2) and to construct the operators $L_i$.

So let us consider the $\bar{\partial}$-problem (2.3.1) with the canonical normalization $\chi \underset{\lambda \to \infty}{\longrightarrow} 1$. We will confine ourselves (similar to [148]) to the operators $\hat{B}_i$ which have singularities at infinity, namely

$$\hat{B}_1(\lambda) \to iA\lambda, \quad \hat{B}_2(\lambda) \to B\lambda^n, \quad \hat{B}_3(\lambda) \to C\lambda^m \tag{2.5.6}$$

as $\lambda \to \infty$, where $A$, $B$, $C$ are commuting matrices and $n$, $m$ are arbitrary integers ($m > n$). A comparison with (2.4.1) indicates that the case (2.5.6) will be closely connected with equations (2.4.8).

The Fourier transform is defined as follows

$$X(s, \bar{s}, x_1, x_2, x_3) = \frac{1}{2\pi} \iint_C d\lambda \wedge d\bar{\lambda} \, \exp(i(\lambda s + \bar{\lambda}\bar{s})) f(\lambda, \bar{\lambda}; x_1, x_2, x_3). \tag{2.5.7}$$

The operators $\hat{B}_i$ ($i = 1, 2, 3$) are converted by the transformation (2.5.7) into the differential operators with respect to $s$ of the orders 1, $n$ and $m$, respectively. The Fourier transform of equations (2.5.2) looks like

$$(\partial_{x_1} - A\partial_s + \tilde{w}_0(s, x_1, x_2, x_3))X_0 = 0, \tag{2.5.8a}$$

$$(\partial_{x_2} - \sum_{k=1}^{n} \tilde{u}_{0k}(s, x_1, x_2, x_3)\partial_s^k)X_0 = 0, \tag{2.5.8b}$$

$$(\partial_{x_3} - \sum_{k=1}^{m} \tilde{v}_{0k}(s, x_1, x_2, x_3)\partial_s^k)X_0 = 0 \tag{2.5.8c}$$

where $\tilde{w}_0$, $\tilde{u}_{0k}$, $\tilde{v}_{0k}$ are matrix-valued functions. The compatibility condition for the system (2.5.8) implies the nonlinear integrable system for the functions $\tilde{w}_0$, $\tilde{u}_{0k}$, and $\tilde{v}_{0k}$.

Having in mind the linear problems (2.4.8) we choose here $\tilde{w}_0 = 0$ and $A = 1$. In this case $X_0 = X_0(x_1 + s, x_2, x_3)$. In virtue of the condition (2.5.5), one also has

$$\tilde{u}_{0k}(s, x_1, x_2, x_3) = \tilde{u}_k(s + x_1, x_2, x_3),$$

$$\tilde{v}_{0k}(s, x_1, x_2, x_3) = \tilde{v}_k(s + x_1, x_2, x_3). \tag{2.5.9}$$

So, with such choice the system (2.5.8) is equivalent to

$$(\partial_{x_2} - \sum_{k=1}^{n} \tilde{u}_k(z, x_2, x_3)\partial_z^k)X_0 = 0,$$

$$(\partial_{x_3} - \sum_{k=1}^{m} \tilde{v}_k(z, x_2, x_3)\partial_z^k)X_0 = 0 \quad (2.5.10)$$

where $z = x_1 + s$.

The system (2.5.10) is nothing but the system (2.4.8). The $\bar{\partial}$-dressing method, as was shown in the previous sections, provides a wide class of exact solutions for the linear system (2.5.10). Note that, in general, the variable $z$ is complex. The reduction to the real $s$ corresponds to the reduction to the nonlocal Riemann—Hilbert problem [148].

With the exact solutions $X_{0i}$ and $Y_{0i}$ of the system (2.5.10) in the hand one constructs the exact solutions $R$ of equation (2.3.6) of the form

$$R(\mu, \bar{\mu}; \lambda, \bar{\lambda}, x) = \sum_{k=1}^{N} f_k(\mu, \bar{\mu}) g_k^*(\lambda, \bar{\lambda}) \quad (2.5.11)$$

where

$$f_k(\mu, \bar{\mu}, x_1, x_2, x_3) = \frac{1}{2\pi} \iint_C ds \wedge d\bar{s} \, \exp(-i(\mu s + \bar{\mu}\bar{s})) X_{0k}(s + x_1, x_2, x_3), \quad (2.5.12)$$

$$g_k^*(\lambda, \bar{\lambda}, x_1, x_2, x_3) = \frac{1}{2\pi} \iint_C ds \wedge d\bar{s} \, \exp(i(\lambda s + \bar{\lambda}\bar{s})) Y_{0k}^*(s + x_1, x_2, x_3). \quad (2.5.13)$$

Together with $X_{0k}$ and $Y_{0k}$ one now has also the explicit expressions for the Fourier transform $\hat{\tilde{B}}_1, \hat{\tilde{B}}_2, \hat{\tilde{B}}_3$ of the operators $\hat{B}_1, \hat{B}_2, \hat{B}_3$:

$$\hat{\tilde{B}}_1 = \partial_s,$$

$$\hat{\tilde{B}}_2 = \sum_{k=1}^{n} \tilde{u}_k(s + x_1, x_2, x_3)\partial_s^k, \quad (2.5.14)$$

$$\hat{\tilde{B}}_3 = \sum_{k=1}^{m} \tilde{v}_k(s + x_1, x_2, x_3)\partial_s^k.$$

Note that in terms of the Fourier transform $\tilde{R}$ of the kernel $R$ defined by

$$\tilde{R}(s, \tilde{s}; x_1 x_2 x_3) \doteq \frac{1}{(2\pi)^2} \iint_C \iint_C d\mu \wedge d\bar{\mu} \, d\lambda \wedge d\bar{\lambda}$$

$$\times R(\mu, \bar{\mu}; \lambda, \bar{\lambda}; x_1, x_2, x_3) \exp(i(\mu s + \bar{\mu}\bar{s}) - i(\lambda \tilde{s} + \bar{\lambda}\bar{\tilde{s}})) \quad (2.5.15)$$

equations (2.3.6) are of the form

$$\frac{\partial}{\partial x_i} \tilde{R}(s, \tilde{s}; x_1, x_2, x_3) = (\hat{\tilde{B}}_i(s) - \hat{\tilde{B}}_i^*(\tilde{s})) \tilde{R}(s, \tilde{s}; x_1, x_2, x_3) \quad (i = 1, 2, 3) \quad (2.5.16)$$

Chapter 2    61

where the operators $\tilde{B}_i$ are given by (2.5.14). Equations (2.5.16) admit, of course, the separation of variables $s$ and $\tilde{s}$

$$\tilde{R}(s,\tilde{s},x) = \sum_{i,k=1}^{N} \rho_{ik} X_{0i}(s,x) Y_{0k}^{*}(\tilde{s},x) \qquad (2.5.17)$$

and the functions $X_{0i}$ and $Y_{0k}$ obey equations (2.5.10), $\rho_{ik}$ are arbitrary constants.

In the original spectral space $\lambda$ the operators $\hat{B}_1$, $\hat{B}_2$ and $\hat{B}_3$ are the convolutive integral operators which act as follows

$$(\hat{B}_1 f)(\lambda) = i\lambda f(\lambda),$$

$$(\hat{B}_2 f)(\lambda) = (B_2 * f)(\lambda) = \sum_{k=1}^{n} \iint_C d\mu \wedge d\bar{\mu}\, \hat{u}_k(\lambda - \mu)(i\mu)^k f(\mu), \qquad (2.5.18)$$

$$(\hat{B}_3 f)(\lambda) = (B_3 * f)(\lambda) = \sum_{k=1}^{m} \iint_C d\mu \wedge d\bar{\mu}\, \hat{v}_k(\lambda - \mu)(i\mu)^k f(\mu), \qquad (2.5.19)$$

where

$$\hat{u}_k(\lambda) \doteq \frac{1}{2\pi} \iint_C ds \wedge d\bar{s} \exp(-i(\lambda s + \bar{\lambda}\bar{s})) \tilde{u}_k(s) \qquad (2.5.20)$$

and similar for $\hat{v}_k(\lambda)$. Note that $\hat{u}_n(\lambda) \sim \delta(\lambda)$ and $\hat{v}_m(\lambda) \sim \delta(\lambda)$. So the operators $\hat{B}_2$ and $\hat{B}_3$ has the singularities $\lambda^n$ and $\lambda^m$ as $\lambda \to \infty$, in agreement with (2.5.6).

Now, using the operators $D_i = \partial_{x_i} + \hat{B}_i^*$ with $\hat{B}_1$, $\hat{B}_2$, $\hat{B}_3$ given by (2.5.18) – (2.5.20) we have to construct the operators $L_i$ which obey the conditions (2.3.15) and (2.3.16). One can use for this exactly the same procedure as it was described in section (2.4). It is not difficult to show, that there exist again the two desired operators $L_1$ and $L_2$ and they are of the form (2.4.4) and (2.4.6). But now $u_k$ and $v_k$ are, in general, the convolutive integral operators. They are expressed via the coefficients of the asymptotic expansion of $\chi$ as $\lambda \to \infty$ and the solutions $\tilde{u}_k$, $\tilde{v}_k$ of the system (2.5.10). So, by this procedure we get the solutions of some new nonlinear system on the background of nonlinear system associated with (2.5.10).

It occurs, due to the properties of the method, that this new nonlinear system coincides with the starting nonlinear system defined by (2.5.10). Thus, the $\bar{\partial}$-dressing method generalized as described above, allows us to construct the exact solutions of the nonlinear integrable equations on nontrivial background. For the first time such extension of the dressing method has been given in [148] with the use of the nonlocal Riemann—Hilbert problem.

For several nonlinear systems (DS-I, Ishimori-I and 2DISG-I equations) there exists very particular almost trivial but important backgrounds. Indeed, considering DS-I and Ishimori-I equations (1.2.7), (1.2.11) as the background equations, associated with (2.5.10), one can choose the solutions $q = 0$, $\Phi \neq 0$ and $\vec{s} = 0$, $\Phi \neq 0$ as the backgrounds. General

$\Phi$ in this case is $(\sigma = 1)$ $\Phi(x,y) = \alpha(x+y) + \beta(x-y)$ where $\alpha$ and $\beta$ are arbitrary functions. The functions $\alpha$ and $\beta$ are the boundary values of the function $\Phi(x,y,t)$ at the proper infinities. So, the generalized $\bar{\partial}$-dressing method described above also provides us the exact solutions of the DS-I, Ishimori-I and 2DISG-I equations with nontrivial boundaries. The exponentially localized solitons of these equations belong to such class of solutions.

In the next chapters we will consider the application of the $\bar{\partial}$-dressing method on nontrivial background for several 2+1-dimensional integrable equations. Further generalizations of the $\bar{\partial}$-dressing method will be discussed in the final chapter.

# Chapter 3

# The Kadomtsev–Petviashvili equation

We proceed now to the consideration of concrete 2+1-dimensional integrable equations. We start with the celebrated Kadomtsev—Petviashvili (KP) equation (1.2.2). This equation has been derived in [82] and it describes a slow variations in $y$ direction of the waves propagating along the $x$ direction. The KP equation is the most studied of nonlinear integrable equations in three independent variables $x$, $y$, $t$.

In the differential form the KP equation looks like

$$u_t + u_{xxx} + 6uu_x + 3\sigma^2 w_y = 0,$$

$$w_x = u_y \tag{3.0.1}$$

where $u$ and $w$ are scalar functions and $\sigma^2 = \pm 1$. The system (3.0.1) is equivalent to the compatibility condition for the following linear system [10, 83]

$$(\sigma \partial_y + \partial_x^2 + u)\psi = 0,$$

$$(\partial_t + 4\partial_x^3 + 6u\partial_x + 3u_x - 3\sigma w + \alpha)\psi = 0, \tag{3.0.2}$$

where $\alpha$ is an arbitrary constant. The function $w$ can of course be easily eliminated from the system (3.0.1). To do this one should fix the operator $\partial_x^{-1}$. For the class of solutions $u$ decaying as $x$, $y \to \infty$ we will choose $(\partial_x^{-1} f)(x) = \int_{-\infty}^{x} dx' f(x')$.

## 3.1. $\bar{\partial}$-dressing

At first we will consider the $\bar{\partial}$-dressing on trivial background.

The KP equation belongs to the class of scalar integrable equations described in the first example in section (2.4). It is associated with the scalar $\bar{\partial}$-problem (2.3.1) with [108, 109]

$$\hat{B}_1 = i\lambda, \quad \hat{B}_2 = \frac{1}{\sigma}\lambda^2, \quad \hat{B}_3 = 4i\lambda^3. \tag{3.1.1}$$

The operators $D_i$ ($i = 1, 2, 3$) are of the form

$$D_1 = \partial_x + i\lambda, \quad D_2 = \partial_y + \frac{1}{\sigma}\lambda^2, \quad D_3 = \partial_t + 4i\lambda^3 \qquad (3.1.2)$$

and the general kernel $R$ of the $\bar{\partial}$-problem is

$$R(\mu, \bar{\mu}; \lambda, \bar{\lambda}; x, y, t) = e^{F(\mu)} R_0(\mu, \bar{\mu}; \lambda, \bar{\lambda}) e^{-F(\lambda)} \qquad (3.1.3)$$

where $R_0$ is an arbitrary function and

$$F(\lambda) = i\lambda x + \frac{1}{\sigma}\lambda^2 y + 4i\lambda^3 t. \qquad (3.1.4)$$

Following the general scheme described in section (2.4) we proceed to the derivation of the basis equations (2.3.18) or (2.4.4) and (2.4.6). The normalization of the solution $\chi$ of the $\bar{\partial}$-problem is assumed to be canonical:

$$\chi = 1 + \frac{1}{\lambda}\chi_{-1} + \frac{1}{\lambda^2}\chi_{-2} + \ldots \quad (\lambda \to \infty). \qquad (3.1.5)$$

The quantity $\sigma D_2 \chi$ has the second order pole as $\lambda \to \infty$. In virtue of (3.1.2) this singularity is obviously absent in the quantity

$$\sigma D_2 \chi + D_1^2 \chi = \sigma \partial_y \chi + \partial_x^2 \chi + 2i\lambda \partial_x \chi. \qquad (3.1.6)$$

Moreover, due to (3.1.5), the quantity (3.1.6) has no singularity as $\lambda \to \infty$ at all, namely

$$\sigma D_2 \chi + D_1^2 \chi \underset{\lambda \to \infty}{-} 2i\partial_x \chi_{-1}. \qquad (3.1.7)$$

Using (3.1.7), one readily concludes that the operator $L_1$ defined by

$$L_1 \chi = (\sigma D_2 + D_1^2 + u)\chi = 0 \qquad (3.1.8)$$

with

$$u(x, y, t) = -2i\partial_x \chi_{-1}(x, y, t) \qquad (3.1.9)$$

obeys the conditions (2.3.15), (2.3.16).

So, equation (3.1.8) is our first desired linear problem.

Substitution of the asymptotic expansion (3.1.5) into equation (3.1.8) also gives

$$\sigma \partial_y \chi_{-1} + \partial_x^2 \chi_{-1} + 2i\partial_x \chi_{-2} + u\chi_{-1} = 0, \qquad (3.1.10)$$

$$\sigma \partial_y \chi_{-2} + \partial_x^2 \chi_{-2} + 2i\partial_x \chi_{-3} + u\chi_{-2} = 0. \qquad (3.1.11)$$

In order to construct the second linear problem one should use the operator $D_3$. The third order singularity in $D_3 \chi$ is obviously absent in the function

$$D_3 \chi + 4D_1^3 \chi = \partial_t \chi + \partial_x^3 \chi + 12i\lambda \partial_x^2 \chi - 12\lambda^2 \partial_x \chi. \qquad (3.1.12)$$

Note that due to (3.1.5) the function (3.1.12) also has no second order singularity as $\lambda \to \infty$. One needs only to annihilate in (3.1.12) the first order singularity. This can be reached by adding the term $v_1 D_1 \chi$ to (3.1.12). The condition of the absence of the first order singularity in the quantity

$$D_3\chi + 4D_1^3\chi + v_1 D_1\chi = \partial_t\chi + \partial_x^3\chi + 12i\lambda\partial_x^2\chi - 12\lambda^2\partial_x\chi + v_1\partial_x\chi + i\lambda v_1\chi \quad (3.1.13)$$

implies

$$iv_1 - 12\partial_x\chi_{-1} = 0 \quad (3.1.14)$$

or, due to (3.1.9),

$$v_1 = 6u. \quad (3.1.15)$$

So, the quantity $D_3\chi + 4D_1^3\chi + 6uD_1\chi$ obeys the condition (2.3.15). In order to satisfy the condition (2.3.16) one should add the term $v_0\chi$ with the proper $v_0$. The condition (2.3.16), i. e.

$$D_3\chi + 4D_1^3\chi + 6uD_1\chi + v_0\chi \to 0 \quad (3.1.16)$$

as $\lambda \to \infty$ gives

$$12i\partial_x^2\chi_{-1} - 12\partial_x\chi_{-2} + 6iu\chi_{-1} + v_0 = 0. \quad (3.1.17)$$

Eliminating $\chi_{-2}$ from (3.1.17) with the use of equation (3.1.10), one gets

$$v_0 = -6i\partial_x^2\chi_{-1} + 6i\sigma\partial_y\chi_{-1}. \quad (3.1.18)$$

Denoting

$$w = -2i\partial_y\chi_{-1}, \quad (3.1.19)$$

one has

$$v_0 = 3u_x - 3\sigma w. \quad (3.1.20)$$

Thus, the second linear problem is of the form

$$(D_3^2 + 4D_1^3 + 6uD_1 + 3u_x - 3\sigma w)\chi = 0, \quad (3.1.21)$$

where

$$w_x = u_y.$$

The linear problems (3.1.8) and (3.1.21) are the desired linear problems which form the basis of linear problems (2.3.18) in the case (3.1.1).

One can also construct another second linear problem. Indeed, starting with the quantity $D_3\chi - 4\sigma D_2 D_1\chi$ which also has no third order singularity as $\lambda \to \infty$ and repeating the above procedure, one obtains the equation

$$D_3\chi - 4\sigma D_2 D_1\chi + 2uD_1\chi - (u_x + 3\sigma w)\chi = 0. \quad (3.1.22)$$

Equation (3.1.22) together with (3.1.8) also form a basis. But it is easy to show that equations (3.1.22) and (3.1.21) are modulo equivalent to equation (3.1.8). In what follows we will consider equations (3.1.8) and (3.1.21).

The linear system (3.1.8) and (3.1.21) strictly implies the KP equation. Indeed, substituting the asymptotic expansion (3.1.5) into equation (3.1.21) and considering the terms of order $\lambda^{-1}$ one gets

$$\partial_t \chi_{-1} + \partial_x^3 \chi_{-1} + 12i\partial_x^2 \chi_{-2} - 12\partial_x \chi_{-3} + 6u\partial_x \chi_{-1} + 6iu\chi_{-2} + (3u_x + 3\sigma w)\chi_{-1} = 0. \quad (3.1.23)$$

Then, using (3.1.9)–(3.1.11) and (3.1.19), one obtains from (3.1.23) the system (3.0.1). The same procedure for equation (3.1.22) leads, of course, to the same system (3.0.1). Note that within this construction the variable $\chi_{-1}$ is the most natural one. In virtue of (3.1.9) $\chi_{-1}$ is the potential for $u$ and equation (3.1.23) is the potential KP equation.

Further it is easy to show that according to the general scheme the transition to the function

$$\psi = \chi \exp\left(i\lambda x + \frac{1}{\sigma}\lambda^2 y + 4i\lambda^3 t\right) \quad (3.1.24)$$

converts the linear problems (3.1.8) and (3.1.21) into the more usual form (3.0.2) with $\alpha = 0$.

Note that if one introduces another function $\tilde{\psi}$ by

$$\tilde{\psi} = \chi \exp\left(i\lambda x + \frac{1}{\sigma}\lambda^2 y\right) \quad (3.1.25)$$

then it obeys the linear system

$$(\sigma \partial_y + \partial_x^2 + u)\tilde{\psi} = 0,$$

$$(\sigma \partial_t + 4\partial_x^3 + 6u\partial_x + 3u_x - 3\sigma w + 4i\lambda^3)\tilde{\psi} = 0 \quad (3.1.26)$$

i. e. the system (3.0.2) with $\alpha = 4i\lambda^3$. The system (3.1.26) is convenient for the analysis of the initial value problem for the KP equation.

Now let us proceed to the $\bar{\partial}$-dressing on nontrivial background.

According to the general scheme presented in section (2.5) one should choose the operators $\hat{B}_i$ as the convolutive operators such that

$$\hat{B}_1(\lambda) \to i\lambda, \quad \hat{B}_2(\lambda) \to \frac{1}{\sigma}\lambda^2, \quad \hat{B}_3(\lambda) \to 4i\lambda^3 \quad (3.1.27)$$

as $\lambda \to \infty$. In the dual space the operators $\tilde{B}_1$, $\tilde{B}_2$, $\tilde{B}_3$ given by (2.5.14) are the differential operators of the first, second and third order in $\partial_s$, respectively.

The system (2.5.10) with $n = 2$, $m = 3$ is the basic system for constructing the background. Choosing

$$\tilde{u}_2 = -\frac{1}{\sigma}, \quad \tilde{u}_1 = 0, \quad \tilde{u}_0 = -\frac{u_0}{\sigma}$$

$$\tilde{v}_3 = -4, \quad \tilde{v}_2 = 0, \quad \tilde{v}_1 = -6u_0, \quad \tilde{v}_0 = -3u_{0z} + 3\sigma w_0 \tag{3.1.28}$$

one has the system ($x_2 = \frac{1}{\sigma}y$, $x_3 = t$)

$$\left(\partial_y + \frac{1}{\sigma}\partial_z^2 + \frac{1}{\sigma}u_0(z,y,t)\right)X(z,y,t) = 0, \tag{3.1.29}$$

$$(\partial_t + 4\partial_z^3 + 6u_0\partial_z + 3u_{0z} - 3\sigma w_0)X(z,y,t) = 0$$

where $w_{0z}(z,y,t) = u_{0y}(z,y,t)$.

The system (3.1.29) is nothing but the linear system (3.0.2) for the KP equation in the variables $z$, $y$, $t$. Thus, $u_0(z,y,t)$ is the solution of the KP equation and $X(z,y,t)$ is the corresponding wave function. The function $u_0(s+x,y,t)$ will be the background for the further dressing.

In the space of original spectral variable $\lambda$ the operators $\hat{B}_i$ act as follows:

$$\hat{B}_1 = i\lambda,$$

$$(\hat{B}_2 f)(\lambda) = -\frac{(i\lambda)^2}{\sigma}f(\lambda) - \frac{1}{\sigma}\iint_C d\mu \wedge d\bar{\mu}\, \hat{u}_0(\lambda - \mu, x, y, t)f(\mu), \tag{3.1.30}$$

$$(\hat{B}_3 f)(\lambda) = -4(i\lambda)^3 f(\lambda) - 6\iint_C d\mu \wedge d\bar{\mu}\, \hat{u}_0(\lambda - \mu, x, y, t)i\mu f(\mu)$$

$$- \iint_C d\mu \wedge d\bar{\mu}\, (3\hat{u}_0(\lambda - \mu, x, y, t)i(\lambda - \mu) - 3\sigma \hat{w}_0(\lambda - \mu, x, y, t))f(\mu)$$

where

$$\hat{u}_0(\lambda, x, y, t) = \frac{1}{2\pi}\iint_S ds \wedge d\bar{s}\, e^{-i(\lambda s + \bar{\lambda}\bar{s})}u_0(s + x, y, t) \tag{3.1.31}$$

and

$$i\lambda \hat{w}_0(\lambda, x, y, t) = \hat{u}_{0y}(\lambda, x, y, t). \tag{3.1.32}$$

Hence, the operators $D_i = \partial_{x_i} + B_i^*$ ($i = 1, 2, 3$) with the nontrivial background act as

$$(D_1\chi)(\lambda) = (\partial_x + i\lambda)\chi(\lambda),$$

$$(D_2\chi)(\lambda) = (\partial_y + \frac{\lambda^2}{\sigma})\chi(\lambda) - \frac{1}{\sigma}\iint_C d\mu \wedge d\bar{\mu}\, \hat{u}_0(\mu - \lambda, x, y, t)\chi(\mu), \tag{3.1.33}$$

$$(D_3\chi)(\lambda) = (\partial_t + 4i\lambda^3)\chi(\lambda) - 3i\lambda \iint_C d\mu \wedge d\bar{\mu}\, \hat{u}_0(\mu - \lambda, x, y, t)\chi(\mu)$$

$$-3\iint_C d\mu \wedge d\bar\mu\,(i\mu\hat u_0(\mu-\lambda,x,y,t)-\sigma\hat w_0(\mu-\lambda,x,y,t))\chi(\mu).$$

Using the obvious identity

$$\iint_C d\mu \wedge d\bar\mu\, u_0(\mu-\lambda)\chi(\lambda) = \iint_C d\mu \wedge d\bar\mu\, u_0(\mu)\chi(\lambda+\mu), \tag{3.1.34}$$

one can represent the operators $D_i$ in the form more convenient for further construction:

$$(D_1\chi)(\lambda) = (\partial_x + i\lambda)\chi(\lambda), \tag{3.1.35a}$$

$$(D_2\chi)(\lambda) = \left(\partial_y + \frac{\lambda^2}{\sigma}\right)\chi(\lambda) - \frac{1}{\sigma}\iint_C d\mu \wedge d\bar\mu\,\hat u_0(\mu,x,y,t)\chi(\lambda+\mu), \tag{3.1.35b}$$

$$(D_3\chi)(\lambda) = (\partial_t + 4i\lambda^3)\chi(\lambda) - 6i\lambda\iint_C d\mu \wedge d\bar\mu\,\hat u_0(\mu,x,y,t)\chi(\lambda+\mu)$$

$$-3\iint_C d\mu \wedge d\bar\mu\,(i\mu\hat u_0(\mu,x,y,t)-\sigma\hat w_0(\mu,x,y,t))\chi(\lambda+\mu). \tag{3.1.35c}$$

Now using these operators $D_i$ one has to construct the operators $L_1$ and $L_2$. Since the most singular parts in $D_i$ ($i=1,2,3$) are the same as those for the trivial background $u_0 = 0$, the new operators $L_1$ and $L_2$ will have the same order and, in general, a similar structure.

Let us start again with the quantity

$$\sigma D_2\chi + D_1^2\chi = \sigma\partial_y\chi + \partial_x^2\chi + 2i\lambda\partial_x\chi - \iint_C d\mu \wedge d\bar\mu\,\hat u_0(\mu,x,y,t)\chi(\lambda+\mu). \tag{3.1.36}$$

It again has no second and first order singularities as $\lambda\to\infty$. Here we assume that the functions $u_0(s,x,y,t)$ and $\hat u_0(\lambda,x,y,t)$ are smooth and bounded. So all convolutive integrals are bounded. Thus, the function

$$\sigma D_2\chi + D_1^2\chi + u(x,y,t)\chi \tag{3.1.37}$$

with the proper choice of $u$ obeys the condition (2.3.16) (i. e. it tends to zero as $\lambda\to\infty$). Using (3.1.35), (3.1.36), (3.1.37), one finds that such $u(x,y,t)$ is given by

$$u(x,y,t) = -2i\partial_x\chi_{-1} + \iint_C d\mu \wedge d\bar\mu\,\hat u_0(\mu,x,y,t). \tag{3.1.38}$$

Taking into account (3.1.31), one finally gets

$$u(x,y,t) = u_0(x,y,t) - 2i\partial_x\chi_{-1}. \tag{3.1.39}$$

Thus, the first linear problem is

$$\sigma D_2\chi + D_1^2\chi + u\chi = 0, \tag{3.1.40}$$

where $u$ is given by (3.1.39) or, equivalently,

$$\sigma\partial_y\chi + \partial_x^2\chi + 2i\lambda\partial_x\chi + u\chi - (\hat{T}\chi) = 0, \qquad (3.1.41)$$

where

$$(\hat{T}\chi)(\lambda) = \frac{1}{2\pi}\iint_C ds\wedge d\bar{s}\iint_C d\mu\wedge d\bar{\mu}\,\exp(-i(\mu s + \bar{\mu}\bar{s}))u_0(s+x,y,t)\chi(\lambda+\mu,x,y,t). \qquad (3.1.42)$$

Equation (3.1.41), in addition to (3.1.39), implies

$$\sigma\partial_y\chi_{-1} + \partial_x^2\chi_{-1} + 2i\partial_x\chi_{-2} + (u-u_0)\chi_{-1} = 0 \qquad (3.1.43)$$

$$\sigma\partial_y\chi_{-2} + \partial_x^2\chi_{-2} + 2i\partial_x\chi_{-3} + (u-u_0)\chi_{-2} - iu_{0x}\chi_{-1} = 0. \qquad (3.1.44)$$

The second linear problem is constructed similar to the previous case. The quantity $D_3\chi + 4D_1^3\chi$ again has no third and second order singularities as $\lambda \to \infty$. So it is rather obvious that the desired second linear problem is of the form

$$D_3\chi + 4D_1^3\chi + \hat{v}_1 D_1\chi + \hat{v}_0\chi = 0, \qquad (3.1.45)$$

where $\hat{v}_1$ and $\hat{v}_0$ are some convolutive operators:

$$(\hat{v}_i f)(\lambda) = \iint_C d\mu\wedge d\bar{\mu}\,v_i(\mu)f(\lambda+\mu) \quad (i=0,1). \qquad (3.1.46)$$

The requirement of the absence of the first order singularity as $\lambda \to \infty$ and the fulfillment of condition (2.3.16) with the use of (3.1.39), (3.1.44) give

$$\hat{v}_1(x,y,t) = \iint_C d\mu\wedge d\bar{\mu}\,v_1(\mu,x,y,t) = 6u(x,y,t), \qquad (3.1.47)$$

$$\hat{v}_0(x,y,t) = \iint_C d\mu\wedge d\bar{\mu}\,v_0(\mu,x,y,t) = 3u_x(x,y,t) - 3\sigma w_0, \qquad (3.1.48)$$

where

$$w = w_0 - 2i\partial_y\chi_{-1} \qquad (3.1.49)$$

or

$$w_x = u_y. \qquad (3.1.50)$$

The formulae (3.1.47) and (3.1.48) implies

$$v_1(\mu) = \frac{1}{2\pi}\iint_C ds\wedge d\bar{s}\,\exp(-i(\mu s + \bar{\mu}\bar{s}))6u(x,y,t) = 6u(x,y,t), \qquad (3.1.51)$$

$$v_0(\mu) = \frac{1}{2\pi}\iint_C ds\wedge d\bar{s}\,\exp(-i(\mu s+\bar{\mu}\bar{s}))\{3u_x(x,y,t)-3\sigma w(x,y,t)\} = 3u_x - 3\sigma w. \qquad (3.1.52)$$

Thus, the second linear problem is

$$\partial_t\chi + 4\partial_x^3\chi + 12i\lambda\partial_x^2\chi - 12\lambda^2\partial_x\chi + 6u(x,y,t)\partial_x\chi$$

$$+i\lambda u\chi + (3u_x - 3\sigma w) - 6i\lambda(\hat{T}\chi) - 3(\hat{F}\chi) = 0, \qquad (3.1.53)$$

where $\hat{T}\chi$ is given by (3.1.42) and

$$(\hat{F}\chi)(\lambda) = \frac{1}{2\pi}\iint_C ds \wedge d\bar{s} \iint_C d\mu \wedge d\bar{\mu}\, \exp(-i(\mu s + \bar{\mu}\bar{s}))$$

$$+\{i\mu u_0(s+x,y,t) - \sigma w_0(s+x,y,t)\}\chi(\lambda+\mu,x,y,t) \qquad (3.1.54)$$

and the function $w$ is related to $u$ by equation (3.1.50).

So, the linear problems (3.1.41) and (3.1.53) are desired linear problems in the case of nontrivial background. Emphasize that in terms of the operators $D_1$, $D_2$, $D_3$ they are, due to (3.1.51) and (3.1.52), exactly of the same form of (3.1.40) and (3.1.45) as those for trivial background. Transition from trivial to nontrivial background is reflected only in the form of the commuting operators $D_i$ ($i = 1, 2, 3$). This is an important feature of the $\bar{\partial}$-dressing method on nontrivial background. The linear problems (3.1.41), (3.1.53) are the integral equations in the space of spectral parameter $\lambda$. Note that in the dual space $s$ they are the purely differential equations in $s, x, y, t$.

Now, one can derive the nonlinear equation associated with the linear problems (3.1.41) and (3.1.53) by substituting the asymptotic expansion (3.1.5) into the problem (3.1.53) and considering the terms of the order $\lambda^{-1}$. One gets

$$\partial_t\chi_{-1} + 4\partial_x^3\chi_{-1} + 12i\partial_x^2\chi_{-2} - 12\partial_x\chi_{-3} + 6u\partial_x\chi_{-1}$$

$$+6i(u-u_0)\chi_{-2} + 3(u_x + u_{0x} - \sigma w + \sigma w_0)\chi_{-1} = 0. \qquad (3.1.55)$$

Eliminating $\chi_{-3}$ and $\chi_{-2}$ with the use of (3.1.43), (3.1.44), and differentiating with respect to $x$ and taking into account that $u_0$ obeys the KP equation, one obtains the function $u(x,y,t)$ obeys the KP equation too.

The result that the $\bar{\partial}$-dressing on nontrivial background leads to the KP equation is also an easy consequence of the form of the linear problems. Indeed, the linear problems (3.1.40) and (3.1.45) with $\tilde{v}_0$ and $\tilde{v}_1$ given by (3.1.51), (3.1.52) are exactly the same as the problems (3.1.8) and (3.1.21) for the trivial background case. Then, the operators $D_i$ ($i = 1, 2, 3$) (3.1.35) commute with each other by construction. Hence, the compatibility condition for the system (3.1.40), (3.1.45) implies the KP equation for $u(x,y,t)$.

Thus, starting with the background $u_0(z,y,t)$ which obey the KP equation, the $\bar{\partial}$-dressing provides us the other solution of the KP equation given by the formula (3.1.39).

For the first time such construction of the solutions of the KP equation on nontrivial background has been proposed in paper [148] based on the nonlocal Riemann—Hilbert problem and in slightly different form. In fact, if we want to start with the background $u_0(s + x, y, t)$ defined for real $s$ then all the formulae derived above clearly indicate that the reduction to real-valued $\lambda$ is required. This forces us to reduce the nonlocal $\bar{\partial}$-problem to the nonlocal Riemann—Hilbert problem with the jump across the real axis.

## 3.2. Exact solutions

The results of the previous section allow us to solve the KP equation by the $\bar{\partial}$-dressing method. Solutions of the KP equation are given by the formula (3.1.39) where $u_0(x, y, t)$ is an arbitrary background and $\chi$ is the solution of the $\bar{\partial}$-problem with the kernel $R$ of the form (2.5.17).

First we will consider the case of trivial background. In virtue of (3.1.5) and (2.2.3) one has

$$u(x, y, t) = \frac{1}{\pi} \frac{\partial}{\partial x} \iint_C d\lambda \wedge d\bar{\lambda} \iint_C d\mu \wedge d\bar{\mu}\, \chi(\mu, \bar{\mu}) R(\mu, \bar{\mu}; \lambda, \bar{\lambda}), \quad (3.2.1)$$

where

$$R(\mu, \bar{\mu}; \lambda, \bar{\lambda}) = e^{F(\mu, x, y, t)} R_0(\mu, \bar{\mu}; \lambda, \bar{\lambda}) e^{-F(\lambda, x, y, t)} \quad (3.2.2)$$

and

$$F = i\lambda x + \frac{\lambda^2}{\sigma} y + 4i\lambda^3 t. \quad (3.2.3)$$

For small $R$ one has $\chi \sim 1$ and hence

$$u = \frac{1}{\pi} \frac{\partial}{\partial x} \iint_C d\lambda \wedge d\bar{\lambda} \iint_C d\mu \wedge d\bar{\mu}\, R_0(\mu, \bar{\mu}; \lambda, \bar{\lambda})$$
$$\times \exp\left(i(\mu - \lambda)x + \frac{1}{\sigma}(\mu^2 - \lambda^2)y + 4i(\mu^3 - \lambda^3)t\right). \quad (3.2.4)$$

This formula allows us to find the necessary conditions on $R_0$ in order for $u$ to be real. These conditions are different for the cases $\sigma = i$ and $\sigma = 1$. For the KP-I equation ($\sigma = i$) this condition is of the form

$$\overline{R_0(\mu, \bar{\mu}; \lambda, \bar{\lambda})} = R_0(\bar{\lambda}, \lambda; \bar{\mu}, \mu) \quad (3.2.5)$$

while for the KP-II equation ($\sigma = 1$) one has

$$\overline{R_0(\mu, \bar{\mu}; \lambda, \bar{\lambda})} = R_0(-\bar{\mu}, -\mu; -\bar{\lambda}, -\lambda). \quad (3.2.6)$$

According to sections (2.2) and (2.3) a wide class of explicit exact solutions correspond to the general degenerate kernel $R_0(\mu, \bar{\mu}; \lambda, \bar{\lambda})$:

$$R_0(\mu, \bar{\mu}; \lambda, \bar{\lambda}) = \sum_{k=1}^{N} f_{0k}(\mu, \bar{\mu}) g_{0k}(\lambda, \bar{\lambda}) \quad (3.2.7)$$

where $f_{0k}$, $g_{0k}$ are arbitrary functions and $N$ is an arbitrary integer. Using the formulae (3.1.39), (2.2.22), (2.2.21) and (2.2.20), one gets the corresponding solutions of the KP equation

$$u(x,y,t) = \frac{1}{\pi}\partial_x \left( \sum_{l,k=1}^{N} \xi_k (1+A)^{-1}_{kl} \eta_l \right). \tag{3.2.8}$$

where

$$\xi_k(x,y,t) = \iint_C d\lambda \wedge d\bar{\lambda}\, f_{0k}(\lambda, \bar{\lambda}) \exp\left(i\lambda x + \frac{1}{\sigma}\lambda^2 y + 4i\lambda^3 t\right) \tag{3.2.9}$$

$$\eta_l(x,y,t) = \iint_C d\lambda \wedge d\bar{\lambda}\, g_{0k}(\lambda, \bar{\lambda}) \exp\left(-i\lambda x - \frac{1}{\sigma}\lambda^2 y - 4i\lambda^3 t\right) \tag{3.2.10}$$

and

$$A_{lk}(x,y,t) = \frac{1}{2\pi i} \iint_C d\mu \wedge d\bar{\mu} \iint_C d\lambda \wedge d\bar{\lambda}\, \frac{g_{0k}(\mu,\bar{\mu}) f_{0k}(\lambda,\bar{\lambda})}{\lambda - \mu}$$

$$\times \exp\left(i(\lambda-\mu)x + \frac{1}{\sigma}(\lambda^2 - \mu^2)y + \varphi_i(\lambda^3 - \mu^3)t\right). \tag{3.2.11}$$

The solutions (3.2.8) of the KP equation are parametrized by $2N$ arbitrary functions of the two independent variables. The r.h.s. of (3.2.8) can be rewritten in a more compact form. Using the formal identity

$$\frac{1}{i\lambda}e^{i\lambda x} = \int^x dx' e^{i\lambda x'}, \tag{3.2.12}$$

the definitions (3.2.9), (3.2.10) and performing the standard transformations in (3.2.8), one gets

$$u(x,y,t) = 2\partial_x^2 \ln \det(1+A), \tag{3.2.13}$$

where

$$A_{kl}(x,y,t) = \frac{1}{2\pi} \partial_x^{-1}(\xi_k \eta_l) \tag{3.2.14}$$

and $\xi_k(x,y,t)$, $\eta_l(x,y,t)$ are given by (3.2.9), (3.2.10). The integration $\partial_x^{-1}$ in (3.2.14) is chosen in a way which guarantee the existence of $A_{kl}(x,y,t)$. For the first time the formula (3.2.13) has been obtained by the original Zakharov—Shabat dressing method (see e. g. [38]).

It is easy to see that the functions $\xi_k$ and $\eta_k$ obey the simple systems of equations

$$(\sigma \partial_y + \partial_x^2)\xi = 0,$$

$$(\partial_t + 4\partial_x^3)\xi = 0 \tag{3.2.15}$$

and

$$(\sigma \partial_y - \partial_x^2)\eta = 0,$$

$$(\partial_t + 4\partial_x^3)\eta = 0. \tag{3.2.16}$$

Consequently $\xi$ and $\eta$ obey also the equation

$$(\partial_t + \partial_x^3 + 3\sigma^2 \partial_x^{-1}\partial_y^2)\xi = 0$$

and the same for $\eta$, which is nothing but the linear part of the KP equation.

So the exact solutions (3.2.13) are parametrized by the general solution of the systems (3.2.15) and (3.2.16) or by arbitrary but specific solutions of the linear part of the KP equation.

In general, the solutions (3.2.13) are complex-valued. The reality conditions (3.2.5) and (3.2.6) are satisfied of

$$\overline{g_{0k}(\lambda,\bar{\lambda})} = f_{0k}(\bar{\lambda},\lambda) \tag{3.2.17}$$

for the KP-I equation and

$$\overline{f_{0k}(\lambda,\bar{\lambda})} = f_{0k}(-\lambda,-\bar{\lambda}),$$

$$\overline{g_{0k}(\lambda,\bar{\lambda})} = g_{0k}(-\lambda,-\bar{\lambda}) \tag{3.2.18}$$

for the KP-II equation.

For the KP-I equation the condition (3.2.17) gives

$$\eta_k(x,y,t) = \overline{\xi_k(x,y,t)}. \tag{3.2.19}$$

Hence, the real valued solutions with functional parameters of the KP-I equation is of the form (3.2.12) with

$$A_{kl} = \frac{1}{2\pi}\partial_x^{-1}(\xi_k \bar{\xi}_l). \tag{3.2.20}$$

The simplest solution of the KP-I equation of this type looks like

$$u(x,y,t) = 2\partial_x^2 \ln\left(1 + \frac{1}{2\pi}\int_{-\infty}^x dx' |\xi(x',y,t)|^2\right), \tag{3.2.21}$$

where $\xi(x,y,t)$ is an arbitrary function of the form (3.2.9).

Now let us proceed to the line solutions. They correspond to the choice of the functions $f_{0k}$, $g_{0k}$ as the delta-functions (2.2.23). In virtue of the reality conditions (3.2.17) and (3.2.18) one has for the KP-I equation

$$R_0(\lambda',\bar{\lambda}';\lambda,\bar{\lambda}) = \sum_{k=1}^{N} |f_{0k}|^2 \delta(\lambda' - \lambda_k)\delta(\lambda - \bar{\lambda}_k), \tag{3.2.22}$$

where $f_{0k}$ and $\lambda_k$ are arbitrary complex constants. For the KP-II equation, respectively,

$$R_0(\lambda',\bar{\lambda}';\lambda,\bar{\lambda}) = \sum_{k=1}^{n} f_{0k}\, g_{0k}\delta(\lambda' + i\lambda_k)\delta(\lambda - i\mu_k), \tag{3.2.23}$$

where $f_{0k}$, $g_{0k}$, $\lambda_k$ and $\mu_k$ are arbitrary real constants.

As a result, the line multisolitons of the KP-I equation are given by the formula (3.2.13) where (see e. g. [38])

$$A_{lk} = \frac{1}{\pi} \frac{2 f_{0k} \bar{f}_{0l}}{i(\lambda_k - \bar{\lambda}_l)} \exp(i(\lambda_k - \bar{\lambda}_l)x - i(\lambda_k^2 - \bar{\lambda}_l^2)y + 4i(\lambda_k^3 - \bar{\lambda}_l^3)t). \quad (3.2.24)$$

The simplest solution of the KP-I equation of such type is of the form

$$u(x,y,t) = 2\rho^2 \cosh^{-2}\{\rho(x + 2\nu y + 2(3\nu^2 - \rho^2)t - x_0)\}, \quad (3.2.25)$$

where

$$x_0 = -\frac{1}{\rho} \ln \frac{|f_{01}|^2}{|\rho|}, \quad \lambda_1 = \nu + i\rho, \quad \operatorname{Im}\nu = \operatorname{Im}\rho = 0, \quad \rho < 0.$$

An interesting solution of the KP-I equation is given by the formula (3.2.21), with the function $\xi$ of the form

$$\xi(x,y,t;k) = C e^{kx + ik^2 y - 4k^3 t} + \int_0^k dk' e^{k'x + ik'^2 y - 4k'^3 t}.$$

It describes the annihilation of the line soliton. Indeed, as $t \to -\infty$ this solution coincides with the usual soliton (3.2.25) with $\nu = 0$, but as $t \to +\infty$ the soliton is absent [87, 38]. This solution illustrates the well-known instability of the plane solitons for the KP-I equation [87, 38].

Other interesting real solution of the KP-I equation corresponds to the case $n = 2$ and $R_0 = \sum_{k=1}^{2} R_k \delta(\lambda' - \mu_k)\delta(\lambda - \eta_k)$ where $\mu_1 = -\eta_2 = i\nu_1$ and $\eta_2 = \mu_1 = i\nu_2$, $\operatorname{Im}\nu_1 = \operatorname{Im}\nu_2 = 0$. In this case [109]

$$u(x,y,t) = 2 \frac{\partial^2}{\partial x^2} \ln\left(1 + ae^{-\delta(x+\nu t)} \cos \delta\nu y + \frac{a^2 e^{-2\delta x}}{4(\nu^2 - \delta^2)}\right),$$

where $\delta = \nu_1 - \nu_2$, $\nu = \nu_1 + \nu_2$. This solution is periodic in $y$ and decreases as $|x| \to \infty$.

Bounded line soliton solutions of the KP-II equation correspond to the kernels $R_0$ (3.2.23) with $\lambda_k, \mu_k > 0$. They are of the form (3.2.13) with a matrix $A$ given by

$$A_{kl} = \frac{f_{0k} g_{0l} \exp\left((\lambda_k + \mu_l)x - (\lambda_k^2 - \mu_l^2)y - 4(\lambda_k^3 + \mu_l^3)t\right)}{\lambda_k + \mu_l}. \quad (3.2.26)$$

The one-soliton solution is (see e. g. [38])

$$u(x,y,t) = \frac{(\lambda+\mu)^2}{2} \cosh^{-2}\left\{\frac{(\lambda+\mu)x - (\lambda^2 - \mu^2)y - 4(\lambda^3 + \mu^3)t + \alpha}{2}\right\} \quad (3.2.27)$$

where $\alpha = \ln \frac{f_{0k} g_{0l}}{\lambda + \mu}$. In the case $\lambda = \mu$ the solution (3.2.27) is nothing but the KdV soliton. At $\lambda \neq \mu$ the soliton (3.2.27) propagates under some angle with respect to the $x$-axis.

Chapter 3   75

General solution (3.2.13), (3.2.26) describes the intersection of $n$ such "skew" solitons and does not decrease along the directions $x/y = \lambda_k - \mu_k$ ($k = 1,\ldots,n$) as $x,y \to \infty$.

The general line-soliton solutions of the KP-II equation can be represented in the form (3.2.12) with another matrix $A$ [114]

$$A_{kl} = \frac{f_{0k} f_{0l} \exp\left((\lambda_k + \lambda_l)x - (\lambda_k^2 - \mu_k^2)y - 4(\lambda_k^3 + \mu_k^3)t\right)}{\lambda_k + \mu_l}. \qquad (3.2.28)$$

The solution (3.2.13), (3.2.26) describes the elastic scattering of $n$ line solitons (3.2.27). The corresponding phase shift can be found in a closed form.

Further, we will consider the solution of the KP-II equation which is close to the soliton (3.2.27). This solution corresponds to the case $n = 1$ where the functions $f_1(\lambda, \bar{\lambda})$ and $g_1(\lambda, \bar{\lambda})$ are of the form [162]

$$f_1(\lambda, \bar{\lambda}) = \frac{A}{2} \theta(i\lambda) \delta(\lambda + \bar{\lambda})[\mathrm{sgn}(-i\lambda - (x_1 - a)) - \mathrm{sgn}(-i\lambda - (\lambda_1 + a))],$$

$$g_1(\lambda, \bar{\lambda}) = \frac{A}{2} \theta(-i\lambda) \delta(\lambda + \bar{\lambda})[\mathrm{sgn}(-i\lambda - (\lambda_1 - a)) - \mathrm{sgn}(-i\lambda - (\lambda_1 + a))] \qquad (3.2.29)$$

with $a \ll \lambda_1$, i. e. these functions $f_1(\lambda, \bar{\lambda})$ and $g_1(\lambda, \bar{\lambda})$ are concentrated on the narrow supports near the point $\lambda_1$. This solution is [162]

$$u(x,y,t) = 2\frac{\partial^2}{\partial x^2} \ln \Delta, \qquad (3.2.30)$$

where

$$\Delta = 1 + 8A^2 \lambda_1 e^{-4\lambda_1(x - 8\lambda_1^2 t)} \Phi(x + 2\lambda_1 y - 12\lambda_1^2 t)\Phi(x - 2\lambda_1 y - 12\lambda_1^2 t) \qquad (3.2.31)$$

and $\Phi(\xi) = \sinh a\xi/\xi$. The solution (3.2.30) has rather interesting asymptotics. At $2\lambda_1 y \gg 8\lambda_1^2 t$ we have

$$\Delta = 1 + \frac{A^2}{y^2} e^{-2\lambda_1(x + 2\lambda_1 a y)}.$$

This asymptotics corresponds to the weakly curved soliton with the amplitude $\lambda_1^2$ the top of which is disposed on the line $x = -2\lambda_1 a y - \frac{1}{2\lambda_1} \ln \frac{A^2}{y^2}$. In the asymptotic region $y \ll 4\lambda_1 t$ one can neglect the $y$-dependence. As a result, in this region, one has

$$\delta \sim 1 + \frac{A^2}{16\lambda_1^4 t^2} e^{-2[(\lambda_1 - a)(x - 4\lambda_1^2 t) + 8\lambda_1^2 a t]}. \qquad (3.2.32)$$

The expression (3.2.32) corresponds to the "erected" soliton with the decreased amplitude $2(\lambda_1 - a)^2$ which propagates back with the velocity $v = \frac{8\lambda_1^2 a}{\lambda_1 - a} + \frac{1}{2\lambda_1 t}$. Thus, the constructed solution (3.2.30) describes the propagation of the straightening wave on the curved soliton. Under the straightening, the soliton loses the energy which is taking off by the sound propagating. So one can also treat such waves as the shock waves of rarefaction accompanied by the bend of the soliton [162].

Many of the solutions of the KP equation described above have been calculated for the first time within the framework of the original version of the dressing method.

Now let us consider the rational solutions. They correspond to the special kernel $R_0$ of the form (2.2.31), i. e.

$$R_0 = \sum_{k=1}^{M} f_{0k}(\mu) g_{0k}(\lambda) \delta(\mu - \lambda_k) \delta(\lambda - \lambda_k). \qquad (3.2.33)$$

The reality conditions (3.2.5) and (3.2.6) imply that for the KP-I equation either $M = 2N$ ($N$ is an arbitrary integer) and

$$\lambda_{k+N} = \bar{\lambda}_k, \quad f_{0k+N} = \bar{f}_{0k}, \quad g_{0k+N} = \bar{g}_{0k} \quad (i = 1, \ldots, N) \qquad (3.2.34)$$

or $M$ is an arbitrary integer and

$$\lambda_k = \bar{\lambda}_k, \quad f_{0k} = \bar{f}_{0k}, \quad g_{0k} = \bar{g}_{0k}. \qquad (3.2.35)$$

For the KP-II equation the corresponding reality conditions are

1. $M = 2N$ ($N$ is an arbitrary integer)

$$\lambda_{k+N} = -\bar{\lambda}_k, \quad f_{0k+N} = \bar{f}_{0k}, \quad g_{0k+N} = \bar{g}_{0k} \quad (k = 1, \ldots, N) \qquad (3.2.36)$$

and

2. $M$ is an arbitrary integer and

$$\bar{\lambda}_k = -\lambda_k, \quad \bar{f}_{0k} = f_{0k}, \quad \bar{g}_{0k} = g_{0k}. \qquad (3.2.37)$$

Let us concentrate on the case (3.2.24). For further purpose it is convenient to redenote the constants $\lambda_k$ in the following manner

$$\lambda_k = \lambda_k^+ \quad (k = 1, \ldots, N), \quad \lambda_{k+N} = \lambda_k^- \quad (k = 1, \ldots, N),$$

$$\chi(\lambda_k) = \chi_k^+ \quad (k = 1, \ldots, N), \quad \chi(\lambda_{k+N}) = \chi_k^- \quad (k = 1, \ldots, N)$$

and choose

$$f_{0k} = -\frac{\pi}{2}, \quad g_{0k}(\lambda_k) = 1. \qquad (3.2.38)$$

Now we use the results of section (2.2). The formula (3.2.1) together with (2.2.32) gives

$$u(x, y, t) = 2\partial_x \sum_{k=1}^{N} (\chi_k^+ + \chi_k^-). \qquad (3.2.39)$$

Then, with the use of (3.2.38), (3.2.3) the system (2.2.35) becomes

$$(x - 2\lambda_i^\pm y + 12(\lambda_i^\pm)^2 t + \gamma_i^\pm(0))\chi_i^\pm + \sum_{k=1, k\neq i}^{N}\left(\frac{\chi_k^\pm}{\lambda_k^+ - \lambda_i^\pm} + \frac{\chi_k^-}{\lambda_k^- - \lambda_i^\pm}\right) = 1, \qquad (3.2.40)$$

where $\gamma_i^\pm(0) = \frac{\partial g_{0i}(\lambda)}{\partial \lambda}|_{\lambda=0}$. The system (3.2.40) is the system of $2N$ equations. Solving this system with respect to $\chi_k^+$ and $\chi_k^-$ and substituting the result into (3.2.39), one can obtain the explicit form of the general $N$-lump solution. Using the known properties of the linear systems, it is not difficult to show that this general $N$-lump solution is representable in the following compact form [114]

$$u(x, y, t) = 2\frac{\partial^2}{\partial x^2}\ln \det B \qquad (3.2.41)$$

where $B$ is the $2N \times 2N$ matrix with the elements

$$B_{kl} = \delta_{kl}(x - 2\lambda_k y + 12\lambda_k^2 t + \gamma_k) + i(1 - \delta_{kl})\frac{1}{\lambda_k - \lambda_l} \quad (k, l = 1, \ldots, 2N). \qquad (3.2.42)$$

Here $\lambda_k = \lambda_k^+$ ($k = 1, \ldots, N$), $\lambda_{N+k} = \lambda_k^-$ ($k = 1, \ldots, N$), $\gamma_k = \gamma_{k0}^+$ ($k = 1, \ldots, N$), $\gamma_{N+k} = \gamma_k^-(0)$ ($k = 1, \ldots, N$) and $\delta_{kl}$ is the Kronecker delta-symbol.

A noncomplicated algebraic argument shows that the solution (3.2.41) has no singularities in the case when all $\lambda_i^\pm$ are distinct.

The simplest one-lump ($N = 1$) solution of the KP-I equation is of the form [114]

$$u(x, y, t) = 4\frac{-(\tilde{x} + a\tilde{y})^2 + b^2\tilde{y}^2 + b^{-2}}{((\tilde{x} + a\tilde{y})^2 + b^2\tilde{y}^2 + b^{-2})^2} \qquad (3.2.43)$$

where $\tilde{x} = x - 3(a^2 + b^2)t - x_0$, $\tilde{y} = y + 6at - y_0$ and $a = -2\lambda_{1R}$, $b = 2\lambda_{1I}$, $x_0 = \frac{\lambda_{1R}\gamma_{1I}}{\lambda_{1I}} - \gamma_{1R}$, $y_0 = \frac{\gamma_{1I}}{2\lambda_{1I}}$.

The lump solution (3.2.43) obviously has no singularities and decreases in all directions as $\frac{1}{x^2+y^2}$ as $x^2 + y^2 \to \infty$. In the general case the lump (3.2.43) moves with the velocity $v = (v_x, v_y)$ where $v_x = 3(a^2 + b^2) = 6|\lambda_1|^2$ and $v_y = -6a = 12\text{Re}\lambda_1$. For the first time the existence of the solution (3.2.43) has been demonstrated in the paper [163] by numerical methods.

Both the one-lump solution (3.2.43) and the general $n$-lump solutions (3.2.41) are the transparent potentials for the one-dimensional nonstationary Schrödinger equation.

It is not difficult to see that in the case of all distinct $\lambda_k$ only the diagonal elements of the matrix $B$ (3.2.41) contribute to $\det B$ as $t \to \pm\infty$. Hence, the general solution (3.2.41) asymptotically is the superposition of $n$ lumps

$$u(x, y, t)\underset{t\to\pm\infty}{\to} \sum_{i=1}^{n} u_i(x - v_{ix}t - x_{0i}, y - v_{iy}t - y_{0i})$$

where $u_i$ $(i = 1, \ldots, n)$ are the one-lump solutions of the form (3.2.43). So, the general solution (3.2.41), (3.2.42) describes the collision of $N$ lumps. Since the asymptotics of the solution (3.2.41) as $t \to +\infty$ and $t \to -\infty$ coincide, the lumps do not interact at all. The phase shift, which is typical in the $1+1$-dimensional case, is absent [114]. Emphasize that the lumps are weakly localized bounded solutions of the KP-I equation.

More general rational solutions of the KP-I equations with the poles can also be described by the formula (3.2.41). But it is more convenient to represent them in the form

$$u(x, y, t) = -2 \sum_{i=1}^{n} \frac{1}{(x - x_i(y, t))^2} \tag{3.2.44}$$

where $x_i(y, t)$ are certain functions on $y$ and $t$. Krichever has demonstrated [164] that the solutions of the KP-I equation of this type possess the remarkable property. Namely, if the function $u(x, y, t)$ evolves according to the KP-I equation, the dynamics of the poles in (3.2.44) is described by the Calogero—Moser system [165, 166], i. e. the system of $n$ points on a line with the Hamiltonian $H = \sum_{i=1}^{n} \frac{p_i^2}{2m} + \sum_{i \neq j} \frac{2}{(x_i - x_j)^2}$. One can check this straightforwardly, substituting (3.2.44) into the KP-I equation.

In a similar manner one can analyze the other types of the rational solutions of the KP-I and the KP-II equations which correspond to (3.2.35), (3.2.36) and (3.2.37). It occurs that all such solutions are singular.

Finally, we will consider some solutions of the KP equation on nontrivial background. One can choose any exact solution presented above as the background $u_0$. In general, it is of the form (3.2.13) with the corresponding matrix $A_0$. To construct a new solution $u$ one should first find $\chi_{-1}$. To do this we should know the kernel $R$ of the $\bar{\partial}$-problem. It obeys the linear equation, e. g. (2.5.16) where the operators $\tilde{B}_i$ are completely defined in terms of the background solution $u_0$.

The most general explicit solution of equation (2.5.16) is of the form (2.5.17) where the functions $X_i, Y_i$ obey the linear equation (2.5.10). It is easy to see that the corresponding kernel is also degenerate

$$R(\mu, \bar{\mu}, \lambda, \bar{\lambda}, x) = \sum_{k=1}^{N} c_{ik} f_i(\mu, \bar{\mu}) g_k^*(\lambda, \bar{\lambda}) \tag{3.2.45}$$

where the functions $f_i$ and $g_k^*$ are given by (2.5.12) and (2.5.13). The functions $f_i$ and $g_k$ due to (2.5.10) are, in fact, the wave functions of the KP equation for the solution $u_0(x, y, t)$. For exact explicit solution $u_0$ the corresponding wavefunction is also found explicitly (see e. g. the formula (2.2.22)). So the functions $f_i$ and $g_k^*$ in (3.2.45) are known too. The kernel (3.2.45) is degenerate. As a result, we are able to write explicitly the corresponding solution $u = u_0 - 2i\partial_x \chi_{-1}$.

Thus, if one takes the explicit exact solution of the KP equation presented above as the background and chooses the degenerate kernel $R$ of the type (3.2.45) then one gets the explicit exact solution $u$ of the KP equation on the background $u_0$. Since in this case both $u_0$ and $u$ are, in general, of the form (3.2.13) we have

$$u = 2\partial_x^2 \ln \det((1+A)(1+A_0)) \qquad (3.2.46)$$

where the matrix $A_0$ corresponds to the background $u_0$ and the matrix $A$ is given by (3.2.14) where $\xi_k$ and $\eta_l$ are calculated via wave functions associated with the background $u_0$. If $u_0$ is the solution with the functional parameters then the matrix $A_0$ is of the form (3.2.14) too.

In particular, if $u_0$ is the line multi-soliton then [148]

$$1 + A = (1 + c(1 + A_0^+)^{-1})^{-1} \qquad (3.2.47)$$

where + denotes the hermitian conjugation, the matrix $A_0$ is given by (3.2.24) and $(c)_{ik} = c_{ik}$ are arbitrary constants in (2.5.17). The solution $u$ in this case is [148]

$$u = 2\partial_x^2 \ln\{(1 + A_0)(1 + c(1 + A_0^+)^{-1})\}. \qquad (3.2.48)$$

At this point we complete the enumeration of the exact solutions of the KP equation. Using the $\bar{\partial}$-dressing method one can easily multiply the list of the explicit solutions. Exact solutions of the KP equation and their properties have been considered in a number of papers (see e. g. [167–191]).

Most of the exact solutions considered in this section do not decay at the infinity $x^2 + y^2 \to \infty$. According to section (2.3) the formula (3.2.4) gives us the necessary condition for boundness and decayness of $u$. The corresponding conditions (2.3.28) look like

$$\operatorname{Im}(\mu - \lambda) = 0, \quad \operatorname{Re}\left(\frac{1}{\sigma}(\mu^2 - \lambda^2)\right) = 0. \qquad (3.2.49)$$

For the KP-I equation ($\sigma = i$) it gives $\operatorname{Im} \mu = \operatorname{Im} \lambda = 0$ while for the KP-II equation one has $\mu = -\bar{\lambda}$.

Thus, the bounded solutions of the KP-I equation correspond to the kernels $R_0$ of the form

$$R_0 = R_0(\mu, \lambda)\delta(\mu - \bar{\mu})\delta(\lambda - \bar{\lambda}) \qquad (3.2.50)$$

and for the KP-II equation the kernels $R_0$ should be as follows

$$R_0 = R_0(\lambda, \bar{\lambda})\delta(\mu + \bar{\lambda}). \qquad (3.2.51)$$

For the kernel (3.2.50) the nonlocal $\bar{\partial}$-problem is reduced to the nonlocal Riemann—Hilbert problem with the jump across the real axis

$$\chi^+(\lambda) - \chi^-(\lambda) = \int_R d\mu\, \chi(\mu) R(\mu, \lambda). \qquad (3.2.52)$$

For the kernel (3.2.51) one gets

$$\frac{\partial \chi(\lambda, \bar{\lambda})}{\partial \bar{\lambda}} = R(\lambda, \bar{\lambda}) \chi(-\bar{\lambda}, -\lambda). \qquad (3.2.53)$$

So the analysis of the bounded and decaying solutions of the KP equation is closely connected with the problems (3.2.52) and (3.2.53). We will see that these problems also play a key role in the study of the initial value problems for the KP-I and KP-II equations.

## 3.3. Initial value problem for the KP-I equation

In principle, the $\bar{\partial}$-dressing method can be used for solution of the initial value problem $u(x, y, 0) \to u(x, y, t)$ for the KP equation. For bounded and decaying initial data one should deal as it was shown in the conclusion of the previous section, with the nonlocal Riemann—Hilbert problem or the quasi-local $\bar{\partial}$-problem. But the problem of the calculation of the kernels $R(\mu, \bar{\mu}; \lambda, \bar{\lambda}, x, y, 0)$ which corresponds to the initial data $u(x, y, 0)$ was not effectively solved within the dressing method.

A standard approach to the solution of the initial value problem for the KP equation consists in the analysis of the forward and inverse spectral problems for the auxiliary linear problem (3.0.2) or, more precisely, the problem (3.1.26):

$$(\sigma \partial_y + \partial_x^2 + u)\tilde{\psi} = 0 \qquad (3.3.1a)$$

$$\left(\partial_t + 4\partial_x^3 + 6u\partial_x + 3u_x - 3\sigma \int_{-\infty}^x dx'\, u_y(x', y, t) + 4i\lambda^3\right) \tilde{\psi} = 0. \qquad (3.3.1b)$$

We will assume that the scalar function $u(x, y, t)$ is bounded and decaying rapidly enough at infinity $x^2 + y^2 \to \infty$ and that it obeys the constraints $\int_{-\infty}^{\infty} dx\, u(x, y, t) = 0$.

A spectral parameter is introduced into the problem (3.3.1) by considering the class of solutions $\tilde{\psi}$ of the form

$$\tilde{\psi}(x, y, t, \lambda) = \mu(x, y, t, \lambda, \bar{\lambda}) e^{i\lambda x + \frac{1}{\sigma}\lambda^2 y} \qquad (3.3.2)$$

where $\lambda$ is a complex parameter and $\mu \to 1$ as $\lambda \to \infty$. Such introduction of the spectral parameter $\lambda$ is in complete agreement with the $\bar{\partial}$-dressing method (see e. g. the formulae (3.1.25), (3.1.26)). Note that the function $\mu$ is differed from the function $\chi$ of the $\bar{\partial}$-dressing method by the factor $e^{4i\lambda^3 t}$.

The function $\mu$ obeys the following linear system

$$(\sigma\partial_y + \partial_x^2 + 2i\lambda\partial_x + u)\mu = 0, \qquad (3.3.3)$$

$$\left(\partial_t + 4\partial_x^3 + 12i\lambda\partial_x^2 - 12\lambda^2\partial_x + 6u\partial_x + 6i\lambda\partial_x + 3u_x - 3\sigma\int_{-\infty}^x dx'\,u_y(x',y,t)\right)\mu = 0. \qquad (3.3.4)$$

The solution of the initial value problems are essentially different for the KP-I and KP-II equations.

Here we will consider the KP-I equation following mainly [118, 128]. In this case the problem (3.3.1a) is the one-dimensional nonstationary Schrödinger equation and equation (3.3.4) is of the form

$$(i\partial_y + \partial_x^2 + 2i\lambda\partial_x + u(x,y,t))\mu = 0. \qquad (3.3.5)$$

An analysis of the forward and inverse spectral problems for equation (3.3.5) is our main goal now.

First we note that the general solution $\mu_0$ of the free equation (3.3.3) ($u \equiv 0$) can be represented in the form

$$\mu_0(x,y,\lambda) = \int_{-\infty}^{+\infty} dk\, A(\lambda,k)\exp(ikx - i(k^2 + 2k\lambda)y)$$

$$= \int_{-\infty}^{+\infty} d\lambda'\, B(\lambda,\lambda')\exp(i(\lambda'-\lambda)x - i(\lambda'^2-\lambda^2)y), \qquad (3.3.5')$$

where $A(\lambda,k)$ and $B(\lambda,\lambda')$ are arbitrary functions. The particular solutions of the type (3.3.5'), namely $\mu_0 = 1$ and $\mu_0(x,y;\lambda,\mu) = e^{i(\mu-\lambda)x - i(\mu^2-\lambda^2)y}$ will arise many times in further constructions.

We will not assume that the potential $u$ is small in any sense. In this case it is convenient to begin the study of the properties of the solutions of the partial differential equations by converting the differential equation into an integral one.

In our case the partial differential equation (3.3.3) is equivalent to the integral equation

$$\mu(x,y,\lambda) = \mu_0(x,y,\lambda) - (Gu\chi)(x,y) \qquad (3.3.6)$$

where, in the general case, the free term $\chi_0$ is of the form (3.3.5),

$$(Gf)(x,y) \doteq \int_{-\infty}^{+\infty}\int_{-\infty}^{+\infty} dx'dy'\, G(x-x',y-y',\lambda)f(x',y') \qquad (3.3.7)$$

and the Green function $G(x-x',y-y',\lambda)$ is defined in the standard manner, namely,

$$L_0 G \doteq (i\partial_y + \partial_x^2 + 2i\lambda\partial_x)G(x-x',y-y',\lambda) = \delta(x-x')\delta(y-y') \qquad (3.3.8)$$

where $\delta(x)$ is the Dirac delta-function. The solutions of the integral equation (3.3.6) which correspond to the different choices of the free term $\mu_0$ and the Green function $G$ determine the different classes of solutions of the partial differential equation (3.3.3).

The formal expression for the Green function $G(x,y,\lambda)$ is given by

$$G(x,y,\lambda) = -\frac{1}{(2\pi)^2} \int_{-\infty}^{+\infty} \int_{-\infty}^{+\infty} dk\, dk'\, \frac{e^{ikx+ik'y}}{k^2 + 2k\lambda + k'}. \quad (3.3.9)$$

It is readily seen that the function (3.3.9) is defined not for all $\lambda = \lambda_R + i\lambda_I$. Indeed, for the real $\lambda$ the integrand in (3.3.9) has the poles on the real axis and, hence, the function $G$ is ambiguous for $\lambda_I = 0$. As a result, the function $G$ has a jump across the real axis $\lambda_I = 0$.

For the complex $\lambda$ ($\lambda_I \neq 0$), integrating in (3.3.9) over $k'$, one obtains

$$G(x,y,\lambda) = \frac{1}{2\pi i} \int_{-\infty}^{+\infty} dk\, \{\theta(y)\theta(-k\lambda_I) - \theta(-y)\theta(k\lambda_i)\} e^{ikx - i(k^2 + 2\lambda k)y} \quad (3.3.10)$$

where $\theta(\xi)$ is the step (Heaviside) function:

$$\theta(\xi) = \begin{cases} 1, & \xi > 0 \\ 0, & \xi < 0 \end{cases}.$$

So, the Green function $G$ is well defined for the complex $\lambda$ at $\lambda_I \neq 0$. The boundary values $G^+, G^-$ of this function at $\lambda_I \to \pm 0$, as it follows from (3.3.10), are

$$G^+(x,y,\lambda) = \frac{1}{2\pi i} \int_{-\infty}^{+\infty} dk (\theta(y)\theta(-k) - \theta(-y)\theta(k)) e^{ikx - i(k^2 + 2\lambda k)y} \quad (3.3.11)$$

and

$$G^-(x,y,\lambda) = \frac{1}{2\pi i} \int_{-\infty}^{+\infty} dk\, (\theta(y)\theta(k) - \theta(-y)\theta(-k)) e^{ikx - i(k^2 + 2\lambda k)y}. \quad (3.3.12)$$

The functions $G^+$ and $G^-$ admit the analytic continuation to the upper ($\lambda_I > 0$) and lower ($\lambda_I < 0$) half planes respectively. On the real axis $\lambda_I = 0$ one has

$$G^+ - G^- = -\frac{1}{2\pi i} \int_{-\infty}^{+\infty} dk\, \text{sgn}(k) e^{ikx - i(k^2 + 2\lambda k)y} \quad (3.3.13)$$

where $\text{sgn}(\xi) \doteq \theta(\xi) - \theta(-\xi)$. Note that the r.h.s. of (3.3.13) is the solution of the type (3.3.5) of equation (3.3.3) with $u = 0$ in the agreement with the obvious equation $L_0(G^+ - G^-) = 0$.

The Green functions $G^+$ and $G^-$ are just the desired Green functions which allow to define the solutions of equation (3.3.3) with good analytic properties.

Let us consider the functions $\mu^+(x,y,\lambda)$ and $\mu^-(x,y,\lambda)$ which are bounded for all complex $\lambda = \lambda_R + i\lambda_I$ tend to 1 as $\lambda \to \infty$ and which are the solutions of the following integral equations

$$\mu^+(x,y,\lambda) = 1 - (G^+ u \mu^+(\cdot,\lambda))(x,y) \qquad (3.3.14)$$

and

$$\mu^-(x,y,\lambda) = 1 - (G^- u(\cdot) \mu^+(\cdot,\lambda))(x,y) \qquad (3.3.15)$$

where the Green functions $G^+$ and $G^-$ are given by the formulae (3.3.11) and (3.3.12). Then let us introduce the function

$$\mu(x,y,\lambda) \doteq \begin{cases} \mu^+(x,y,\lambda), & \lambda_I > 0, \\ \mu^-(x,y,\lambda), & \lambda_I < 0. \end{cases}$$

This function is, of course, the solution of equation (3.3.3) bounded for all $\lambda$ and $\mu \to 1$ as $\lambda \to \infty$.

To ascertain the properties of the function $\mu$ one should investigate the integral equations (3.3.14) and (3.3.15). These equations are Fredholm integral equations of the second type with the parameter dependence on $\lambda$. We will assume that the kernels $G^\pm(x,y,\lambda)u(x,y)$ of equations (3.3.14) and (3.3.15) are of the Fredholm type and so the standard Fredholm theory can be applied.

For convenience we recall some elements of the theory of the Fredholm integral equations

$$\mu(\xi) = \mu_0(\xi) - (\tilde{\mathcal{F}}\mu)(\xi) \qquad (3.3.16)$$

of the second type (see e. g. [150, 151]). The Fredholm determinant $\Delta$ is one of the most important notion of this theory. Formally the Fredholm determinant $\Delta$ defines by the formula $\ln \Delta = \text{tr}(\ln(1 + \tilde{\mathcal{F}}) - \tilde{\mathcal{F}})$ where tr denotes the operator trace. If $\Delta \neq 0$ then equation (3.3.16) is solvable for all $\mu_0$. In the case $\Delta = 0$ the homogeneous equation (3.3.16) may have nontrivial solutions and the properties of the inhomogeneous equation (3.3.16) are the subject of the well-known Fredholm theorems [150, 151]. In particular, for the solvability of equation (3.3.16) in the case $\Delta = 0$ the free term $\mu_0$ in (3.3.16) should be orthogonal to all solutions $\tilde{\mu}_i$ of the equation which is adjoint to the homogeneous equation (3.3.16).

In our case $\mu^\pm(\lambda) = (1 + \hat{G}^\pm u)^{-1} \cdot 1$ formally. The functions $G^\pm$ are analytic at $\lambda_I > 0$ and $\lambda_I < 0$. Hence, the functions $\mu^+(\lambda)$ and $\mu^-(\lambda)$ are analytic in the upper- and lower-half planes, respectively, too, with the exception of the points where the Fredholm determinant of the operator $1 + G^\pm u$ vanishes. At these points $\lambda_i$ $(\Delta^\pm(\lambda_i) = 0)$ the functions $\mu^+$ and $\mu^-$ have the poles. So, the existence of the nontrivial solutions for the homogeneous equations (3.3.14) and (3.3.15) is connected with the presence of the poles in the functions $\mu^+$ and $\mu^-$. An investigation of the properties of the Fredholm determinant is a rather complicated problem. Here we will assume that the Fredholm determinants $\Delta^\pm(\lambda)$ for equations (3.3.14) and (3.3.15) have a finite number of simple zeros at the distinct points $\lambda_i^\pm$ and none of them are situated on the real axis $\lambda_I = 0$. The corresponding solutions of the homogeneous equations (3.3.14) and (3.3.15) will be denoted by $\mu_i^+(x,y)$ and $\mu_i^-(x,y)$.

Accordingly the Fredholm theory implies that the solutions of equations (3.3.14) and (3.3.15) are of the form

$$\mu^\pm(x,y,\lambda) = 1 + i \sum_{i=1}^n \frac{\mu_i^\pm(x,y)}{\lambda - \lambda_i^\pm} + \tilde{\mu}^\pm(x,y,\lambda) \qquad (3.3.17)$$

where $\tilde{\mu}^+(x,y,\lambda)$ and $\tilde{\mu}^-(x,y,\lambda)$ are analytic functions in $\text{Im}\lambda > 0$ and $\text{Im}\lambda < 0$ respectively. The functions $\mu_i^\pm(x,y)$ are normalized such that $\mu_i^\pm(x - 2\lambda_i^\pm y) \to 1$ as $x^2 + y^2 \to \infty$. For real potential $u(x,y)$ one has $\lambda_i^- = \bar{\lambda}_i^+$.

The functions $\mu_i^\pm$ obey the following important relations [128]

$$\lim_{\lambda \to \lambda_i^\pm} \left( \mu^\pm - i \frac{\mu_i^\pm}{\lambda - \lambda_i} \right) = (x - 2\lambda_i^\pm y + \gamma_i^\pm(t))\mu_i \qquad (3.3.18)$$

and

$$1 \mp \frac{1}{2\pi} \int_{-\infty}^{+\infty} \int_{-\infty}^{+\infty} dx\, dy\, u(x,y) \mu_i^\pm(x,y) = 0 \qquad (3.3.19)$$

where $\gamma_i^\pm$ are time-independent constants.

To prove these relations, for instance, for the function $\mu^-$ we introduce the auxiliary functions

$$\tilde{\mu}_i^-(x,y,\lambda) \doteq \left( \mu^- - \frac{i\mu_i^-}{\lambda - \lambda_i^-} \right) \exp(i\lambda x - i\lambda^2 y) \qquad (3.3.20)$$

and

$$\tilde{\mu}_i(x,y,\lambda) \doteq \mu_i^-(x,y) \exp(i\lambda x - i\lambda^2 y). \qquad (3.3.21)$$

It follows from equation (3.3.15) that

$$((1 + \tilde{G}^- u)\tilde{\mu}_i^-(\cdot,\lambda))(x,y) = e^{i\lambda^2 x - i\lambda^2 y} - \frac{i}{\lambda - \lambda_i^-}((1 + \tilde{G}^- u)\tilde{\mu}_i(\cdot,\lambda))(x,y) \qquad (3.3.22)$$

where the function $\tilde{G}^-$ is of the form

$$\tilde{G}^-(\xi,\eta,\lambda) \doteq e^{i\lambda\xi - i\lambda^2 \eta} G^-(\xi,\eta,\lambda)$$

$$= \frac{1}{2\pi i} \int_{-\infty}^{+\infty} dk\, (\theta(y)\theta(k - \lambda) - \theta(-y)\theta(\lambda - k)) e^{ik\xi - ik^2 \eta}. \qquad (3.3.23)$$

Taking the limit $\lambda \to \lambda_i^-$ in (3.3.22) and taking into account that

$$(1 + \tilde{G}^-(\cdot, \lambda_i^-)u(\cdot))\tilde{\mu}_i^-(\cdot, \lambda_i^-) = 0, \tag{3.3.24}$$

one obtains

$$\left\{(1 + \tilde{G}^-(\cdot, \lambda_i^-)u)\left(\tilde{\mu}_i^-(\cdot, \lambda_i^-) + i\frac{\partial \tilde{\mu}_i}{\partial \lambda}(\cdot, \lambda_i^-)\right)\right\}(x, y)$$
$$= \left\{1 + \frac{1}{2\pi}\int_{-\infty}^{+\infty}\int_{-\infty}^{+\infty} d\xi d\eta\, u(\xi, \eta)\mu_i^-(\xi, \eta)\right\} e^{i\lambda_i^- x - i\lambda_i^{-2} y}. \tag{3.3.25}$$

The Fredholm alternative now implies that

$$1 + \frac{1}{2\pi}\int_{-\infty}^{+\infty}\int_{-\infty}^{+\infty} d\xi\, d\eta\, u(\xi, \eta)\mu_i^-(\xi, \eta) = 0$$

and

$$\tilde{\mu}_i^-(x, y, \lambda_i^-) + i\frac{\partial \tilde{\mu}_i(x, y, \lambda_i^-)}{\partial \lambda} = \gamma_i \mu_i^-(x, y, \lambda_i^-) \tag{3.3.26}$$

where $\gamma_i$ are constants. At last, it is not difficult to show that equation (3.3.26) is equivalent to equation (3.3.18) for $\mu_i^-$. The relations (3.3.18) and (3.3.19) for the function $\mu^+$ are proved analogously.

Thus, in view of (3.3.17), the problem of constructing the solution of equation (3.3.4) with good analytic properties is solved: the function $\mu(\lambda) = \begin{cases} \mu^+, & \lambda_I > 0 \\ \mu^-, & \lambda_I < 0 \end{cases}$ is the analytic function on the entire complex plane $\lambda$, with the exception of a finite number of simple poles and it has a jump across the real axis $\lambda_I = 0$.

Just the existence of this jump allows us to formulate the corresponding conjugation problem and introduce the inverse problem data.

With this in mind let us calculate the quantity $K(x, y; \lambda) \doteq \mu^+(x, y, \lambda) - \mu^-(x, y, \lambda)$ for real $\lambda$. Subtracting equations (3.3.14) and (3.3.15), we obtain

$$K(x, y; \lambda) = ((G^- - G^+)u\mu^+)(x, y, \lambda) - (G^-(\cdot, \lambda)u(\cdot)K(\cdot; \lambda))(x, y). \tag{3.3.27}$$

Then, the use of (3.3.13) gives

$$K(x, y; \lambda) = \int_{-\infty}^{+\infty} d\lambda'\, T(\lambda, \lambda') \exp(\omega(\lambda', \lambda)) - (G^-(\cdot; \lambda)u(\cdot)K(\cdot; \lambda))(x, y) \tag{3.3.28}$$

where

$$\omega(\lambda', \lambda) \doteq i(\lambda' - \lambda)x - i(\lambda'^2 - \lambda^2)y \tag{3.3.29}$$

and

$$T(\lambda, \lambda') = \frac{i}{2\pi}\operatorname{sgn}(\lambda - \lambda')\int_{-\infty}^{+\infty}\int_{-\infty}^{+\infty} d\xi\, d\eta\, u(\xi, \eta)\chi^+(\xi, \eta, \lambda)e^{\omega(\lambda, \lambda', \xi, \eta)}. \tag{3.3.30}$$

So the quantity $K(x,y;\lambda)$ is the solution of the integral equation of the type (3.3.15) with the free term $\int_{-\infty}^{+\infty} d\lambda' T(\lambda,\lambda')e^{\omega(\lambda',\lambda,x,y)}$. In view of this, it is natural to consider the solutions $N(x,y,\lambda',\lambda)$ of equation (2.1.4) which are simultaneously the solutions of the integral equation

$$N(x,y,\lambda',\lambda) = \exp(i(\lambda' - \lambda)x - i(\lambda'^2 - \lambda^2)y) - (G^-(\cdot,\lambda)u(\cdot)N(\cdot,\lambda',\lambda))(x,y) \quad (3.3.31)$$

where $\lambda_I = \lambda'_I = 0$. The solutions $N(x,y,\lambda',\lambda)$ with the different $\lambda'$ form the one-parametric family of solutions of equation (3.3.4).

Then we multiply equation (3.3.31) by $T(\lambda,\lambda')$, integrate over $\lambda'$ and compare the equation obtained with equation (3.3.28). As a result, we obtain

$$K(x,y;\lambda) = \mu^+(x,y,\lambda) - \mu^-(x,y,\lambda) = \int_{-\infty}^{+\infty} d\lambda' T(\lambda,\lambda')N(x,y,\lambda',\lambda). \quad (3.3.32)$$

It is assumed that the homogeneous equation (3.3.15) has no nontrivial solutions for real $\lambda$.

Further, in order to be able to treat (3.3.32) as the relation between the boundary values of the meromorphic function one must find the interrelation between $N(x,y,\lambda',\lambda)$ and $\chi^-(x,y,\lambda)$.

To do this, we introduce the functions

$$\tilde{N}(x,y,\lambda',\lambda) \doteq N(x,y,\lambda',\lambda)e^{i\lambda x - i\lambda^2 y}, \quad (3.3.33)$$

$$\tilde{\mu}^\pm(x,y,\lambda) \doteq \mu^\pm(x,y,\lambda)e^{i\lambda x - i\lambda^2 y}. \quad (3.3.34)$$

These functions, as follows from (3.3.31) and (3.3.15) are the solutions of the integral equations

$$((1 + \tilde{G}u)\tilde{N}(\cdot,\lambda',\lambda))(x,y) = e^{i\lambda' x - i\lambda'^2 y} \quad (3.3.35)$$

and

$$((1 + \tilde{G}u)\tilde{\mu}^-(\cdot,\lambda))(x,y) = e^{i\lambda x - i\lambda^2 y} \quad (3.3.36)$$

where the function $\tilde{G}(x,y,\lambda)$ is given by (3.3.23). Note that $\tilde{G}(x,y,\lambda)$ is nothing but the Green function of the operator $i\partial_y + \partial_x^2$ and $\tilde{N}(x,y,\lambda',\lambda)$ and $\tilde{\mu}^\pm(x,y,\lambda)$ are the solutions of the original equation (3.3.3).

Differentiating (3.3.35) over $\lambda$, one gets

$$\left((1 + \tilde{G}u)\frac{\partial \tilde{N}(\cdot,\lambda',\lambda)}{\partial \lambda}\right)(x,y) = F(\lambda',\lambda)e^{i\lambda x - i\lambda^2 y} \quad (3.3.37)$$

where

$$F(\lambda',\lambda) = \frac{1}{2\pi i}\int_{-\infty}^{+\infty}\int_{-\infty}^{+\infty} d\xi\, d\eta\, u(\xi,\eta)N(\xi,\eta,\lambda',\lambda). \quad (3.3.38)$$

Comparing (3.3.37) with (3.3.36) and assuming that the equation $(1 + \tilde{G}u)M = 0$ has no nontrivial solutions for real $\lambda$, we obtain

$$\frac{\partial}{\partial \lambda}\tilde{N}(x,y,\lambda',\lambda) = F(\lambda',\lambda)\mu^-(x,y,\lambda). \tag{3.3.39}$$

Finally, the integration of the equality (3.3.39) over $\lambda$, with the use of (3.3.39), (3.3.34) and the equality $N(x,y,\lambda',\lambda) = \mu^-(x,y,\lambda)$, gives

$$N(x,y,\lambda',\lambda) = \mu^-(x,y,\lambda')e^{i(\lambda'-\lambda)x - i(\lambda'^2-\lambda^2)y}$$

$$+ \int_{\lambda'}^{\lambda} d\mu\, F(\lambda',\mu)\mu^-(x,y,\mu)e^{i(\mu-\lambda)x - i(\mu^2-\lambda^2)y}. \tag{3.3.40}$$

The formula (3.3.40) is the crucial one for the subsequent constructions. With the use of this formula the relation (3.3.32) can be rewritten in the form [128]

$$\mu^+(x,y,\lambda) = \mu^-(x,y,\lambda) + \int_{-\infty}^{+\infty} d\lambda'\, \mu^-(x,y,\lambda')f(\lambda',\lambda)e^{i(\lambda'-\lambda)x - i(\lambda'^2-\lambda^2)y} \tag{3.3.41}$$

where

$$f(\lambda',\lambda) \doteq T(\lambda',\lambda)$$

$$+ \theta(\lambda - \lambda')\int_{-\infty}^{\lambda'} d\mu\, T(\lambda,\mu)F(\mu,\lambda') - \theta(\lambda' - \lambda)\int_{\lambda'}^{\infty} d\mu\, T(\lambda,\mu)F(\mu,\lambda'). \tag{3.3.42}$$

The latter expression for the function $f(\lambda',\lambda)$ can be simplified if one takes into account the formulae (3.3.30) and (3.3.38). One gets

$$f(\lambda',\lambda) = \frac{\text{sgn}(\lambda - \lambda')}{2\pi i}\int_{-\infty}^{+\infty}\int_{-\infty}^{+\infty} dx\, dy\, u(x,y)N(x,y,\lambda',\lambda) \tag{3.3.43}$$

where $\text{Im}\lambda' = \text{Im}\lambda = 0$. The formula (3.3.43) manifestly demonstrates the significance of the function $N(x,y,\lambda',\lambda)$. Note that this function does not admit the continuation from the real $\lambda$ and $\lambda'$. Note also that for the real potential $u(x,y)$ one has $\bar{f}(\lambda',\lambda) = f(\lambda,\lambda')$.

Thus, our intermediate purpose has been achieved. For equation (3.3.3) we have constructed the solution $\mu(x,y,\lambda)$ which is the meromorphic function on $\lambda$ in the upper and lower half planes and the boundary values $\mu^+, \mu^-$ of which are connected by the relation (3.3.41). So, we have arrived at the nonlocal Riemann—Hilbert problem. In our case the contour $\Gamma$ is the real axis, the function $\mu$ has the canonical normalization $(\mu \underset{\lambda \to \infty}{\to} 1)$ and the function $f(\lambda',\lambda)$ plays a role of the inverse problem data.

The solution of the nonlocal Riemann—Hilbert problem is given by the reduction of the general formula (2.2.3), i. e. by

$$\mu(x,y,\lambda) = 1 + i\sum_{i=1}^{n}\left(\frac{\mu_i^+}{\lambda - \lambda_i^+} + \frac{\mu_i^-}{\lambda - \lambda_i^-}\right)$$

$$+\frac{1}{2\pi i}\int_{-\infty}^{+\infty}\int_{-\infty}^{+\infty}d\mu\,d\lambda'\,\frac{\mu^-(x,y,\lambda')f(\lambda',\mu)e^{\omega(x,y,\lambda',\mu)}}{\mu-\lambda}. \quad (3.3.44)$$

The equations of the inverse problem which we are looking for just follow from this formula. Passing in (3.3.44) to the limit $\lambda \to \lambda_R - i0$, one obtains

$$\mu^-(x,y,\lambda) = 1 + i\sum_{i=1}^{n}\left(\frac{\mu_i^+}{\lambda-\lambda_i^+} + \frac{\mu_i^-}{\lambda-\lambda_i^-}\right)$$

$$+\frac{1}{2\pi i}\int_{-\infty}^{+\infty}\int_{-\infty}^{+\infty}d\mu\,d\lambda'\,\frac{\mu^-(x,y,\lambda')f(\lambda',\mu)e^{\omega(x,y,\lambda',\mu)}}{\mu-\lambda+i0} \quad (3.3.45)$$

where $\text{Im}\,\lambda = 0$. In the limits $\lambda \to \lambda_i^\pm$ the formula (3.3.44), together with (3.3.18), also yields

$$(x - 2\lambda_i^\pm y + \gamma_i(t))\mu_i^\pm = 1 + i\sum_{k=1, k\neq i}^{n}\left(\frac{\mu_k^\pm}{\lambda_i^\pm - \lambda_k^+} + \frac{\mu_k^-}{\lambda_i^\pm - \lambda_k^-}\right)$$

$$+\frac{1}{2\pi i}\int_{-\infty}^{+\infty}\int_{-\infty}^{+\infty}d\mu\,d\lambda'\,\frac{\mu^-(x,y,\lambda')f(\lambda',\mu,t)e^{\omega(\lambda',\mu,x,y)}}{\mu-\lambda_i^\pm} \quad (i=1,\ldots,n). \quad (3.3.46)$$

Using (2.3.9) and (3.3.44), one gets

$$u(x,y,t) = \frac{\partial}{\partial x}\Bigg\{2\sum_{i=1}^{n}(\mu_i^+ + \mu_i^-)$$

$$+\frac{1}{\pi}\int_{-\infty}^{+\infty}\int_{-\infty}^{+\infty}d\lambda d\lambda'\,f(\lambda',\lambda,t)\mu^-(x,y,\lambda')e^{i(\lambda'-\lambda)x - i(\lambda'^2-\lambda^2)y}\Bigg\}. \quad (3.3.47)$$

Equations (3.3.45), (3.3.46) and the formula (3.3.47) is the system of equations which solve the inverse problem for equation (3.3.3) [128]. The set $\{f(\lambda',\lambda,t)\,(\text{Im}\,\lambda' = \text{Im}\,\lambda = 0)$, $\lambda_i^+, \lambda_i^-, \gamma_i^+(t), \gamma_i^-(t), i = 1,\ldots,n\}$ is the data of the inverse problem. Given these inverse problem data one can calculate the functions $\mu^-$ and $\mu_i^\pm$ ($i = 1,\ldots,n$) with the help of equations (3.3.45) and (3.3.46). Then one reconstructs the potential $u(x,y,t)$ by the formula (3.3.47).

With the solution of the inverse problem for the spectral problem (3.3.3) we are now in a better position to address the initial value problem of the KP-I equation. For this one should find the dependence of the inverse problem data $\{f(\lambda',\lambda,t), \lambda_i^+, \lambda_i^-, \gamma_i^+(t), \gamma_i^-(t)\}$ in time $t$. This dependence is determined by equation (3.3.4) with $\sigma = i$. Indeed, substituting (3.3.41) into equation (3.3.4) we obtain

$$\frac{\partial f(\lambda',\lambda,t)}{\partial t} = 4i(\lambda'^3 - \lambda^3)f(\lambda',\lambda,t). \quad (3.3.48)$$

To ascertain the time-independence of $\lambda_i^\pm$ and $\gamma_i^\pm$ it is sufficient to consider equation (3.3.4) as $x, y \to \infty$, i.e. the equation

$$\mu_t + 4\mu_{xxx} + 12i\lambda\mu_{xx} - 12\lambda^2\mu_x = 0.$$

Substituting the expression (3.3.17) into this equation, passing to the limits $\lambda \to \lambda_i^{\pm}$ and using (3.3.18), one gets

$$\frac{\partial \lambda_i^{\pm}}{\partial t} = 0, \quad \frac{\partial \gamma_i^{\pm}}{\partial t} = 12(\lambda_i^{\pm})^2. \tag{3.3.49}$$

As a result

$$f(\lambda', \lambda, t) = e^{4i(\lambda'^3 - \lambda^3)t} f(\lambda', \lambda, 0),$$

$$\lambda_i^{\pm}(t) = \lambda_i^{\pm}(0), \quad \gamma_i^{\pm}(t) = 12(\lambda_i^{\pm})^2 t + \gamma_i^{\pm}(0). \tag{3.3.50}$$

The inverse problem equations (3.3.45), (3.3.46), (3.3.48) and the formulae (3.3.50) allow us to solve the initial value problem for the KP-I equation by the standard for the IST method scheme

$$u(x, y, 0) \xrightarrow{I} \{f(\lambda', \lambda, 0), \lambda_i^+(0), \lambda_i^-(0), \gamma_i^+(0), \gamma_i^-(0)\}$$

$$\xrightarrow{II} \{f(\lambda', \lambda, t), \lambda_i^+(0), \lambda_i^-(0), \gamma_i^+(t), \gamma_i^-(t)\} \xrightarrow{III} u(x, y, t). \tag{3.3.51}$$

Indeed, let we have $u(x, y, 0)$. Solving the forward problem for (3.3.3), we find the data $\{f(\lambda', \lambda, 0), \lambda_i^{\pm}(0), \gamma_i^{\pm}(0)\}$. For example,

$$f(\lambda', \lambda, 0) = \frac{1}{2\pi i} \text{sgn}(\lambda' - \lambda) \int_{-\infty}^{+\infty} \int_{-\infty}^{+\infty} d\xi \, d\eta \, u(\xi, \eta, 0) N(\xi, \eta, 0, \lambda', \lambda).$$

Then, the formulae (3.3.50) give the inverse problem data for an arbitrary time $t$. Using these expression and equations (3.3.45) and (3.3.46), we calculate (in principle) the functions $\mu^-(x, y, t, \lambda)$ and $\mu_i^+(x, y, t)$, $\mu_i^-(x, y, t)$. Finally, the formula (3.3.48) gives us the potential $u(x, y, t)$ at arbitrary $t$.

The procedure described above gives the solution of the general initial value problem for the KP-I equation in the class of decreasing potentials $u$. This procedure is implicit since it includes the singular integral equations (3.3.45) and (3.3.46) at the third step. Nevertheless it allow us to investigate the KP-I equation rather completely. In particular, using these formulae one can construct the wide class of the exact solutions of the KP-I equation.

Indeed, choosing the pure continuous degenerate data

$$f(\lambda, \mu) = \sum_{i=1}^{N} f_k(\lambda) g_k(\mu) \tag{3.3.52}$$

where $f_k(\lambda)$ and $g_k(\mu)$ are arbitrary functions and $\lambda_i^{\pm} = 0$, $\gamma_i^{\pm} = 0$, one gets the solutions with functional parameters. They are the particular case of the solutions (3.2.12), (3.2.20).

Then, choosing the pure discrete data $(f(\lambda, \mu) \equiv 0)$, one gets, in fact, the system (3.2.40). As a result, one has the multilump solutions (3.2.41).

In addition, the formulae (3.3.45), (3.3.46) and (3.3.48) are very useful for the calculation of the asymptotical behavior of the solutions of the KP-I equation as $t \to \pm\infty$ [192]. The use of these asymptotics allows to construct the action-angle type variables for the KP-I equation [193].

## 3.4. Initial value problem for the KP-II equation

Here we discuss the solution of the initial value problem for the KP-II equation following mainly the paper [119]. As in the case of the KP-I equation we assume that $u(x,y,t)$ decreases sufficiently rapidly as $x^2 + y^2 \to \infty$. Instead of (3.3.3) one has

$$(\partial_y + \partial_x^2 + 2i\lambda\partial_x + u(x,y,t))\mu = 0. \quad (3.4.1)$$

Bounded solutions $\mu_0$ of equation (3.4.1) with $u \equiv 0$ are of the form

$$\mu_0(x,y,t) = A(\lambda,\bar{\lambda})e^{-i(\bar{\lambda}+\lambda)x+(\bar{\lambda}^2-\lambda^2)y} = B(p,q)e^{ipx+iqy} \quad (3.4.2)$$

where $p = -(\lambda+\bar{\lambda})$, $q = i(\lambda^2 - \bar{\lambda}^2)$ and $A(\lambda,\bar{\lambda})$, $B(p,q)$ are arbitrary bounded functions.

The solutions of the problem (3.4.1) bounded for all $\lambda$ and canonically normalized ($\mu \underset{\lambda \to \infty}{\to} 1$) are defined by the integral equation

$$\mu(x,y,\lambda) = 1 - (G(\cdot)u(\cdot)\mu(\cdot,\lambda))(x,y). \quad (3.4.3)$$

The formal expression for the Green function $G(x,y,\lambda)$ of the operator $L_0 = \partial_y + \partial_x^2 + 2i\lambda\partial_x$ is given, similar to (3.3.9), by

$$G(x,y,\lambda) = -\frac{1}{(2\pi)^2} \iint dk'dk \frac{e^{ikx+ik'y}}{k^2 + 2k\lambda - ik'}. \quad (3.4.4)$$

In contrast to the KP-I case, the integrand in (3.4.4) has poles in the points $k = -2\lambda_R$, $k' = -4\lambda_R\lambda_I$ and $k' = k = 0$. Integrating in (3.4.4) over $k'$, one obtains

$$G(x,y,\lambda) = -\frac{1}{2\pi}\left\{\theta(\lambda_R)\left[-\theta(y)\int_{-2\lambda_R}^0 dk + \theta(-y)\left(\int_0^\infty dk + \int_{-\infty}^{-2\lambda_R} dk\right)\right]\right.$$

$$\left. + \theta(-\lambda_R)\left[-\theta(y)\int_0^{-2\lambda_R} dk + \theta(-y)\left(\int_{-\infty}^0 dk + \int_{-2\lambda_R}^\infty dk\right)\right]\right\} e^{ikx+(k^2+2\lambda k)y}. \quad (3.4.5)$$

The Green function (3.4.5) explicitly depends on $\lambda_R = \frac{1}{2}(\lambda + \bar{\lambda})$. The integral equation (3.4.3) likewise has such a dependence. For example, at $\lambda_R > 0$

$$\mu(x,y,\lambda,\bar{\lambda}) = 1 - \int_{-\infty}^{+\infty}\int_{-\infty}^{+\infty} dx'dy'\, G(x-x',y-y';\lambda,\bar{\lambda})u(x',y')\mu(x',y';\lambda,\bar{\lambda}) \quad (3.4.6)$$

where

$$G(\xi,\eta;\lambda,\bar{\lambda}) = -\frac{1}{2\pi}\left[-\theta(y)\int_{-2\lambda_R}^{0}dk + \theta(-y)\left(\int_{0}^{\infty}dk + \int_{-\infty}^{-2\lambda_R}dk\right)\right]e^{ik\xi+(k^2+\lambda k)\eta}.$$

So neither the Green function $G(x,y;\lambda,\bar{\lambda})$ (3.4.5) nor the solutions of the integral equation (3.4.3) are analytic functions on the whole complex plane of $\lambda$. Moreover the Green function $G(x,y;\lambda,\bar{\lambda})$ has no jump across the real axis $\lambda_I = 0$. This indicates a cardinal difference between the problems (3.4.1) and (3.3.4). Thus, one is unable to formulate any conjugation problem of the Riemann—Hilbert problem type for the spectral problem (3.4.1). The KP-II equation is the first example of the integrable equation where such situation has arisen.

The method which, nevertheless, permits to formulate and solve the inverse problem for equation (3.4.1) has been discovered by Ablowitz, Bar Yaakov and Fokas in the paper [119]. The main idea of this approach is to calculate $\partial\mu/\partial\bar{\lambda}$ and then use the generalized Cauchy formula (2.1.14).

Following this idea, we differentiate equation (3.4.3) with respect to $\bar{\lambda}$:

$$\frac{\partial\mu(x,y;\lambda,\bar{\lambda})}{\partial\bar{\lambda}} = -\int_{-\infty}^{+\infty}\int_{-\infty}^{+\infty}dx'dy'\,\frac{\partial G}{\partial\bar{\lambda}}(x-x',y-y';\lambda,\bar{\lambda})u(x',y')\mu(x',y';\lambda,\bar{\lambda})$$

$$-\int_{-\infty}^{+\infty}\int_{-\infty}^{+\infty}dx'dy'\,G(x-x',y-y';\lambda,\bar{\lambda})u(x',y')\frac{\partial\mu(x',y';\lambda,\bar{\lambda})}{\partial\bar{\lambda}}. \qquad (3.4.7)$$

It is not difficult to show, using the formula (3.4.5), that

$$\frac{\partial G(x,y;\lambda,\bar{\lambda})}{\partial\bar{\lambda}} = -\frac{\sigma_0}{2\pi}e^{ipx+iqy}, \qquad (3.4.8)$$

where $\sigma_0 = \text{sgn}(-\lambda_R)$, $p = -2\lambda_R = -(\lambda+\bar{\lambda})$, $q = -4\lambda_I\lambda_R = i(\lambda^2 - \bar{\lambda}^2)$. Note that the proportionality of $\partial G/\partial\bar{\lambda}$ to the exponent $e^{ipx+iqy}$ is the trivial consequence of the obvious equation $L_0(\partial G/\partial\bar{\lambda}) = 0$ and the formula (3.4.2).

Substituting the expression for $\partial G/\partial\bar{\lambda}$ given in (3.4.8) into (3.4.7), we obtain

$$\frac{\partial\mu(x,y,\lambda,\bar{\lambda})}{\partial\bar{\lambda}} = F(\lambda,\bar{\lambda})e^{ipx+iqy}$$

$$-\int_{-\infty}^{+\infty}\int_{-\infty}^{+\infty}dx'dy'\,G(x-x',y-y';\lambda,\bar{\lambda})u(x',y')\frac{\partial\mu(x',y';\lambda,\bar{\lambda})}{\partial\bar{\lambda}}, \qquad (3.4.9)$$

where

$$F(\lambda,\bar{\lambda}) = \frac{\sigma_0}{2\pi}\int_{-\infty}^{+\infty}\int_{-\infty}^{+\infty}dx\,dy\,e^{-ipx-iqy}u(x,y)\mu(x,y,\lambda,\bar{\lambda}). \qquad (3.4.10)$$

Thus, $\partial\mu/\partial\bar{\lambda}$ is the solution of the integral equation of the type (3.4.3) with the free term $e^{ipx+iqy}F(\lambda,\bar{\lambda})$ instead of 1. Emphasize that the fact $\partial\mu/\partial\bar{\lambda}$ must obey the integral equation of the type (3.4.9) with some free term immediately follows from the observation that $\partial\mu/\partial\bar\lambda$ obeys equation (3.4.1). The calculation presented above only gives the explicit form of the free term in (3.4.9).

In view of (3.4.9) it is quite natural to consider the function $N(x,y,\lambda,\bar\lambda)$ which obeys the integral equation

$$N(x,y,\lambda,\bar\lambda) = e^{ipx+iqy} - (G(\cdot,\lambda,\bar\lambda)u(\cdot)N(\cdot,\lambda,\bar\lambda))(x,y). \qquad (3.4.11)$$

Multiplying (3.4.11) by $F(\lambda,\bar\lambda)$, comparing the equation obtained with equation (3.4.9) and assuming that the homogeneous equation (3.4.3) has no nontrivial solutions, we obtain

$$\frac{\partial\mu(x,y,\lambda,\bar\lambda)}{\partial\bar\lambda} = F(\lambda,\bar\lambda)N(x,y,\lambda,\bar\lambda). \qquad (3.4.12)$$

The assumption about the absence of the nontrivial solutions for the homogeneous equation (3.4.3) is a very important one. For the KP-I case the corresponding homogeneous integral equations (3.3.14) and (3.3.15) have the nontrivial solutions and the KP-I equation possesses the lump type solutions.

Now we must express $N(x,y,\lambda,\bar\lambda)$ via $\mu(x,y,\lambda,\bar\lambda)$. Taking into account the symmetry property of the Green function

$$G(x,y,-\bar\lambda,-\lambda) = G(x,y,\lambda,\bar\lambda)e^{-ipx-iqy} \qquad (3.4.13)$$

and comparing (3.4.3) with (3.4.11), we obtain

$$N(x,y,\lambda,\bar\lambda) = \mu(x,y,-\bar\lambda,-\lambda)e^{ipx+iqy}. \qquad (3.4.14)$$

As a result, by substituting (3.4.14) into (3.4.12) we arrive at the following linear $\bar\partial$-problem [119]

$$\frac{\partial\mu(x,y,\lambda,\bar\lambda)}{\partial\bar\lambda} = F(\lambda,\bar\lambda)e^{ipx+iqy}\mu(x,y,\bar\lambda,\lambda), \qquad (3.4.15)$$

where $F(\lambda,\bar\lambda)$ is given by (3.4.10) or by the formula

$$F(\lambda,\bar\lambda) = \frac{\sigma_0}{2\pi}\int_{-\infty}^{+\infty}\int_{-\infty}^{+\infty} dx\, dy\, u(x,y)N(x,y,-\bar\lambda,-\lambda). \qquad (3.4.16)$$

Note that in terms of the function $\tilde\psi = \mu e^{i\lambda x + \lambda^2 y}$ the $\bar\partial$-equation (3.4.15) looks like

$$\frac{\partial\tilde\psi(x,y,\lambda,\bar\lambda)}{\partial\bar\lambda} = F(\lambda,\bar\lambda)\tilde\psi(x,y,-\bar\lambda,-\lambda). \qquad (3.4.17)$$

The fact that the bounded in $\lambda$ solutions of equation (3.4.1) obey the linear $\bar\partial$-equation (3.4.15) is one of the most important properties of the problem (3.4.1). Just this $\bar\partial$-equation generates the inverse problem equations for the problem (3.4.1).

Indeed, using the generalized Cauchy formula (2.1.14), substituting the expression for $\partial \mu / \partial \bar{\lambda}$ given by (3.4.15) into this formula and taking into account that $\mu \to 1$ as $\lambda \to \infty$, we obtain

$$\mu(x,y,\lambda,\bar{\lambda}) = 1 + \frac{1}{2\pi i} \iint_C d\lambda' \wedge d\bar{\lambda}' \frac{F(\lambda',\bar{\lambda}')e^{ip'x+iq'y}\mu(x,y,-\bar{\lambda}',\lambda')}{\lambda' - \lambda} \qquad (3.4.18)$$

where $p' = -(\lambda' + \bar{\lambda}')$, $q' = i(\lambda'^2 - \bar{\lambda}'^2)$. Then using (3.1.9) and (3.4.18), one gets

$$u(x,y,t) = \frac{1}{\pi} \frac{\partial}{\partial x} \iint_C d\lambda \wedge d\bar{\lambda} \, F(\lambda,\bar{\lambda},t)e^{ipx+iqy}\mu(x,y,-\bar{\lambda},-\lambda). \qquad (3.4.19)$$

The integral equation (3.4.18) and formula (3.4.19) solve the inverse problem for the problem (3.4.1) [119]. The function $F(\lambda,\bar{\lambda})$ of two real variables $\lambda_R, \lambda_I$ is the inverse problem data. Given $F(\lambda,\bar{\lambda})$, one can calculate $\mu(x,y,\lambda,\bar{\lambda})$ with the use of equation (3.4.18) and then reconstruct the potential $u(x,y,t)$ by the formula (3.4.19).

The potential $u(x,y,t)$ defined by the formula (3.4.19) is complex-valued in the general case. It is a real function if the function $\mu$ obeys the condition

$$\mu(x,y,-\bar{\lambda},-\lambda) = \bar{\mu}(x,y,\lambda,\bar{\lambda}) \qquad (3.4.20)$$

and the inverse problem data obeys the constraint

$$F(-\lambda,-\bar{\lambda}) = \bar{F}(\lambda,\bar{\lambda}). \qquad (3.4.21)$$

In order to use the inverse problem equations (3.4.18) and (3.4.19) for the integration of the KP-II equation it is necessary, as usual, to find the time evolution of the inverse problem data. To do this, firstly note that $\partial \mu / \partial \bar{\lambda}$ obeys the same equation (3.4.1) ($\sigma = 1$) as $\mu$. Considering this equation in the asymptotic region $x^2 + y^2 \to \infty$ and taking into account (3.4.15), we obtain

$$\frac{\partial F(\lambda,\bar{\lambda},t)}{\partial t} = -4i(\lambda^3 + \bar{\lambda}^3)F(\lambda,\bar{\lambda},t). \qquad (3.4.22)$$

Hence

$$F(\lambda,\bar{\lambda},t) = e^{-4i(\lambda^3+\bar{\lambda}^3)t}F(\lambda,\bar{\lambda},0). \qquad (3.4.23)$$

Thus, the initial value problem for the KP-II equation is solved by the standard IST scheme [119]

$$u(x,y,0) \stackrel{(3.4.10)}{\to} F(\lambda,\bar{\lambda},0) \stackrel{(3.4.23)}{\to} F(\lambda,\bar{\lambda},t) \stackrel{(3.4.18),(3.4.19)}{\to} u(x,y,t). \qquad (3.4.24)$$

In general, the solution of the initial value problem of the KP-II equation has turned out to be extremely instructive. The discovery of the $\bar{\partial}$-method has been a very important step in the development of the IST method.

The inverse spectral transform for the KP-II equation is closely connected with the theory of generalized analytic functions. Indeed, in the case of the real potential $u(x,y,t)$ the $\bar{\partial}$-equation, in virtue of (3.4.20), is equivalent to the following equation

$$\frac{\partial \mu(x,y,\lambda,\bar{\lambda})}{\partial \bar{\lambda}} = B(\lambda,\bar{\lambda})\bar{\mu}(x,y,\lambda,\bar{\lambda}) \qquad (3.4.25)$$

where $B(\lambda,\bar{\lambda}) \doteq F(\lambda,\bar{\lambda})\exp(ipx+iqy)$. But, as we already mentioned in the Introduction (section 1.3), the solutions of equation (3.4.25) are nothing but the generalized analytic functions.

Using the properties of the generalized analytic functions (see e. g. [140–143]), one can prove the unique solvability of the inverse problem for equation (3.4.1) with an arbitrary, decreasing as $x^2+y^2 \to \infty$, real potential $u(x,y,t)$. One can prove the smoothness of the potential $u(x,y,t)$ and also some other properties [137]. So the use of the theory of generalized analytic functions allows us to give the rigorous and complete solution of the inverse problem for equation (3.4.1) and correspondingly of the initial value problem for the KP-II equation within the class of decreasing at the infinity solutions.

The rigorous investigation of the inverse problem for (3.4.1) and of the properties of the solutions of the KP equation has also been given in papers [194–196].

Emphasize that the results presented in this and previous sections are concerned with the solution of the Cauchy problem for the KP-I and KP-II equations within the class of decreasing as $x^2+y^2 \to \infty$ solutions $u(x,y,t)$. A solution of the initial value problem for the KP equation for the class of potentials $u(x,y,t)$ which includes the line solitons has been discussed in papers [197–199].

# Chapter 4

# The modified Kadomtsev–Petviashvili equation

Our second example is given by equation (1.2.4) or by the system

$$V_t + V_{xxx} - \frac{3}{2}V^2V_x + 3\sigma^2 W_y - 3\sigma V_x W = 0, \qquad (4.0.1)$$

$$W_x = V_y$$

where $\sigma^2 = \pm 1$. Equation (4.0.1) has been introduced in [84] within the framework of the gauge-invariant description of the KP equation. In [85] it has appeared as the first member of the 1-st modified KP hierarchy. In the one-dimensional limit $V_y = 0$ equation (4.0.1) is reduced to the mKdV equation. Equation (4.0.1) is equivalent to the compatibility condition for the linear system [199]

$$\sigma \Psi_y + \Psi_{xx} + V\Psi_x = 0, \qquad (4.0.2a)$$

$$\Psi_t + 4\Psi_{xxx} + 6V\Psi_{xx} + \left(3V_x - 3\sigma W_y + \frac{3}{2}V^2\right)\Psi_x + \alpha \Psi = 0 \qquad (4.0.2b)$$

where $\alpha$ is an arbitrary constant.

Generally, solutions of equation (4.0.1) are the complex-valued functions. But in the case $\sigma^2 = -1$ ($\sigma = i$) it admits the reduction to the pure imaginary $V$ while at $\sigma^2 = 1$ ($\sigma=1$) there is an obvious reduction to real $V$. In view of this it is natural to introduce a new dependent variable $u$ defined by $V = \sigma u$. In terms of $u$ equation (4.0.1) looks like

$$u_t + u_{xxx} - 3\sigma^2\left(\frac{1}{2}u^2 u_x - w_y + u_x w\right) = 0, \qquad (4.0.3)$$

$$w_x = \sigma u_y.$$

In what follows we will refer to equation (4.0.3) as the mKP equation, namely, as the mKP-I equation at $\sigma = i$ ($\sigma^2 = -1$) and as the mKP-II equation at $\sigma = 1$. Both the mKP-I and the mKP-II equations obviously admit the reduction to real $u$.

The mKP equation is of great interest by several reasons. Firstly, this equation may be relevant to the description of the water waves on the plane $(x, y)$ in a situation when similar to the mKdV case one should take into account the cubic nonlinearity. Secondly, there is a close algebraic interrelation between the mKP and KP equations similar to the KdV case. In particular, they are related by the two-dimensional Miura transformation $u_{KP} = -\frac{1}{2}\sigma\partial_x^{-1}V_y - \frac{1}{2}V_x - \frac{1}{4}V^2$ [84, 85]. Thirdly, as we shall see in the mKP equation, more exactly, the problem of finding the exact solutions of the linear equation (4.0.2a) arises within the problem of construction of the exponentially localized solitons for the Ishimori equation. Lastly, the case of the mKP equation is an interesting one from the point of view of the IST method itself since it has some features essentially different from those for the KP equation.

We will follow in part to the paper [200].

## 4.1. $\bar{\partial}$-dressing

We start with the $\bar{\partial}$-dressing on trivial background. The mKP equation belongs to the same class of the scalar integrable equations as the KP equation but with one essential difference. It is associated with the scalar $\bar{\partial}$-problem (2.3.1) with the canonical normalization $\chi \to 1$ as $\lambda \to \infty$ and [200]

$$\hat{B}_1(\lambda) = \frac{i}{\lambda}, \quad \hat{B}_2(\lambda) = \frac{1}{\sigma\lambda^2}, \quad \hat{B}_3(\lambda) = 4i\frac{1}{\lambda^3}. \tag{4.1.1}$$

The operators $D_i$ $(i = 1, 2, 3)$ are of the form

$$D_1 = \partial_x + \frac{i}{\lambda}, \quad D_2 = \partial_y + \frac{1}{\sigma}\frac{1}{\lambda^2}, \quad D_3 = \partial_t + 4i\frac{1}{\lambda^3}. \tag{4.1.2}$$

The kernel $R$ of the $\bar{\partial}$-problem, due to (2.3.21), is of the form

$$R(\lambda', \bar{\lambda}'; \lambda, \bar{\lambda}; x, y, t) = e^{F(\lambda')}R_0(\lambda', \bar{\lambda}'; \lambda, \bar{\lambda})e^{-F(\lambda)} \tag{4.1.3}$$

where $R_0$ is an arbitrary function and

$$F(\lambda) \doteq \frac{i}{\lambda}x + \frac{1}{\sigma\lambda^2}y + 4i\frac{1}{\lambda^3}t. \tag{4.1.4}$$

The canonical normalization of $\chi$ implies that at $\lambda \to \infty$

$$\chi = 1 + \frac{1}{\lambda}\chi_{-1} + \frac{1}{\lambda^2}\chi_{-2} + \ldots . \tag{4.1.5}$$

On the other hand, one assumed that near the origin $\lambda = 0$

$$\chi(\lambda, \bar{\lambda}; x, y, t) = \chi_0(x, y, t) + \lambda\chi_1 + \lambda^2\chi_2 + \ldots . \tag{4.1.6}$$

In construction of the linear problems we will follow the general scheme described in section (2.4). In contrast to the KP case now the quantities $D_i\chi$ have the singularities as $\lambda \to 0$.

The second order pole is, obviously, absent in the quantity

$$\sigma D_2\chi + D_1^2\chi = \sigma\partial_y\chi + \partial_x^2\chi + \frac{2i}{\lambda}\partial_x\chi. \tag{4.1.7}$$

But, due to (4.1.6) the quantity $\sigma D_2\chi + D_1^2\chi$ has the first order pole at $\lambda = 0$. To annihilate this pole we add the term $VD_1\chi$ to (4.1.7). The function

$$\sigma D_2\chi + D_1^2\chi + \sigma u D_1\chi = \sigma\partial_y\chi + \partial_x^2\chi + \frac{2i}{\lambda}\partial_x\chi + \sigma u\partial_x\chi + \frac{i}{\lambda}\sigma u\chi \tag{4.1.8}$$

has no singularity as $\lambda \to 0$ if

$$u(x,y,t) = -\frac{2}{\sigma}(\ln\chi_0(x,y,t))_x \tag{4.1.9}$$

where $\chi_0 = \chi(\lambda = 0, x, y, t)$. So the quantity (4.1.8) obeys the condition (2.3.15). Now, let us check the condition (2.3.16). Using (4.1.5), one readily shows that

$$\sigma D_2\chi + D_1^2\chi + \sigma u D_1\chi \to 0 \tag{4.1.10}$$

as $\lambda \to \infty$, i. e. condition (2.3.16) is satisfied.

Thus, the first linear problem is

$$(\sigma D_2 + D_1^2 + \sigma u D_1)\chi = 0. \tag{4.1.11}$$

Emphasize that the absence of the term $p(x,y,t)\chi$ in problem (4.1.11) is the direct consequence of the choice of (4.1.2) and canonical normalization of $\chi$. This is the main reason for the choice $1/\lambda$ in (4.1.2) instead of $\lambda$ as in the KP case. Note that for the KP equation the choice (3.1.1) guarantees the absence of the term $VD_1\chi$ in problem (3.1.8).

In order to construct the second linear problem we start with the quantity

$$D_3\chi + 4D_1^3\chi = \partial_t\chi + 4\partial_x^3\chi + 12i\frac{1}{\lambda}\partial_x^2\chi - 12\frac{1}{\lambda^2}\partial_x\chi \tag{4.1.12}$$

which has no third order pole at $\lambda = 0$. The residue of the quantity (4.1.12) at the second order pole at $\lambda = 0$ is equal to $-12\partial_x\chi_0$. Hence the quantity

$$D_3\chi + 4D_1^3\chi + 6\sigma u D_1^2\chi$$

$$= \partial_t\chi + 4\partial_x^3\chi + 6\sigma u\partial_x^2\chi - \frac{1}{\lambda^2}(6\sigma u\chi + 12\partial_x\chi) + \frac{1}{\lambda}(12i\partial_x^2\chi + 6\sigma iu\partial_x\chi) \tag{4.1.13}$$

has no second order singularity at $\lambda = 0$. To compensate the first order pole contribution in (4.1.13) one should add to (4.1.13) the term $V_1 D_1\chi$ with the proper function $V_1(x,y,t)$. The condition of absence of the first order singularity at $\lambda = 0$ in the quantity

$$D_3\chi + 4D_1^4\chi + 6\sigma u D_1^2\chi + V_1 D_1\chi \tag{4.1.14}$$

gives
$$iV_1\chi_0 - 6\sigma u\chi_1 - 12\partial_x\chi_1 + 12i\partial_x^2\chi_0 + 6\sigma iu\partial_x\chi_0 = 0. \tag{4.1.15}$$

Equation (4.1.4) implies
$$\sigma\partial_y\chi_0 + \partial_x^2\chi_0 + \sigma u\partial_x\chi_0 + i(\sigma u\chi_1 + 2\partial_x\chi_1) = 0, \tag{4.1.16}$$
$$\sigma\partial_y\chi_1 + \partial_x^2\chi_1 + \sigma u\partial_x\chi_1 + i(\sigma u\chi_2 + 2\partial_x\chi_2) = 0. \tag{4.1.17}$$

Using (4.1.16), one can eliminate $\chi_1$ from (4.1.15). As a result, one gets
$$V_1 = 3\sigma u_x + \frac{3}{2}\sigma^2 u^2 - 3\sigma^2 w \tag{4.1.18}$$

where
$$w = -\frac{2}{\sigma}\partial_y \ln \chi_0, \tag{4.1.19}$$

i. e.
$$w_x = u_y. \tag{4.1.20}$$

Further, it is not difficult to show that the expression (4.1.14) obeys the condition (2.3.16). Hence
$$D_3\chi + 4D_1^3\chi + 6\sigma u D_1^2\chi + \left(3\sigma u_x + \frac{3}{2}\sigma^2 u^2 - 3\sigma^2 w\right) D_1\chi = 0 \tag{4.1.21}$$

which is our second linear problem.

In terms of the function
$$\psi = \chi \exp\left(\frac{ix}{\lambda} + \frac{y}{\sigma\lambda^2} + 4i\frac{t}{\lambda^3}\right) \tag{4.1.22}$$

the linear problems (4.1.11) and (4.1.21) are of the form
$$(\sigma\partial_y + \partial_x^2 + \sigma u\partial_x)\psi = 0,$$
$$\left(\partial_t + 4\partial_x^3 + 6\sigma u\partial_x^2 + \left(3\sigma u_x + \frac{3}{2}\sigma^2 u^2 - 3\sigma^2 w\right)\partial_x\right)\psi = 0 \tag{4.1.23}$$

which is nothing but (4.0.2). The compatibility condition for (4.1.23) is the mKP equation.

The mKP equation for $u$ also follows directly from the second linear problem (4.1.21), i. e. the equation

$$\partial_t\chi + 4\partial_x^3\chi + 6\sigma u\partial_x^2\chi - \frac{1}{\lambda^2}(6\sigma u\chi + 12\partial_x\chi) + \frac{1}{\lambda}(12i\partial_x^2\chi + 6i\sigma u\partial_x\chi)$$
$$+ \left(3\sigma u_x + \frac{3}{2}\sigma^2 u^2 - 3\sigma^2 w\right)\partial_x\chi + \frac{i}{\chi}\left(3\sigma u_x + \frac{3}{2}\sigma^2 u^2 - 3\sigma^2 w\right)\chi = 0. \tag{4.1.24}$$

Chapter 4

Evaluating the l.h.s. of (4.1.24) at $\lambda = 0$, one gets

$$\chi_{0t} + 4\chi_{0xxx} + 6\sigma u\chi_{0xx} - (6\sigma u\chi_2 + 12\chi_{2x}) + (12i\chi_{1xx} + 6i\sigma u\chi_{1x})$$
$$+ \left(3\sigma u_x + \frac{3}{2}\sigma^2 u^2 - 3\sigma^2 w\right)\chi_{0x} + i\left(3\sigma u_x + \frac{3}{2}\sigma^2 u^2 - 3\sigma^2 w\right)\chi_1 = 0, \quad (4.1.25)$$

where $\chi_1$ and $\chi_2$ are defined by (4.1.6). Eliminating $\chi_1$ and $\chi_2$ from (4.1.25) with the use of equations (4.1.16) and (4.1.17), we obtain

$$\chi_{0t} + \chi_{0xxx} - \frac{3}{2}\frac{\chi_{0xx}\chi_{0x}}{\chi_0} - 3\sigma\chi_{0xy} - 3\sigma\frac{\chi_{0x}\chi_{0y}}{\chi_0} - 3\sigma w_y \chi_0 = 0, \quad (4.1.26)$$

$$w_x + \sigma\frac{\chi_{0y}}{\chi_0} + \frac{\chi_{0xx}}{\chi_0} = 0.$$

Finally, for $u = -\frac{2}{\sigma}(\ln \chi_0)_x$ one gets the mKP equation (4.0.3).

Note the primary role of the function $\chi_0(x,y,t)$ in this derivation. In fact, equation (4.1.26) is the equation for the KP wave function $\psi$ which arises after elimination of $u$ from the system (3.0.2) [104].

Now let us consider the $\bar{\partial}$-dressing on nontrivial background. According to section (2.5) we choose the operators $\hat{B}_i$ to be the integral operators with the same main singularities as the bare $\hat{B}_i$, i.e. such that

$$\hat{B}_1(\lambda) \to \frac{i}{\lambda}, \quad \hat{B}_2(\lambda) \to \frac{1}{\sigma\lambda^2}, \quad \hat{B}_3(\lambda) \to \frac{4i}{\lambda^3} \quad (4.1.27)$$

as $\lambda \to 0$.

The dual space is defined by the Fourier transform

$$\chi(s,x,y,t) = \frac{1}{2\pi}\iint_C \frac{d\lambda \wedge d\bar{\lambda}}{|\lambda|^4} \exp\left(i\left(\frac{s}{\lambda} + \frac{\bar{s}}{\bar{\lambda}}\right)\right) f(\lambda, \bar{\lambda}, x, y, t). \quad (4.1.28)$$

In the dual space the operators $\hat{B}_1$, $\hat{B}_2$ and $\hat{B}_3$, are the differential operators of the first, second and third orders, respectively.

Choosing in (2.5.14)

$$\tilde{u}_2 = -\frac{1}{\sigma}, \quad \tilde{u}_1 = \sigma u_0(z,y,t), \quad \tilde{u}_0 = 0,$$

$$\tilde{V}_3 = -4, \quad \tilde{V}_2 = -6\sigma u_0, \quad \tilde{V}_1 = -\left(3\sigma u_{0x} + \frac{3}{2}\sigma^2 u_0^2 - 3\sigma^2 w_0\right), \quad \tilde{V}_0 = 0 \quad (4.1.29)$$

one has in the dual space the system

$$\left(\partial_y + \frac{1}{\sigma}\partial_z^2 + u_0(z,y,t)\partial_z\right)X_0 = 0,$$

$$\left(\partial_t + 4\partial_z^3 + 6\sigma u_0 \partial_z^2 + \left(3\sigma u_{0z} + \frac{3}{2}\sigma^2 u_0^2 - 3\sigma^2 w_0\right)\partial_z\right) X_0 = 0$$

where $w_{0z}(z,y,t) = u_{0y}(z,y,t)$. The system (4.1.30) defines the mKP equation (4.0.3) in the dual space $(z,y,t)$.

So the background $u_0$ is given by the solution of the mKP equation and the functions $X_0$ and $Y_0$ are the corresponding mKP wave functions. Note that for the transformation (4.1.28) one has

$$(V(s)f(s))(\lambda) = \iint_C \frac{d\mu \wedge d\bar{\mu}}{|\mu|^4} \iint_C \frac{dv \wedge d\bar{v}}{|v|^4} \delta\left(\frac{1}{\lambda} - \frac{1}{\mu} - \frac{1}{v}\right) V(\mu)f(v)$$

$$= \iint_C \frac{d\mu \wedge d\bar{\mu}}{|\mu|^4} V\left(\frac{\lambda\mu}{\mu - \lambda}\right) f(\mu). \tag{4.1.30}$$

Hence, in the original spectral space the operators $\hat{B}_i$ act as follows

$$(\hat{B}_1 f)(\lambda) = \frac{i}{\lambda}f(\lambda),$$

$$(\hat{B}_2 f)(\lambda) = \frac{1}{\sigma}\frac{1}{\lambda^2}f(\lambda) - i\iint_C \frac{d\mu \wedge d\bar{\mu}}{|\mu|^4} \hat{u}_0\left(\frac{\lambda\mu}{\mu-\lambda}, x, y, t\right)\frac{1}{\mu}f(\mu), \tag{4.1.31}$$

$$(\hat{B}_3 f)(\lambda) = 4i\frac{1}{\lambda^3}f(\lambda) + 6\sigma \iint_C \frac{d\mu \wedge d\bar{\mu}}{|\mu|^4} \hat{u}_0\left(\frac{\lambda\mu}{\mu-\lambda}\right)\frac{1}{\mu^2}f(\mu)$$

$$-\iint_C \frac{d\mu \wedge d\bar{\mu}}{|\mu|^4}\left\{3\sigma\hat{u}_0\left(\frac{\lambda\mu}{\mu-\lambda}\right)i\left(\frac{1}{\lambda}-\frac{1}{\mu}\right) + \frac{3}{2}\sigma^2\hat{u}_0^2\left(\frac{\lambda\mu}{\mu-\lambda}\right) - 3\sigma^2\hat{w}_0\left(\frac{\lambda\mu}{\mu-\lambda}\right)\right\}f(\mu)$$

where

$$\hat{u}_0(\lambda, x, y, t) \doteq \frac{1}{2\pi}\iint_C ds \wedge d\bar{s}\exp\left(-i\left(\frac{s}{\lambda} + \frac{\bar{s}}{\bar{\lambda}}\right)\right)u_0(s+x, y, t),$$

$$\hat{w}_0(\lambda, x, y, t) \doteq \frac{1}{2\pi}\iint_C ds \wedge d\bar{s}\exp\left(-i\left(\frac{s}{\lambda} + \frac{\bar{s}}{\bar{\lambda}}\right)\right)w_0(s+x, y, t). \tag{4.1.32}$$

Thus, the operators $D_i = \partial_{x_i} + \hat{B}_i^*$ are defined as

$$(D_1\chi)(\lambda) = \left(\partial_x + \frac{i}{\lambda}\right)\chi(\lambda),$$

$$(D_2\chi)(\lambda) = \left(\partial_y + \frac{1}{\sigma\lambda^2}\right)\chi(\lambda) - \frac{1}{\lambda}\iint_C \frac{d\mu \wedge d\bar{\mu}}{|\mu|^4}\hat{u}_0(\mu)\chi\left(\frac{\lambda\mu}{\lambda+\mu}\right), \tag{4.1.33}$$

$$(D_3\chi)(\lambda) = (\partial_t + \frac{4i}{\lambda^3})\chi(\lambda) + \frac{6\sigma}{\lambda^2}\iint_C \frac{d\mu \wedge d\bar{\mu}}{|\mu|^4}\hat{u}_0(\mu)\chi\left(\frac{\lambda\mu}{\lambda+\mu}\right)$$

$$-\frac{i}{\lambda}\iint_C \frac{d\mu \wedge d\bar{\mu}}{|\mu|^4}\left(\frac{3\sigma i}{\mu}\hat{u}_0(\mu) + \frac{3}{2}\sigma^2\hat{u}_0^2(\mu) - 3\sigma^2\hat{w}_0(\mu)\right)\chi\left(\frac{\lambda\mu}{\lambda+\mu}\right).$$

Note that the operators $D_i$ are not the convolutive operators in contrast to the KP case.

Now we are in a position to construct the dressed operators $L_1$ and $L_2$.

First, we note that the quantity

$$\sigma D_2\chi + D_1^2\chi = \sigma\partial_y\chi + \partial_x^2\chi + \frac{2i}{\chi}\partial_x\chi - \frac{\sigma}{\lambda}\iint_C \frac{d\mu \wedge d\bar{\mu}}{|\mu|^4}\hat{u}_0(\mu)\chi\left(\frac{\lambda\mu}{\lambda+\mu}\right) \quad (4.1.34)$$

has no second order pole at $\lambda = 0$. But it has the first order pole. Its contribution can be compensated by the term $V_1 D_1\chi$. The requirement that the residue of the quantity

$$\sigma D_2\chi + D_1^2 + V_1 D_1\chi \quad (4.1.35)$$

at the first order pole at $\lambda = 0$ is equal to zero gives

$$V_1\chi(0) + 2\partial_x\chi(0) - \sigma\iint_C \frac{d\mu \wedge d\bar{\mu}}{|\mu|^4}\hat{u}_0(\mu)\chi(0) = 0. \quad (4.1.36)$$

Using (4.1.32) and the property of the delta-function $\delta(1/\mu)$, one gets

$$V_1 = \sigma u(x,y,t) = \sigma u_0(x,y,t) - \frac{2i}{\sigma}\partial_x \ln\chi(0). \quad (4.1.37)$$

Further, one can readily show that

$$\sigma D_2\chi + D_1^2\chi + \sigma u D_1\chi \to 0$$

as $\lambda \to \infty$. Thus the first linear problem in the case of nontrivial background is

$$\sigma D_2\chi + D_1^2\chi + \sigma u D_1\chi = 0 \quad (4.1.38)$$

where $u$ is given by (4.1.37) and the operators $D_2$ and $D_1$ act according to (4.1.33).

The second linear problem can be constructed similar to the KP case. It is not difficult to show that the second problem is of the form

$$D_3\chi + 4D_1^3\chi + 6\sigma u D_1^2\chi + \left(3\sigma u_x + \frac{3}{2}\sigma^2 u^2 - 3\sigma^2 w\right)D_1\chi = 0 \quad (4.1.39)$$

where $u$ is given by (4.1.37) and the action of operators $D_i$ defined in (4.1.33).

Again similar to the KP equation the linear problems (4.1.38), (4.1.39) are exactly the same as those of (4.1.11) and (4.1.21) for trivial background. Only the operators $D_i$ are different and the dressing formula (4.1.37) is the generalization of (4.1.9) to nontrivial background $u_0$.

The evaluation of equation (4.1.39) at $\lambda = 0$ gives rise to the mKP equation for $u$. This result readily follows from the compatibility of the problems (4.1.38), (4.1.39) and commutativity of the operators $D_i$ ($i = 1, 2, 3$).

## 4.2. Exact solutions

Now we will construct the explicit exact solutions of the mKP equation using the exact solutions of the $\bar{\partial}$-problem given in section (2.2). They was first obtained in [200].

Using the formulae (4.1.9) and (2.2.3), one gets for trivial background

$$u(x,y,t) = -\frac{2}{\sigma}\partial_x \ln\left(1 + \frac{1}{2\pi i}\iint_C \frac{d\lambda \wedge d\bar{\lambda}}{\lambda} \iint_C d\mu \wedge d\bar{\mu}\, \chi(\mu,\bar{\mu})R(\mu,\bar{\mu};\lambda,\bar{\lambda})\right) \quad (4.2.1)$$

where

$$R(\mu,\bar{\mu};\lambda,\bar{\lambda}) = e^{F(\mu)}R_0(\mu,\bar{\mu};\lambda,\bar{\lambda})e^{-F(\lambda)} \quad (4.2.2)$$

and

$$F(\lambda) = \frac{ix}{\lambda} + \frac{y}{\sigma\lambda^2} + \frac{4it}{\lambda^3}. \quad (4.2.3)$$

For small data the formula (4.2.1) is reduced to

$$u(x,y,t) = -\frac{1}{\pi i \sigma}\partial_x\left(\iint_C \frac{d\lambda \wedge d\bar{\lambda}}{\lambda} \iint_C d\mu \wedge d\bar{\mu}\, R_0(\mu,\bar{\mu};\lambda,\bar{\lambda})\right.$$

$$\left. \times \exp\left\{i\left(\frac{1}{\mu} - \frac{1}{\lambda}\right)x + \frac{1}{\sigma}\left(\frac{1}{\mu^2} - \frac{1}{\lambda^2}\right)y + 4i\left(\frac{1}{\mu^3} - \frac{1}{\lambda^3}\right)t\right\}\right). \quad (4.2.4)$$

In virtue of (4.1.9) for the reality of $u$ it is sufficient that (for general $u$)

$$|\chi_0| = 1 \quad (4.2.5)$$

at $\sigma = i$ and

$$\chi_0 = \bar{\chi}_0 \quad (4.2.6)$$

at $\sigma = 1$.

Using (4.2.4), one gets the following reality constraints in terms of $R_0$ for mKP-I

$$R_0(\lambda',\bar{\lambda}';\lambda,\bar{\lambda})\lambda' = \overline{R_0(\bar{\lambda},\lambda;\bar{\lambda}',\lambda')}\lambda \quad (4.2.7)$$

and for mKP-II

$$R_0(\lambda',\bar{\lambda}';\lambda,\bar{\lambda}) = \overline{R_0(-\bar{\lambda}',-\lambda';-\bar{\lambda},-\lambda)}. \quad (4.2.8)$$

Iterating the formula (4.2.1), it is not difficult to show that the conditions (4.2.7), (4.2.8) are the reality conditions for nonsmall $u(x,y,t)$ too. The reality constraints (4.2.7) and (4.2.8) are preserved in time due to (4.2.2).

**Solutions with functional parameters.** They correspond to the degenerate kernel $R_0$

$$R_0(\mu,\bar{\mu};\lambda,\bar{\lambda}) = \sum_{k=1}^{N} f_{0k}(\mu,\bar{\mu})g_{0k}(\lambda,\bar{\lambda}) \quad (4.2.9)$$

where $f_{0k}$ and $g_{0k}$ are arbitrary functions. Using the general formula (2.2.22) with $n = 1$ and (2.2.20), (2.2.21) and the reconstruction formula (4.1.9), one gets

$$u = -\frac{2}{\sigma}\partial_x \ln\left(1 + \frac{1}{2i}\sum_{l,k=1}^{N}\eta_k(1+A)_{kl}^{-1}\xi_l\right) \quad (4.2.10)$$

where

$$\xi_l(x,y,t) = \iint_C d\lambda \wedge d\bar\lambda\, f_{0l}(\lambda,\bar\lambda)\exp\left(\frac{i}{\lambda}x + \frac{1}{\sigma\lambda^2}y + 4i\frac{1}{\lambda^3}t\right), \quad (4.2.11)$$

$$\eta_k(x,y,t) = \iint_C \frac{d\lambda \wedge d\bar\lambda}{\lambda} g_{0k}(\lambda,\bar\lambda)\exp\left(-\frac{i}{\lambda}x - \frac{1}{\sigma\lambda^2}y - 4i\frac{t}{\lambda^3}\right) \quad (4.2.12)$$

and

$$A_{lk}(t) = \frac{1}{2\pi i}\iint_C d\lambda \wedge d\bar\lambda \iint_C d\mu \wedge d\bar\mu\, \frac{g_{0l}(\mu,\bar\mu)f_{0k}(\lambda,\bar\lambda)}{\lambda - \mu}$$

$$\times \exp\left(i\left(\frac{1}{\lambda} - \frac{1}{\mu}\right)x + \frac{1}{\sigma}\left(\frac{1}{\lambda^2} - \frac{1}{\mu^2}\right)y + 4i\left(\frac{1}{\lambda^3} - \frac{1}{\mu^3}\right)t\right). \quad (4.2.13)$$

The solutions (4.2.10) are parametrized by $2n$ arbitrary complex functions $f_k$ and $g_k$ ($k = 1,\ldots,N$).

One can represent these solutions in a form which contains only the functions $\xi_k$ and $\eta_k$. First, we note that

$$\chi_0 = 1 + \operatorname{tr} M \quad (4.2.14)$$

where

$$M_{kl} \doteq \frac{1}{2i}\eta_k(1+A)_{kl}^{-1}\xi_l. \quad (4.2.15)$$

The matrix $M$ has rank one. Hence, $\det(1+M) = 1 + \operatorname{tr} M$ and, consequently, $\chi_0 = \det(1+M)$. Then, using the identity

$$\frac{1}{\lambda(\mu-\lambda)} = \frac{1}{\lambda\mu} - \frac{1}{\mu(\lambda-\mu)}, \quad (4.2.16)$$

one can show that

$$\det(1+M) = \frac{\det(1+A+B)}{\det(1+A)} \quad (4.2.17)$$

where $B_{kl} = 1/(2i\pi)\xi_k\eta_l$. Further, using the formal identity (3.2.12) with the change of variable $\lambda \to \bar\lambda'$, one gets

$$A_{kl} = -\frac{1}{2i\pi}\partial_x^{-1}(\eta_l\xi_{kx}) \quad (4.2.18)$$

where the integration $\partial_x^{-1}$ in (4.2.18) is chosen in a way which guarantees the existence of $A_{kl}$. Using (4.2.17) and (4.2.18), we finally obtain

$$u(x,y,t) = -\frac{2}{\sigma}\partial_x \ln\frac{\det(1+\tilde A)}{\det(1+A)} \quad (4.2.19)$$

where
$$\tilde{A}_{kl} = \frac{1}{2i\pi}\partial_x^{-1}(\xi_k\eta_{lx}), \tag{4.2.20}$$

$$A_{kl} = -\frac{1}{2i\pi}\partial_x^{-1}(\xi_{kx}\eta_l). \tag{4.2.21}$$

In virtue of (4.2.11) and (4.2.12) the functions $\xi_k$ and $\eta_k$ are the general solutions of the systems
$$\sigma\xi_y + \xi_{xx} = 0, \quad \xi_t + 4\xi_{xxx} = 0 \tag{4.2.22}$$

and
$$\sigma\eta_y - \eta_{xx} = 0, \quad \eta_t + 4\eta_{xxx} = 0. \tag{4.2.23}$$

Consequently, $\xi$ and $\eta$ obey the equation
$$\xi_t + \xi_{xxx} + 3\sigma^2\partial_x^{-1}\xi_{yy} = 0 \tag{4.2.24}$$

which is the linear part of the mKP equation. Note that it coincides with that for the KP equation.

So the exact solutions (4.2.19) of the mKP equation are parametrized by arbitrary general solutions of the systems (4.2.22) and (4.2.23) or by the special solutions (4.2.11), (4.2.12) of equation (4.2.24).

The reality conditions (4.2.7) and (4.2.8) imply certain constraints on the functions $f_{0k}$ and $g_{0k}$ ($k = 1, \ldots, n$). They are satisfied, in particular, if
$$R_k g_{0k}(\lambda, \bar{\lambda}) = \overline{f_{0k}(\lambda, \bar{\lambda})}\lambda \tag{4.2.25}$$

for the mKP-I equation where $R_k$ are arbitrary real constants and
$$\overline{f_{0k}(\lambda, \bar{\lambda})} = f_{0k}(-\bar{\lambda}, -\lambda),$$
$$\overline{g_{0k}(\lambda, \bar{\lambda})} = g_{0k}(-\bar{\lambda}, -\lambda) \tag{4.2.26}$$

for the mKP-II equation. The condition (4.2.25) implies
$$\bar{\xi}_k = -R_k\eta_k. \tag{4.2.27}$$

As a result, one can show using (4.2.19), that the real-valued solutions of the mKP-I equation can be represented as
$$u = 4\frac{\partial}{\partial x}\arg\det C \tag{4.2.28}$$

where
$$C_{kl} = A_{kl}R_l = R_k\delta_{kl} + \frac{\pi}{2i}\partial_x^{-1}(\xi_{kx}\bar{\xi}_l). \tag{4.2.29}$$

The simplest solution of the mKP-I equation of this type is

$$u(x,y,t) = 4\partial_x \text{arctg} \frac{R|\xi(x,y,t)|^2/2}{1 + \text{Re}(i\int_\infty^x dx' \xi_{x'}(x',y,t)\bar{\xi}(x',y,t))}. \qquad (4.2.30)$$

Correspondingly, the condition (4.2.26) for the mKP-II equation means

$$\bar{\xi}_k = -\xi_k, \quad \bar{\eta}_k = \eta_k. \qquad (4.2.31)$$

**Line solitons and breathers.** The class of exact solutions with functional parameters constructed above contains as the particular cases the line solitons and breathers of the mKP equation.

The real-valued line solitons of the mKP-I equation correspond to the choice

$$f_{0k}(\lambda, \bar{\lambda}) = R_k \delta(\lambda - \lambda_k),$$

$$g_{0k}(\lambda, \bar{\lambda}) = \lambda \delta(\lambda - \bar{\lambda}_k) \qquad (4.2.32)$$

where $R_k$ are arbitrary real constants, i. e.

$$\xi_l(x,y,t) = -2iR_l \exp(F(\lambda_l)),$$

$$\eta_k(x,y,t) = -2i \exp(-F(\bar{\lambda}_k)). \qquad (4.2.33)$$

In this case the solutions are of the form

$$u(x,y,t) = 4\frac{\partial}{\partial x} \arg \det(1 + A) \qquad (4.2.34)$$

where

$$A_{lk} = \frac{2i}{\pi} \sum_{\substack{k=1 \\ k \neq l}}^{n} \frac{R_k \bar{\lambda}_k}{\lambda_l - \bar{\lambda}_k} \exp(F(\lambda_l) - F(\bar{\lambda}_k)) \qquad (4.2.35)$$

and

$$F(\lambda_k) \doteq i\left(\frac{x}{\lambda_k} - \frac{2y}{\lambda_k^2} + \frac{4t}{\lambda_k^3}\right).$$

The simplest soliton looks like ($n = 1$, $\lambda_1 = \lambda_R + i\lambda_I$)

$$u(x,y,t) = -4\frac{2\frac{\lambda_I}{|\lambda|^2} \text{sgn} R_1}{e^{2f} + \left(e^{-f} + \frac{\lambda_R}{\lambda_I}(\text{sgn} R_1)e^f\right)^2} \qquad (4.2.36)$$

where

$$f = \frac{\lambda_I}{|\lambda|^2}\left(x - \frac{2\lambda_R}{|\lambda|^2}y - \frac{4(\lambda_I^2 - 3\lambda_R^2)t}{|\lambda|^4} + \frac{|\lambda|^2}{\lambda_I}\ln|R_1|\right). \qquad (4.2.37)$$

This solution is evidently regular in $x$ and $y$ and is a constant along the direction, $x - \frac{2\lambda_R}{|\lambda|^2}y = $ const, i. e. it is the line soliton. General solution (4.2.34) describes the scattering of $N$ line solitons of the form (4.2.37). At $\lambda_R = 0$ the solutions (4.2.34) are reduced to the mKdV solitons (see e.g. [39]).

The line soliton's eigenfunctions in the points $\lambda_k$ can be found from the system

$$\chi(\lambda_k) + \frac{2i}{\pi}\sum_{l=1}^{n} \frac{R_l \bar{\lambda}_l}{\lambda_k - \bar{\lambda}_l} \exp(F(\lambda_l) - F(\bar{\lambda}_l))\chi(\lambda_l) = 1 \quad (k=1,\ldots,n) \quad (4.2.38)$$

which follows from (2.3.2).

For the mKP-II equation the real line solitons, in virtue of (4.2.26), correspond to the kernel $R_0$ of the form

$$R_0 = \sum_k R_k \delta(\lambda - i\alpha_k)\delta(\mu - i\beta_k) \quad (4.2.39)$$

where $R_k$, $\alpha_k$ and $\beta_k$ are arbitrary real constants. So one has

$$\xi_l = -2iR_l \exp(F(i\alpha_l)),$$

$$\eta_l = -2\beta_l^{-1} \exp(-F(i\beta_l)) \quad (4.2.40)$$

in (4.2.20) and (4.2.21). These solutions of the mKP-II equation are

$$u(x,y,t) = -2\frac{\partial}{\partial x} \ln \frac{\det(1+A+B)}{\det(1+A)} \quad (4.2.41)$$

where

$$A_{km} = \delta_{km} + \frac{R_m}{\alpha_k - \beta_m} \exp(F(i\alpha_k) - F(i\beta_m)) \quad (4.2.42)$$

and

$$B_{km} = 2R_m \beta_m^{-1} \exp(F(i\alpha_k) - F(i\beta_m)). \quad (4.2.43)$$

The simplest line soliton of the mKP-II equation is (N=1)

$$u(x,y,t) = -\frac{2(\alpha-\beta)^2}{\alpha\beta^2} \frac{\varepsilon}{(e^{-f} - \frac{\alpha}{\beta}\varepsilon e^f)(e^{-f} - \varepsilon e^f)} \quad (4.2.44)$$

where

$$2f = \left(\frac{1}{\alpha} - \frac{1}{\beta}\right)x - \left(\frac{1}{\alpha^2} - \frac{1}{\beta^2}\right)y - 4\left(\frac{1}{\alpha^3} - \frac{1}{\beta^3}\right)t + \ln 2\left|\frac{R}{\beta-\alpha}\right|$$

and $\varepsilon = \text{sgn}\left(\frac{R}{\beta-\alpha}\right)$.

The solution (4.2.44) is nonsingular if $\varepsilon < 0$ and $\alpha, \beta^{-1} > 0$.

The real-valued solutions of the mKP equation of the breather type correspond to the kernel $R_0$ with an even number of delta-function contributions. For the mKP-I equation, in virtue of (4.2.7) the suitable kernel $R_0$ is

$$R_0(\lambda', \bar{\lambda}'; \lambda, \bar{\lambda}) = \pi\lambda \sum_{k=1}^{n} [R_k \delta(\lambda' - \lambda_k^+)\delta(\lambda - \lambda_k^-) + \bar{R}_k \delta(\lambda' - \overline{\lambda_k^-})\delta(\lambda - \overline{\lambda_k^+})] \quad (4.2.45)$$

where $R_k$, $\lambda_k^+$ and $\lambda_k^-$ ($k = 1, \ldots, N$) are arbitrary complex parameters. The corresponding real solutions of the mKP-I equation are of the form

$$u(x,y,t) = 2i\frac{\partial}{\partial x} \ln \frac{\overline{\det(1+A)}}{\det(1+A)} = 4\frac{\partial}{\partial x} \arg\ \det(1+A) \quad (4.2.46)$$

where $A$ is the $2N \times 2N$ matrix:

$$A_{km} = \delta_{km} + 2i\frac{\tilde{R}_m \tilde{\lambda}_m}{\lambda_k - \tilde{\lambda}_m} \exp(F(\lambda_k) - F(\tilde{\lambda}_m)) \quad (k,m = 1, \ldots, 2n) \quad (4.2.47)$$

where

$$(\lambda_1, \ldots, \lambda_{2n}) \doteq (\lambda_1^+, \ldots, \lambda_n^+; \overline{\lambda_1^-}, \ldots, \overline{\lambda_n^-}),$$

$$(\tilde{\lambda}_1, \ldots, \tilde{\lambda}_{2n}) \doteq (\lambda_1^-, \ldots, \lambda_n^-; \overline{\lambda_1^+}, \ldots, \overline{\lambda_n^+}), \quad (4.2.48)$$

$$(\tilde{R}_1, \ldots, \tilde{R}_{2n}) \doteq (R_1, \ldots, R_n; \bar{R}_1, \ldots, \bar{R}_n).$$

The solutions (4.2.46) are rather complicated. But it is not difficult to see that they are of the breather type. For instance, in the simplest case $N = 1$ the $\det(1+A)$ looks like

$$\det(1+A) = 1 + e^f \sum_{k=1}^{2} a_k e^{i(\varphi+\delta_k)} + be^{2f} \quad (4.2.49)$$

where $a_1, a_2, b$ are certain constants and

$$f = \left(\frac{\lambda_{1I}}{|\lambda_1|^2} - \frac{\lambda_{2I}}{|\lambda_2|^2}\right)x + 4\left(\frac{\lambda_{2R}\lambda_{2I}}{|\lambda_2|^4} - \frac{\lambda_{1R}\lambda_{1I}}{|\lambda_1|^4}\right)y$$

$$+ 4\left(\frac{\lambda_{1I}^3 + 3\lambda_{1R}^2\lambda_{1I}}{|\lambda_1|^6} - \frac{\lambda_{2I}^3 + 3\lambda_{2R}^2\lambda_{2I}}{|\lambda_2|^6}\right)t,$$

$$\varphi = \left(\frac{\lambda_{1R}}{|\lambda_1|^2} - \frac{\lambda_{2R}}{|\lambda_2|^2}\right)x - 2\left(\frac{\lambda_{1R}^2 - \lambda_{1I}^2}{|\lambda_1|^4} - \frac{\lambda_{2R}^2 - \lambda_{2I}^2}{|\lambda_2|^4}\right)y \quad (4.2.50)$$

$$+ 4\left(\frac{\lambda_{1R}^3 - 3\lambda_{1R}\lambda_{1I}^2}{|\lambda_1|^6} - \frac{\lambda_{2R}^3 - 3\lambda_{2R}\lambda_{2I}^2}{|\lambda_2|^6}\right)t.$$

So, the solution (4.2.46) contains both oscillation and decreasing terms. This simplest solution at $\lambda_1^+ = -\lambda_1^-$ is reduced to the well-known breather of the mKdV equation (see e.g. [39]).

We will also present two particular cases of the solutions (4.2.46) of the mKP-I equation. The first one corresponds to $\lambda^+ = i\nu_1$, $\lambda^- = i\nu_2$ where $\nu_1$ and $\nu_2$ are arbitrary real constants and $R_1 = \bar{R}_2 = b$. It is of the form

$$u(x,y,t) = 8\left(\frac{1}{\nu_2} - \frac{1}{\nu_1}\right) \frac{e^{-f}\cos\varphi(1 - \left(\frac{\nu_1+\nu_2}{\nu_1-\nu_2}\right)^2 e^{2f})}{\left(e^{-f} - \frac{2(\nu_1+\nu_2)}{\nu_1-\nu_2}\sin\varphi + \left(\frac{\nu_1+\nu_2}{\nu_1-\nu_2}\right)^2 e^f\right)^2 + 4\cos^2\varphi} \quad (4.2.51)$$

where

$$f(x,t) = \left(\frac{1}{\nu_1} - \frac{1}{\nu_2}\right)x - 4\left(\frac{1}{\nu_1^3} - \frac{1}{\nu_2^3}\right)t + \ln|b|,$$

$$\varphi(y) = \left(\frac{1}{\nu_1^2} - \frac{1}{\nu_2^2}\right)y + \delta, \quad b = |b|e^{i\delta}. \quad (4.2.52)$$

The solution (4.2.51) is periodic in $y$ and has a soliton behaviour along the coordinate $x$.

Another special breather type solution corresponds to $\lambda^+ = -\overline{\lambda^-} = \lambda$, $R_1 = \bar{R}_2 = b$ and looks like

$$u(x,y,t) = 4\lambda_R \frac{e^f \sin\varphi(1 - \frac{|\lambda|^4}{\lambda_R^2 \lambda_I^2} e^{2f})}{\left(1 - \frac{2\lambda_I}{\lambda_R}e^f\cos\varphi - \frac{(\lambda_R^2 - \lambda_I^2)}{\lambda_R^2 \lambda_I^2}|\lambda|^2 e^{2f}\right)^2 + 4\left(\frac{|\lambda|^2}{\lambda_R \lambda_I} e^{2f} - e^f\cos\varphi\right)^2} \quad (4.2.53)$$

where

$$f(y) = -4\frac{\lambda_R \lambda_I}{|\lambda|^4}y + \ln|b|,$$

$$\varphi(x,t) = \frac{2\lambda_R}{|\lambda|^2}x + \frac{8(\lambda_R^3 - 3\lambda_R\lambda_I^2)t}{|\lambda|^6} + \delta, \quad b = |b|e^{i\delta}. \quad (4.2.54)$$

This solution decreases exponentially as $|y| \to \infty$, it is periodic in $x$ and move along the axis $x$ with the velocity $4\frac{3\lambda_I^2 - \lambda_R^2}{|\lambda|^4}$.

In a similar manner one can construct the breathers for the mKP-II equation. So, the structure of the breather type solutions of the mKP equation is rather rich.

Note that the solutions (4.2.51) and (4.2.53) are not bounded on the whole plane $(x,y)$. The singularity of the solution (4.2.53), as it is not difficult to see, is of the type $\varepsilon^{-1/2}$ ($\varepsilon \to 0$) at the discrete points.

**Lumps.** Now let us proceed to rational solutions of the mKP equation. We choose the kernel $R_0$ of the $\bar{\partial}$-problem in the form (see (2.2.31))

$$R_0(\mu, \bar{\mu}; \lambda, \bar{\lambda}) = \frac{\pi}{2i}\sum_{k=1}^n S_k(\mu, \lambda)\delta(\mu - \lambda_k)\delta(\lambda - \lambda_k) \quad (4.2.55)$$

where $S_k$ are some functions and $\lambda_1, \ldots, \lambda_N$ is the set of isolated points distinct from the origin $\lambda = 0$. The function $\chi$ is given by the formula

$$\chi(\lambda, \bar{\lambda}) = 1 + \sum_{k=1}^{N} \frac{\chi_k S_k}{\lambda - \lambda_k} \qquad (4.2.56)$$

where $\chi_k = \chi(\lambda_k)$. The quantities $\chi_k$ obey the system of equations (2.2.35) that is

$$\chi_i(1 + S'_i - S_i F'(\lambda_i)) + \sum_{k \neq i} \frac{\chi_k S_k}{\lambda_k - \lambda_i} = 1 \quad (i = 1, \ldots, N) \qquad (4.2.57)$$

where

$$S_i \equiv \left. \frac{\partial S_i(\lambda_i, \lambda)}{\partial \lambda} \right|_{\lambda = \lambda_i}$$

and

$$F'(\lambda_i) = -\frac{ix}{\lambda_i^2} - \frac{2y}{\sigma \lambda_i^3} - \frac{12it}{\lambda_i^4}.$$

Solving the system (4.2.57) with respect to $\chi_i$, one then finds the solution of the mKP equation by the formula (4.1.9) with

$$\chi_0 = 1 - \sum_{k=1}^{N} \frac{\chi_k S_k}{\lambda_k}. \qquad (4.2.58)$$

Since $F'(\lambda_i)$ is a the linear function on $x, y, t$ then the solutions constructed are the rational functions on $x, y$ and $t$.

These rational solutions are representable in the following compact form

$$u = -2\sigma^{-1} \frac{\partial}{\partial x} \ln \frac{\det(A + B)}{\det A} \qquad (4.2.59)$$

where the $N \times N$ matrices $A$ and $B$ defined by

$$A_{ik} = \delta_{ik}(1 + S'_i - S_i F'(\lambda_i)) - (1 - \delta_{ik}) \frac{S_k}{\lambda_i - \lambda_k}, \qquad (4.2.60)$$

$$B_{ik} = -S_k \lambda_l^{-1}. \qquad (4.2.61)$$

The reality condition of $u$ implies certain constraints on $\lambda_1, \ldots, \lambda_N$ and $S_1, \ldots, S_N$. We will consider these constraints for the particular choice of $S_i$, namely, $S_i = -i\lambda_i^2$. In this case

$$A_{ik} = \delta_{ik}\left(x - i\frac{2y}{\sigma \lambda_i} + \frac{12t}{\lambda_i^2} + \gamma_i\right) + (1 - \delta_{ik})\frac{i\lambda_k^2}{\lambda_i - \lambda_k}, \qquad (4.2.62)$$

$$B_{ik} = i\lambda_k$$

where we denote $\gamma_i \doteq 1 + S'_i$.

For the mKP-I equation ($\sigma = i$) the solutions (4.2.59) are real and bounded in the two cases.

The first case: $N = 2n$ and

$$\lambda_{n+i} = \bar{\lambda}_i, \quad \gamma_i = -\frac{i\lambda_i}{2} + C_i, \quad \gamma_{n+i} = -\frac{i\bar{\lambda}_i}{2} + \bar{C}_i \quad (i = 1, \ldots, n) \tag{4.2.63}$$

where $\lambda_i$ $(i = 1, \ldots, n)$ are arbitrary isolated points outside the real axis and $C_i$ are arbitrary constants.

The second case: arbitrary $N$ and

$$\text{Im}\,\lambda_i = 0, \quad \gamma_i = -\frac{i\lambda_i}{2} + C_i, \quad \text{Im}\,C_i = 0. \tag{4.2.64}$$

Let us first consider the case (4.2.63). In virtue of (4.2.63), one has

$$\det(A + B) = \overline{\det A}. \tag{4.2.65}$$

Hence

$$u(x, y, t) = 4\frac{\partial}{\partial x}\text{Im}\ln\det A = 4\frac{\partial}{\partial x}\text{arctg}\frac{\text{Im}\,\det A}{\text{Re}\,\det A} \tag{4.2.66}$$

where $A$ is given by (4.2.62). One can also show that under the constraint (4.2.63) the function $u(x, y, t)$ is bounded on the whole plane $(x, y)$. So, the formula (4.2.66) gives us the rational nonsingular solutions, i. e. the lumps of the mKP-I equation. The simplest, one-lump solution is of the form ($N = 1$)

$$u(x, y, t) = 4\frac{\alpha\tilde{x}^2 + \beta\tilde{y}^2 + 2\gamma\tilde{x}\tilde{y} - \alpha c^2}{((\tilde{x} - a\tilde{y})^2 + b^2\tilde{y}^2 + c^2)^2 + (\alpha\tilde{x} + 2\gamma\tilde{y})^2} \tag{4.2.67}$$

where

$$\alpha = \lambda_R, \quad \beta + 4|\lambda|^2\text{Re}\left(\frac{1}{\lambda^3}\right), \quad \gamma = 2|\lambda|^2\text{Re}\left(\frac{1}{\lambda^2}\right),$$

$$a = \frac{2\lambda_R}{|\lambda|^2}, \quad b = \frac{2\lambda_I}{|\lambda|^2}, \quad c = \frac{\lambda_R|\lambda|}{2\lambda_I}. \tag{4.2.68}$$

$$\tilde{x} = x - 3(a^2 + b^2)t + x_0, \quad \tilde{y} = y - 6at + y_0 \tag{4.2.69}$$

and $\delta^+ = \delta_R + i\delta_I$; $x_0 = \delta_R + \delta_I\frac{\lambda_R}{\lambda_I}$, $y_0 = \delta_I\frac{|\lambda|^2}{2\lambda_I}$.

The lump (4.2.67) moves with the velocity $V = (V_x, V_y) = \left(\frac{12}{|\lambda|^2}, \frac{12\lambda_R}{|\lambda|^2}\right)$ and decays as $(x^2 + y^2)^{-1}$ in all directions on the plane $(x, y)$.

The general solution (4.2.66) describes the scattering of lumps of the form (4.2.67). It follows from (4.2.66) and (4.2.60) that as $t \to \pm\infty$:

$$u(x, y, t) = \sum_{\alpha=1}^{n} u_\alpha(x - V_{\alpha x}t, y - V_{\alpha y}t) \tag{4.2.70}$$

where $u_\alpha$ are the one-lump solutions (4.2.67). Hence, the scattering of the lumps is completely trivial similar to the KP-I case: the phase shift is absent.

The lump solutions (4.2.66) are the transparent "potentials" for the problem (4.0.2a). Now let us consider the second case. The simplest solution of this type is ($N = 1$)

$$u(x,y,t) = \frac{2\lambda_1}{\left(x - \frac{2y}{\lambda_1} + \frac{12t}{\lambda_1^2} + \delta_1\right)^2 + \frac{\lambda_1^2}{4}}. \tag{4.2.71}$$

This solution does not decrease in the direction $x - \frac{2y}{\lambda_1} =$ const, i. e. it is the line lump of the mKP-I equation.

The next solution of this type is of the form ($N = 2$)

$$u(x,y,t) = \frac{2\lambda_1 X_2^2 + 2\lambda_2 X_1^2 + \frac{\lambda_1 \lambda_2}{2}\left(\frac{\lambda_1+\lambda_2}{\lambda_1-\lambda_2}\right)^2}{\left[X_1 X_2 - \frac{\lambda_1 \lambda_2}{4}\left(\frac{\lambda_1+\lambda_2}{\lambda_1-\lambda_2}\right)^2\right]^2 + \frac{1}{4}(\lambda_1 X_2 + \lambda_2 X_1)^2} \tag{4.2.72}$$

where

$$X_i = x - \frac{2y}{\lambda_i} + \frac{12t}{\lambda_i^2} + \delta_i \quad (i = 1, 2)$$

and $\delta_i$ are arbitrary real constants. It is easy to see that the solution (4.2.72) describes the scattering of the two lumps of the form (4.2.71). This solution is nonsingular at $\lambda_1 + \lambda_2 \neq 0$.

General solution (4.2.59) in the case (4.2.64) is bounded if all $\lambda_i > 0$ ($i = 1, \ldots, n$), does not decrease in the directions $x - \frac{2y}{\lambda_i} =$ const and describes the scattering of $N$ line lumps (4.2.71). It is easy to see that the scattering of the line lumps is completely trivial: the phase shift is absent.

Emphasize that the line lumps have not been found for the KP equation. Their existence for the mKP-I equation is the novelty for the 2+1-dimensional soliton equations.

In addition to the pure cases (4.2.63) and (4.2.64) one can also consider the general mixed case in which one has $2n$ points of the type (4.2.63) and $N$ points of the type (4.2.64). Such solutions of the mKP-I equation are given by the formula (4.2.59), where $A$ is the $(2n + N) \times (2n + N)$ matrix of the form (4.2.62) with $(\lambda_1, \ldots, \lambda_{2n+N}) \equiv$
$\equiv (\lambda_1, \ldots, \lambda_n, \bar{\lambda}_1, \ldots, \bar{\lambda}_n; \alpha_1, \ldots, \alpha_N)$, (Im $\alpha_k = 0$). They describe the scattering of $n$ decaying lumps (4.2.67) and $N$ line lumps (4.2.71).

For the mKP-II equation the situation is completely different. Similar to the mKP-I case the real-valued $u$ arises from the two cases:

$$1) \quad N = 2n; \quad \lambda_{k+n} = \bar{\lambda}_k, \quad \gamma_{k+n} = \bar{\gamma}_k, \quad (k = 1, \ldots, n);$$

$$2) \text{ arbitrary } N, \quad \lambda_k = i\alpha_k \quad (\text{Im } \alpha_k = 0), \quad \gamma_k = \bar{\gamma}_k, \quad (k = 1, \ldots, N). \tag{4.2.73}$$

But, as it is not difficult to see, all these rational solutions of the mKP-II equation are singular. For instance, the analog of solution (4.2.71) looks like

$$u(x,y,t) = \frac{2\alpha_1}{\frac{\alpha_1^2}{4} - \left(x + \frac{2y}{\alpha_1} - \frac{12t}{\alpha_1^2}\right)^2}. \qquad (4.2.74)$$

This solution describes the uniform motion of the two simple pole, line, opposite sign singularities (located along the lines $x + \frac{2y}{\alpha_1} = $ const) which are parallel to each other (with the distance $\alpha_1$) and move with the same velocity $12\alpha_1^{-2}$.

In addition to the solutions of the mKP equation enumerated above one can also construct the general solutions of the mixed type which includes the solutions of the different type (lumps, line solitons etc.). For instance, the solutions of the mKP-I equation which contains both line lumps and line solitons correspond to the kernel

$$R_0(\lambda, \bar{\lambda}; \mu, \bar{\mu})$$
$$= \pi\mu \sum_{k=1}^{n_1} S_k(\lambda, \mu)\delta(\lambda - \alpha_k)\delta(\mu - \alpha_k) + \pi\mu \sum_{l=1}^{n_2} R_l \delta(\lambda - \lambda_l)\delta(\mu - \bar{\lambda}_l) \qquad (4.2.75)$$

where $\alpha_k$ ($k = 1, \ldots, n_1$) are real constants, $R_l$ are real constants and

$$S_k(\lambda, \mu) = \overline{S_k(\bar{\mu}, \bar{\lambda})}.$$

These solutions are of the form

$$u(x,y,t) = 4\frac{\partial}{\partial x} \arg \det A \qquad (4.2.76)$$

where $A$ is the $(n_1 + n_2) \times (n_1 + n_2)$ matrix with elements

$$A = \begin{pmatrix} A_{kj} & A_{kN} \\ A_{Mj} & A_{MN} \end{pmatrix} \qquad (4.2.77)$$

where

$$A_{kj} = \delta_{kj}\left(x - \frac{2y}{\alpha_k} + \frac{12t}{\alpha_k^2} - \frac{i\alpha_k}{2} + \delta_k\right) + i(1-\delta_{kj})\frac{\alpha_j^2}{\alpha_k - \alpha_j}, \quad k,j = 1,\ldots,n_1;$$

$$A_{kN} = \frac{2iR_N\bar{\lambda}_N}{\alpha_k - \bar{\lambda}_N}\exp(F(\lambda_N) - F(\bar{\lambda}_N)), \quad k = 1,\ldots,n_1, \ N = 1,\ldots,n_2;$$

$$A_{Mj} = \frac{i\alpha_j^2}{\lambda_M - \alpha_j}, \quad M = 1,\ldots,n_2, \ j = 1,\ldots,n_1;$$

$$A_{MN} = \delta_{NM} + \frac{2iR_N\bar{\lambda}_N}{\lambda_M - \bar{\lambda}_N}\exp(F(\lambda_N) - F(\bar{\lambda}_N)), \quad M,N = 1,\ldots,n_2. \qquad (4.2.78)$$

The simplest solution of this type describes the scattering of the line lump (4.2.71) and line soliton (4.2.36):

$$u(x,y,t) = 4\frac{\frac{\alpha}{2} + Re^{2g}\left[\frac{\alpha\lambda_R}{\lambda_I} - \frac{\alpha^2\lambda_I}{2|\lambda|^2}a + \frac{\alpha R}{2}\frac{|\lambda|^2}{\lambda_I^2}e^{2g} - \frac{2\lambda_I}{|\lambda|^2}X_\alpha^2 + \frac{4\alpha^2}{|\lambda|^2}\frac{\lambda_I^2}{|\lambda-\alpha|^2}X_\alpha\right]}{\left(\left(X_\alpha + \frac{ab}{2}\right)Re^{2g} + \frac{\alpha}{2}\right)^2 + \left(X_\alpha\left(1 + \frac{R\lambda_R}{\lambda_I}e^{2g}\right) - \frac{\alpha R}{2}a\,e^{2g}\right)^2} \quad (4.2.79)$$

where

$$a = \left|\frac{\lambda+\alpha}{\lambda-\alpha}\right|^2, \quad b = \frac{\lambda_R}{\lambda_I} - \frac{4\alpha\lambda_I}{|\lambda-\alpha|^2},$$

$$g(x,y,t) = \frac{x\lambda_I}{|\lambda|^2} - \frac{2y\lambda_R\lambda_I}{|\lambda|^4} + \frac{4t(3\lambda_R^2\lambda_I - \lambda_I^3)}{|\lambda|^6},$$

$$X_\alpha = x - \frac{2y}{\alpha} + \frac{12t}{\alpha^2} + \delta.$$

Note that the general multiline soliton solutions of the mKP equation has been constructed also in [201, 202] by the Hirota method.

In conclusion let us consider the case of solutions $u$ decaying at infinity. The formula (4.2.4) provides necessary conditions for boundedness and decayness of solutions. They are of the form

$$\text{Im}\left(\frac{1}{\mu} - \frac{1}{\lambda}\right) = 0, \quad \text{Re}\left(\frac{1}{\sigma}\left(\frac{1}{\mu^2} - \frac{1}{\lambda^2}\right)\right) = 0. \quad (4.2.80)$$

So, for the mKP-I equation the kernel $R_0$ should be of the form

$$R_0 = \tilde{R}_0(\mu,\lambda)\delta(\mu - \bar{\mu})\delta(\lambda - \bar{\lambda}) \quad (4.2.81)$$

while for the mKP-II equation

$$R_0 = \tilde{R}_0(\lambda,\bar{\lambda})\delta(\mu + \bar{\lambda}). \quad (4.2.82)$$

Under these constraints the nonlocal $\bar{\partial}$-problem is reduced, similar to the KP equation, to the nonlocal Riemann—Hilbert problem (3.2.52) and the quasilocal $\bar{\partial}$-problem (3.2.53) respectively.

## 4.3. Initial value problem for the mKP-I equation

In order to analyze the initial value problem we, similar to the KP equation, must study the forward and inverse spectral problems for the corresponding linear problems (4.0.2).

We assume that $u(x,y,t)$ decay sufficiently rapidly as $x^2 + y^2 \to \infty$ and

$$\int_{-\infty}^{+\infty} dx'u(x',y,t) = 0.$$

The linear problem (4.0.2) is converted into the spectral problem by considering the class of solutions defined by

$$\psi = \mu(x, y, t, \lambda, \bar{\lambda}) \exp\left(i\frac{1}{\lambda}x + \frac{1}{\sigma\lambda^2}y\right) \qquad (4.3.1)$$

where $\lambda$ is arbitrary complex parameter and $\mu \to 1$ as $\lambda \to \infty$.

The function $\mu$ obeys the following linear system ($\alpha = 4i\lambda^{-3}$)

$$\sigma\mu_y + \mu_{xxx} + \frac{2i}{\lambda}\mu_x + \sigma u\mu_x + \frac{i\sigma}{\lambda}u\mu = 0 \qquad (4.3.2a)$$

$$\mu_t + 4\mu_{xxx} + 6\sigma u\mu_{xx} + \left(3\sigma u_x + \frac{3}{2}\sigma^2 u^2 - 3\sigma^2 w\right)\mu_x - \frac{1}{\lambda^2}(6\sigma u\mu + 12\mu_x)$$

$$+ \frac{3i}{\lambda}\left(4\mu_{xx} + 2\sigma u\mu_x + \sigma u_x\mu + \frac{1}{2}\sigma^2 u^2\mu - \sigma^2 w\mu\right) + 4i\frac{1}{\lambda^3}\mu = 0 \qquad (4.3.2b)$$

where $w_x = u_y$.

In this section we will consider the mKP-I equation ($\sigma = i$). In this case the spectral problem is

$$i\mu_y + \mu_{xx} + \frac{2i}{\lambda}\mu_x + u\left(i\partial_x - \frac{1}{\lambda}\right)\mu = 0. \qquad (4.3.3)$$

The reconstruction formula is of the form

$$u(x, y, t) = 2i\frac{\mu_x(x, y, t; \lambda = 0)}{\mu(x, y, t; \lambda = 0)}. \qquad (4.3.4)$$

So, we are looking for the solutions of the linear problem (4.3.3) canonically normalized and bounded for all $\lambda$ except, may be, a finite number of points. Such solutions of (4.3.3) can be found as the solutions of the linear integral equation

$$\mu(x, y, t; \lambda)$$
$$= 1 + \iint_{-\infty}^{+\infty} dx' dy'\, G(x - x', y - y'; \lambda) u(x', y', t) \left(i\frac{\partial}{\partial x'} - \lambda\right) \mu(x', y', t; \lambda) \qquad (4.3.5)$$

where $G(x - x', y - y'; \lambda)$ is the Green function for the operator $L_0 = i\partial_y + \partial_x^2 + \frac{2i}{\lambda}\partial_x$. The operator $L_0$ and the Green function $G$ are exactly the same (up to the substitution $\lambda \to \frac{1}{\lambda}$) as for the KP equation. Hence, similar to the KP case, one can construct the two Green functions $G^+$ and $G^-$ given by the following formulae

$$G^{\pm}(x, y, t; \lambda) = \frac{i}{2\pi}\int_{-\infty}^{+\infty}\frac{d\mu}{\mu}\left[\Theta\left(\pm\left(\frac{1}{\mu} - \frac{1}{\lambda}\right)\right)\Theta(y)\right.$$

$$\left. - \Theta\left(\pm\left(\frac{1}{\lambda} - \frac{1}{\mu}\right)\right)\Theta(-y)\right] e^{(\omega(x,y;\mu,\lambda))} \qquad (4.3.6)$$

where $\Theta(\xi)$ is the Heaviside (step) function:

$$\Theta(\xi) = \begin{cases} 1, & \xi > 0 \\ 0, & \xi < 0 \end{cases}$$

and

$$\omega(x,y;\mu,\lambda) \doteq ix\left(\frac{1}{\mu} - \frac{1}{\lambda}\right) - iy\left(\frac{1}{\mu^2} - \frac{1}{\lambda^2}\right). \tag{4.3.7}$$

The Green functions $G^{\pm}(x,y;\lambda)$ are analytic and bounded in the upper and lower half-planes of $\lambda$, respectively.

They allow us to define the two solutions of the problem (4.3.3) via the integral equations

$$\mu^{\pm}(x,y;\lambda) = 1 + \left[G^{\pm}(\cdot;\lambda)u(\cdot)\left(i\partial - \frac{1}{\lambda}\right)\mu^{\pm}(\cdot,\lambda)\right](x,y). \tag{4.3.8}$$

As far as the Green functions $G^+$ and $G^-$ and the solutions $\mu^+$ and $\mu^-$ of equations (4.3.8) are bounded and analytic in the upper and lower half-planes of $\lambda$, respectively, except the points $\lambda_i$ where the homogeneous equations (4.3.8) have nontrivial solutions. Further, since $G^+ - G^- \neq 0$ at Im $\lambda = 0$, then $\mu^+ - \mu^- \neq 0$ at Im $\lambda = 0$ too. Thus, one can define the function

$$\mu = \begin{cases} \mu^+, & \text{Im } \lambda > 0 \\ \mu^-, & \text{Im } \lambda < 0 \end{cases}$$

which is analytic and bounded on the entire complex plane (except a finite number of points) and has a jump across the real axis.

So, we arrive at the standard singular Riemann—Hilbert problem. The nature of this problem can be analyzed exactly in the same manner as for the KP equation. To do this one have to express the jump $\Delta(x,y,t;\lambda) \doteq \mu^+(x,y,t;\lambda) - \mu^-(x,y,t;\lambda)$ at Im $\lambda = 0$ via $\mu^-$.

Subtraction of equations (4.3.8) gives

$$\mu^+(\lambda) - \mu^-(\lambda)$$
$$= (G^+(\lambda) - G^-(\lambda))u\left(i\partial - \frac{1}{\lambda}\right)\mu^+(\lambda) + G^-(\lambda)u\left(i\partial - \frac{1}{\lambda}\right)(\mu^+(\lambda) - \mu^-(\lambda)). \tag{4.3.9}$$

Using the expression for $G^+(\lambda) - G^-(\lambda)$ at Im $\lambda = 0$

$$(G^+ - G^-)(x,y;\lambda) = \frac{i}{2\pi}\int_{-\infty}^{+\infty}\frac{d\mu}{\mu^2}\,\text{sgn}\left(\frac{1}{\mu} - \frac{1}{\lambda}\right)e^{(\omega(x,y;\mu,\lambda))} \tag{4.3.10}$$

which follows from (4.3.6) and easily verified identity

$$\left(i\partial_\xi - \frac{1}{\lambda}\right)(\phi(\xi,\eta)\exp(\omega(\xi,\eta;\mu,\lambda))) = e^{(\omega(\xi,\eta;\mu,\lambda))}\left(i\partial_\xi - \frac{1}{\mu}\right)\phi(\xi,\eta) \tag{4.3.11}$$

one gets

$$\Delta(x,y;\lambda) = \int_{-\infty}^{+\infty} \frac{d\mu}{\mu^2} T(\lambda,\mu) e^{\omega(x,y;\mu,\lambda)} + \left[G^-(\cdot;\lambda)u(\cdot)\left(i\partial - \frac{1}{\lambda}\right)\Delta(\cdot,\lambda)\right](x,y) \quad (4.3.12)$$

where

$$T(\lambda,\mu) = \frac{i}{2\pi}\operatorname{sgn}\left(\frac{1}{\mu} - \frac{1}{\lambda}\right)\iint_{-\infty}^{+\infty} d\xi d\eta\, u\left(i\partial_\xi - \frac{1}{\mu}\right)\mu^+(\xi,\eta)e^{\omega(\xi,\eta;\lambda,\mu)}. \quad (4.3.13)$$

Then we introduce the solution $N(x,y;\mu,\lambda)$ of the problem (4.3.3) which is also the solution of the integral equation

$$N(x,y;\mu,\lambda) = e^{\omega(x,y;\mu,\lambda)} + \left[G^+(\cdot,\lambda)u(\cdot)\left(i\partial - \frac{1}{\lambda}\right)N(\cdot;\mu,\lambda)\right](x,y) \quad (4.3.14)$$

where $\omega$ is given by (4.3.7). Comparing the integral equations (4.3.12) and (4.3.14) and assuming that the homogeneous equation (4.3.8) (and (4.3.12)) has no nontrivial solution for real $\lambda$, one concludes

$$\Delta(x,y;\lambda) = \int_{-\infty}^{+\infty} \frac{d\mu}{\mu^2} T(\lambda,\mu) N(x,y;\mu,\lambda). \quad (4.3.15)$$

Now one needs to express $N$ via $\chi^-$. It follows from (4.3.8) and (4.3.14) that

$$\mu^-(x,y;\lambda)e^{\tilde\omega(x,y;\lambda)} - \left[\tilde G^-(\cdot,\lambda)u(\cdot)\left(i\partial - \frac{1}{\lambda}\right)\mu^-(\cdot,\lambda)e^{\tilde\omega(\cdot;\lambda)}\right](x,y) = e^{\tilde\omega(x,y;\lambda)}, \quad (4.3.16)$$

$$\frac{\partial}{\partial(\lambda^{-1})}(N(x,y;\mu,\lambda)e^{\tilde\omega(x,y,\lambda)}) - \left[\tilde G^-(\cdot,\lambda)u(\cdot)\left(i\partial - \frac{1}{\lambda}\right)\frac{\partial(N(\cdot;\mu,\lambda)e^{\tilde\omega(\cdot,\lambda)})}{\partial(\lambda^{-1})}\right](x,y)$$
$$= e^{\tilde\omega(x,y;\lambda)} F(\mu,\lambda) \quad (4.3.17)$$

where

$$\tilde\omega(x,y;\lambda) = i\left(\frac{x}{\lambda} - \frac{y}{\lambda^2}\right),$$

the Green function $\tilde G^-$ acts as

$$(\tilde G^- \phi)(x,y) = e^{\tilde\omega(x,y;\lambda)}(G^-(\cdot;\lambda)e^{-\tilde\omega(\cdot;\lambda)}\phi(\cdot))(x,y) \quad (4.3.18)$$

and

$$F(\mu,\lambda) = \frac{i}{2\pi}\iint_{-\infty}^{+\infty} d\xi d\eta\, u(\xi,\eta)\left(i\partial_\xi - \frac{1}{\lambda}\right)N(\xi,\eta;\mu,\lambda). \quad (4.3.19)$$

Comparison of equations (4.3.16) and (4.3.17) gives

$$\frac{\partial}{\partial(\lambda^{-1})}(N(x,y;\mu,\lambda)e^{\tilde\omega(x,y;\lambda)}) = F(\mu,\lambda)\mu^-(x,y;\lambda)e^{\tilde\omega(x,y;\lambda)}. \quad (4.3.20)$$

Since $N(x,y;\lambda,\lambda) = \mu^-(x,y;\lambda)$ at Im $\lambda = 0$, equation (4.3.20) implies

$$N(x,y;\mu,\lambda) = \mu^-(x,y;\mu)e^{\omega(x,y;\mu,\lambda)} - \int_\mu^\lambda \frac{d\rho}{\rho^2} F(\mu,\rho)\mu^-(x,y;\rho)e^{\omega(x,y;\rho,\lambda)}. \tag{4.3.21}$$

Substituting (4.3.21) into (4.3.15), one obtains the expression for the jump $\mu^+ - \mu^-$ via $\mu^-$ we are interesting in:

$$\mu^+(x,y;\lambda) - \mu^-(x,y;\lambda) = \int_{-\infty}^{+\infty} \frac{d\mu}{\mu^2} T(\lambda,\mu)\mu^-(x,y;\mu)e^{\omega(x,y;\mu,\lambda)}$$

$$- \int_{-\infty}^{+\infty} \frac{d\mu}{\mu^2} T(\lambda,\mu) \int_\mu^\lambda \frac{d\rho}{\rho^2} F(\mu,\rho)\mu^-(x,y;\rho)e^{\omega(x,y;\rho,\lambda)} \tag{4.3.22}$$

where the functions $T(\lambda,\mu)$ and $F(\mu,\lambda)$ are given by (4.3.13) and (4.3.19) respectively. Changing the order of integration on the r.h.s. of (4.3.22), one finally gets [200]

$$\mu^+(x,y;\lambda) - \mu^-(x,y;\lambda) = \int_{-\infty}^{+\infty} \frac{d\mu}{\mu^2} f(\lambda,\mu)\mu^-(x,y;\mu)e^{\omega(x,y;\mu,\lambda)} \tag{4.3.23}$$

where

$$f(\lambda,\mu) = T(\lambda,\mu)$$

$$-\Theta(\lambda-\mu)\int_{-\infty}^{+\infty} \frac{d\rho}{\rho^2} T(\lambda,\rho)F(\rho,\mu) + \Theta(\mu-\lambda)\int_\mu^{+\infty} \frac{d\rho}{\rho^2} T(\lambda,\rho)F(\rho,\mu) \tag{4.3.24}$$

and $\omega(x,y;\mu,\lambda)$ is given by (4.3.7).

The relation (4.3.23) demonstrates that the Riemann—Hilbert problem we are dealing with is the nonlocal Riemann—Hilbert problem. Similar to the KP equation the nonlocal Riemann—Hilbert problem (4.3.23) generates the inverse problem equations for the problem (4.3.3).

To obtain the complete set of the inverse problem equations one has to take into account the possible singularities of the functions $\mu^+$ and $\mu^-$. In our case equations (4.3.3) and (4.3.8) are not self-adjoint. As a result, the structure of singularities for the functions $\mu^+$ and $\mu^-$ may be rather complicated. Here we will restrict ourselves by considering the simplest, the simple pole singularities, i. e. we will consider the functions $\mu^+$ and $\mu^-$ of the form

$$\mu^\pm(x,y;\lambda) = 1 + \sum_{k=1}^{n_\pm} \frac{C_k^\pm \mu_k^\pm(x,y)}{\lambda - \lambda_k^\pm} + \tilde{\mu}^\pm(x,y;\lambda) \tag{4.3.25}$$

where $\tilde{\mu}^+$ and $\tilde{\mu}^-$ are the functions analytic in the upper and lower half-planes, $C_k^\pm$ are normalization constants and, as usual, $\mu_k^\pm(x,y)$ are the solutions of the homogeneous equations (4.3.8)

$$\mu_k^\pm(x,y) = \left[G^\pm(\cdot,\lambda_k^\pm)u(\cdot)\left(i\partial - \frac{1}{\lambda_k^\pm}\right)\mu_k^\pm(\cdot)\right](x,y) \quad (k=1,\ldots,n_\pm). \tag{4.3.26}$$

We will normalize the functions $\mu_k^\pm(x,y)$ as follows

$$\lim_{(x^2+y^2)^{1/2}\to\infty}\left(x-\frac{2y}{\lambda_k^\pm}\right)\mu_k^\pm(x,y)=1. \tag{4.3.27}$$

For the functions $\mu_k^\pm(x,y)$ the following important relations also hold:

$$\lim_{\lambda\to\lambda_k^\pm}\left(\mu^\pm(x,y;\lambda)-\frac{(-i\lambda_k^{\pm 2})\mu_k^\pm(x,y)}{\lambda-\lambda_k^\pm}\right)=\left(x-\frac{2y}{\lambda_k^\pm}+\gamma_k^\pm\right)\mu_k^\pm(x,y) \tag{4.3.28}$$

where $\gamma_k^\pm(t)$ are some functions independent of $x$ and $y$.

To derive the relation (4.3.28) we, similar to the KP case, introduce the functions

$$\hat{\mu}^\pm(x,y;\lambda) \doteq \mu^\pm(x,y;\lambda)e^{\tilde{\omega}(x,y,\lambda)} \tag{4.3.29}$$

and

$$\hat{\mu}_k^\pm(x,y) \doteq \mu_k^\pm(x,y)e^{\tilde{\omega}(x,y,\lambda_k^\pm)}. \tag{4.3.30}$$

They obey the integral equations

$$\hat{\mu}^\pm(x,y;\lambda)-\left[\tilde{G}^\pm(\cdot,\lambda)u(\cdot)\left(i\partial'-\frac{1}{\lambda}\right)\hat{\mu}^\pm(\cdot;\lambda)\right](x,y)=0 \tag{4.3.31}$$

and

$$\hat{\mu}_k^\pm(x,y)-\left[\tilde{G}^\pm(\cdot,\lambda_k^\pm)u(\cdot)\left(i\partial'-\frac{1}{\lambda_k^\pm}\right)\hat{\mu}_k^\pm(\cdot)\right](x,y)=0 \tag{4.3.32}$$

where

$$(\tilde{G}^\pm(\cdot,\lambda)f)(x,y) \doteq e^{\tilde{\omega}(x,y;\lambda)}(G^\pm(\cdot,\lambda)e^{-\tilde{\omega}(\cdot,\lambda)}f)(x,y) \tag{4.3.33}$$

and the functions $G^\pm$ are given by (4.3.6).

Equations (4.3.31) and (4.3.32) imply that the function

$$\hat{\phi}_k^\pm(x,y,\lambda) \doteq \left(\mu^\pm(x,y;\lambda)-\frac{C_k^\pm\mu_k^\pm(x,y)}{\lambda-\lambda_k^\pm}\right)e^{\tilde{\omega}(x,y,\lambda)} \tag{4.3.34}$$

obeys the following integral equation:

$$\hat{\phi}_k^\pm(x,y;\lambda)-\left[\tilde{G}^\pm(\cdot,\lambda)u(\cdot)\left(i\partial'-\frac{1}{\lambda}\right)\hat{\phi}_k^\pm(\cdot,\lambda)\right](x,y)$$

$$=e^{\tilde{\omega}(x,y,\lambda)}-\frac{C_k^\pm}{\lambda-\lambda_k^\pm}\left[\hat{\mu}_k^\pm-\tilde{G}^\pm(\cdot,\lambda)u\left(i\partial'-\frac{1}{\lambda}\right)\hat{\mu}^\pm\right](x,y). \tag{4.3.35}$$

Proceeding in (4.3.35) to the limit $\lambda\to\lambda_k^\pm$, one obtains

$$\hat{\Phi}_k^\pm(x,y;\lambda_k^\pm)-\left[\tilde{G}^\pm(\cdot,\lambda_k^\pm)u(\cdot)\left(i\partial'-\frac{1}{\lambda_k^\pm}\right)\hat{\phi}_k^\pm(\cdot,\lambda_k^\pm)\right](x,y)$$

$$=e^{\tilde{\omega}(x,y,\lambda_k^\pm)}-C_k^\pm\left[\frac{\partial}{\partial\lambda}\left(\tilde{G}^\pm(\cdot,\lambda)u(\cdot)\left(i\partial'-\frac{1}{\lambda}\right)\right)\bigg|_{\lambda=\lambda_k^\pm}\hat{\mu}_k^\pm(\cdot)\right](x,y)$$

$$-C_k^\pm\left[\frac{\partial\hat{\mu}^\pm}{\partial\lambda}\bigg|_{\lambda=\lambda_k^\pm}-\tilde{G}^\pm(\cdot,\lambda_k^\pm)u(\cdot)0\left(i\partial'-\frac{1}{\lambda_k^\pm}\right)\frac{\partial\hat{\mu}^\pm}{\partial\lambda}\bigg|_{\lambda=\lambda_k^\pm}\right](x,y). \tag{4.3.36}$$

*Chapter 4* 119

It follows from (4.3.33) that

$$\left[\left.\frac{\partial\left(\tilde{G}^{\pm}(\cdot,\lambda)u(\cdot)\left(i\partial'-\frac{1}{\lambda}\right)\right)}{\partial\lambda}\right|_{\lambda=\lambda_k^{\pm}} f(\cdot)\right](x,y)$$

$$= \mp\frac{1}{2\pi\lambda_k^{\pm 2}}\iint_{-\infty}^{+\infty} d\xi d\eta\, e^{\tilde{\omega}(x-\xi,y-\eta,\lambda_k^{\pm})}u(\xi,\eta)\partial_\xi f(\xi,\eta). \tag{4.3.37}$$

In virtue of (4.3.37) equation (4.3.36) is equivalent to the following

$$\left[\left(1-\tilde{G}^{\pm}(\cdot,\lambda_k^{\pm})u(\cdot)\left(i\partial'-\frac{1}{\lambda_k^{\pm}}\right)\right)\left(\left.\hat{\phi}_k^{\pm}\right|_{\lambda=\lambda_k^{\pm}}+C_k^{\pm}\left.\frac{\partial\hat{\mu}^{\pm}}{\partial\lambda}\right|_{\lambda=\lambda_k^{\pm}}\right)\right](x,y)$$

$$= e^{\tilde{\omega}(x,y,\lambda_k^{\pm})}\left[1\pm\frac{C_k^{\pm}}{2\pi\lambda_k^{\pm 2}}\iint_{-\infty}^{+\infty} d\xi d\eta\, u(\xi,\eta)\left(\partial_\xi+\frac{i}{\lambda_k^{\pm}}\right)\mu_k^{\pm}(\xi,\eta)\right]. \tag{4.3.38}$$

We assume that the singular points for equation (4.3.8) are nondegenerate. As a result, the Fredholm alternative implies

$$1\pm\frac{C_k^{\pm}}{2\pi\lambda_k^{\pm 2}}\iint_{-\infty}^{+\infty} d\xi d\eta\, u(\xi,\eta)\left(\partial_\xi+\frac{i}{\lambda_k^{\pm}}\right)\mu_k^{\pm}(\xi,\eta) = 0 \tag{4.3.39}$$

and

$$\left.\hat{\phi}_k^{\pm}\right|_{\lambda=\lambda_k^{\pm}} + C_k^{\pm}\left.\frac{\partial\hat{\mu}^{\pm}}{\partial\lambda}\right|_{\lambda=\lambda_k^{\pm}} = \gamma_k\hat{\mu}_k^{\pm}(x,y) \tag{4.3.40}$$

where $\gamma_k$ are some constants. It is easy to see that (4.3.40) is equivalent to the relation

$$\lim_{\lambda\to\lambda_k^{\pm}}\left(\mu^{\pm}(x,y;\lambda)-\frac{C_k^{\pm}\mu_k^{\pm}(x,y)}{\lambda-\lambda_k^{\pm}}\right) = \left(\frac{iC_k^{\pm}}{\lambda_k^{\pm 2}}\left(x-\frac{2y}{\lambda_k^{\pm}}\right)+\gamma_k\right)\mu_k^{\pm}. \tag{4.3.41}$$

The constants $C_k^{\pm}$ are fixed by the normalization of $\mu_k^{\pm}(x,y)$. Integral equations (4.3.31) and (4.3.32) imply that

$$\mu^{\pm}(x,y;\lambda)\xrightarrow[(x^2+y^2)^{1/2}\to\infty]{} 1, \tag{4.3.42}$$

and

$$\mu_k^{\pm}(x,y)\xrightarrow[(x^2+y^2)^{1/2}\to\infty]{} \frac{\alpha}{x-\frac{2y}{\lambda_k^{\pm}}} \tag{4.3.43}$$

where $\alpha$ is some constant. So, the functions $\mu_k^{\pm}(x,y)$ admit the normalization

$$\lim_{(x^2+y^2)^{1/2}\to\infty}\left(x-\frac{2y}{\lambda_k^{\pm}}\right)\mu_k^{\pm}(x,y) = 1. \tag{4.3.44}$$

Considering the relation (4.3.41) as $(x^2+y^2)^{1/2} \to \infty$ and taking into account (4.3.42) and (4.3.44) one gets

$$C_k^\pm = -i\lambda_k^{\pm 2}. \qquad (4.3.45)$$

With such $C_k^\pm$ the relation (4.3.41) gives (4.3.28).

Thus, we have the nonlocal Riemann—Hilbert problem with the simple poles. Using (4.3.25), (4.3.23), one gets

$$\mu(x,y;\lambda) = 1 + \sum_{k=1}^{n_+} \frac{(-i\lambda_k^{+2})\mu_k^+(x,y)}{\lambda - \lambda_k^+} + \sum_{k=1}^{n_-} \frac{(-i\lambda_k^{-2})\mu_k^-(x,y)}{\lambda - \lambda_k^-}$$
$$+ \frac{1}{2\pi i} \int_{-\infty}^{+\infty} \frac{d\mu}{\mu - \lambda} \int_{-\infty}^{+\infty} \frac{d\rho}{\rho^2} f(\mu,\rho)\mu^-(x,y;\rho)e^{\omega(x,y;\rho,\mu)}. \qquad (4.3.46)$$

Proceeding in (4.3.46) to the limits Im $\lambda \to -0$, $\lambda \to \lambda_k^+$, $\lambda \to \lambda_k^-$ and using (4.3.28), we obtain the system of equations

$$\mu^-(x,y;\lambda) + i\sum_{k=1}^{n_+} \frac{\lambda_k^{+2}\mu_k^+}{\lambda - \lambda_k^+} + i\sum_{k=1}^{n_-} \frac{\lambda_k^{-2}\mu_k^-}{\lambda - \lambda_k^-}$$
$$-\frac{1}{2\pi i}\int_{-\infty}^{+\infty} \frac{d\mu}{\mu - (\lambda - i0)} \int_{-\infty}^{+\infty} \frac{d\rho}{\rho^2} f(\mu,\rho)\mu^-(x,y;\rho)e^{\omega(x,y;\rho,\mu)} = 1, \quad (\text{Im } \lambda = 0) \qquad (4.3.47)$$

$$\left(x - \frac{2y}{\lambda_j^\pm} + \gamma_j^\pm\right)\mu_j^\pm + i\sum_{\substack{k=1\\k\neq j}}^{n_+} \frac{\lambda_k^{+2}\mu_k^+}{\lambda_j^\pm - \lambda_k^+} + i\sum_{\substack{k=1\\k\neq j}}^{n_-} \frac{\lambda_k^{-2}\mu_k^-}{\lambda_j^\pm - \lambda_k^-}$$
$$-\frac{1}{2\pi i}\int_{-\infty}^{+\infty} \frac{d\mu}{\mu - \lambda_j^\pm} \int_{-\infty}^{+\infty} \frac{d\rho}{\rho^2} f(\mu,\rho)\mu^-(\rho)e^{\omega(x,y;\rho,\mu)} = 1 \quad (j = 1,\ldots,n_\pm). \qquad (4.3.48)$$

Together with the reconstruction formula for $u$

$$u(x,y,t) = 2i\dot\partial_x \ln\left(1 + i\sum_{k=1}^{n_+} \lambda_k^+ \mu_k^+ + i\sum_{k=1}^{n_-} \lambda_k^- \mu_k^-\right.$$
$$\left. + \frac{1}{2\pi i}\int_{-\infty}^{+\infty} \frac{d\mu}{\mu} \int_{-\infty}^{+\infty} \frac{d\rho}{\rho^2} f(\mu,\rho)\mu^-(\rho)e^{\omega(x,y;\rho,\mu)}\right) \qquad (4.3.49)$$

the system of equations (4.3.47), (4.3.48) form the inverse problem equations for the problem (4.3.3). The set $\mathcal{F} \doteq \{f(\mu,\rho), -\infty < \mu,\rho < +\infty; \lambda_k^+, \gamma_k^+ \ (k=1,\ldots,n_k^+);$ $\lambda_k^-, \gamma_k^- \ (k=1,\ldots,n_-)\}$ is the inverse problem data. The inverse problem equations (4.3.47), (4.3.48) are uniquely solvable at least for small data. Solving equations (4.3.47), (4.3.48) for a given $\mathcal{F}$, one reconstructs the potential $u(x,y)$ via (4.3.49). Note that this set of inverse data is not complete, since we are restricted only by the simple pole contribution to the discrete spectrum.

To apply the inverse problem equations obtained above for the linearization of the initial value problem for the mKP-I equation one has to find, as usual, the time dependence of the inverse problem data. This can be done in a standard manner by consideration of the linear equation (4.2.26) with $\int_{-\infty}^{+\infty} dx\, u(x,y,0) = 0$ at the limit $x^2 + y^2 \to \infty$. One gets

$$\frac{\partial f(\lambda,\mu,t)}{\partial t} = 4i\left(\frac{1}{\mu^3} - \frac{1}{\lambda^3}\right) f(\lambda,\mu,t),$$

$$\frac{\partial \lambda_k^{\pm}}{\partial t} = 0, \quad \frac{\partial \gamma_k^{\pm}(t)}{\partial t} = \frac{12}{\lambda_k^{\pm 2}} \quad (k = 1,\ldots,n_{\pm}). \tag{4.3.50}$$

Hence

$$f(\lambda,\mu,t) = f(\lambda,\mu,0) e^{4i\left(\frac{1}{\mu^3} - \frac{1}{\lambda^3}\right)t},$$

$$\gamma_k^{\pm}(t) = \frac{12}{\lambda_k^{\pm 2}} t + \gamma_k^{\pm}(0), \quad \lambda_k^{\pm}(t) = \lambda_k^{\pm}(0) \tag{4.3.51}$$

where $f(\lambda,\mu,0)$ and $\gamma_k^{\pm}(0)$ are arbitrary function and constants, respectively.

The formulae (4.3.51) and the inverse problem equations (4.3.47), (4.3.48) allow us to solve the initial value problem for the mKP-I equation by the standard IST procedure

$$u(x,y,0) \to \mathcal{F}(0) \to \mathcal{F}(t) \to u(x,y,t). \tag{4.3.52}$$

The solution $u(x,y,t)$ of the mKP-I equation reconstructed from the generic inverse problem data $\mathcal{F}$ is the complex-valued one in general. Of course, the reduction to the real valued $u$ is of the main interest. Unfortunately one is unable to describe such a reduction as an involution for the function $\mu$. The constraints on the inverse problem data $\mathcal{F}$ which lead to the reality of $u$ can be obtained in a different manner, namely, by the analysis of the weak $u$ limit of the inverse problem equations. Indeed, for small (in a suitable sense) $u(x,y,t)$ one has

$$\chi^{\pm} \sim 1, \quad N \sim e^{\omega(x,y;\mu,\lambda)},$$

$$F(\mu,\lambda) \simeq -\frac{i}{2\pi\mu} \iint_{-\infty}^{+\infty} d\xi d\eta\, u(\xi,\eta,t) e^{\omega(\xi,\eta;\mu,\lambda)}. \tag{4.3.53}$$

Hence for small $u$

$$f(\lambda,\mu) \simeq \frac{i}{2\pi\lambda} \operatorname{sgn}\left(\frac{1}{\lambda} - \frac{1}{\mu}\right)$$

$$\times \iint_{-\infty}^{+\infty} d\xi d\eta\, u(\xi,\eta) \exp\left\{i\xi\left(\frac{1}{\lambda} - \frac{1}{\mu}\right) - i\eta\left(\frac{1}{\lambda^2} - \frac{1}{\mu^2}\right)\right\}. \tag{4.3.54}$$

Now it is easy to see that the condition $\bar{u} = u$ implies

$$\overline{f(\mu,\lambda)}\mu = f(\lambda,\mu)\lambda \quad (\text{Im } \lambda = \text{Im } \mu = 0). \tag{4.3.55}$$

The reality constraints can also be derived for the discrete part of the inverse problem data: $n_+ = n_-$, $\lambda_k^+ = \overline{\lambda_k^-}$, $\gamma_k^\pm = -\frac{i\lambda_k^\pm}{2} + \delta_k^\pm$, $\overline{\delta_k^+} = \delta_k^-$.

Inverse problem equations allows us, similar to the KP equation, to construct classes of exact explicit solutions of the KP-I equation which decay as $x^2 + y^2 \to \infty$. The first class corresponds to the pure discrete inverse problem data ($f(\lambda,\mu) \equiv 0$). The corresponding algebraic system (4.3.48), as it is easy to see, is nothing but the system (4.2.57) in the case (4.2.63) with the redefinition

$$\lambda_i = \lambda_i^+ \quad (i = 1,\ldots,n), \quad \lambda_{n+i} = \lambda_i^- \quad (i = 1,\ldots,n).$$

The corresponding solutions are the multilump solutions (4.2.66).

Another class corresponds to the pure continuous and degenerate data $f(\lambda,\mu)$

$$f(\lambda,\mu) = \sum_{k=1}^{N} R_k f_k(\lambda) \bar{f}_k(\lambda) \lambda.$$

The corresponding solutions of the mKP-I equation are the particular case of the solutions (4.2.28) with functional parameters for which

$$f(\lambda,\bar{\lambda}) = f(\lambda)\delta(\lambda - \bar{\lambda}).$$

## 4.4. Initial value problem for the mKP-II equation

For the mKP-II equation ($\sigma = 1$) the spectral problem (4.3.2a) looks like

$$\mu_y + \mu_{xx} + \frac{2i}{\lambda}\mu_x + u\chi_x + \frac{i}{\lambda}u\chi = 0 \tag{4.4.1}$$

and the reconstruction formula is

$$u(x,y,t) = -2(\ln \mu_0)_x \tag{4.4.2}$$

where $\mu_0(x,y,t) = \mu(x,y,t,\lambda = 0)$.

So, we will search for the solutions of the problem (4.4.1) bounded on the whole complex plane $\lambda$ and normalize as $\mu \underset{\lambda\to\infty}{\to} 1$. Such a solutions can be constructed as the solutions of the integral equation

$$\mu(x,y;\lambda) = 1 + \left[G(\cdot,\lambda)u\left(\partial + \frac{i}{\lambda}\right)\mu(\cdot)\right](x,y) \tag{4.4.3}$$

*Chapter 4* 123

where $G(x, y; \lambda)$ is the bounded Green function of the operator $L_0 = \partial_y + \partial_x^2 + \frac{2i}{\lambda}\partial_x$. This is exactly the same Green function (up to the change of variable $\lambda \to \lambda^{-1}$) as for the KP-II equation

$$G(x,y;\lambda) = \frac{1}{2\pi}\left\{\Theta(\lambda_R)\left[\Theta(-y)\left(\int_0^\infty \frac{d\mu}{\mu^2} + \int_{-\frac{|\lambda|^2}{2\lambda_R}}^0 \frac{d\mu}{\mu^2}\right) - \Theta(y)\int_{-\infty}^{-\frac{|\lambda|^2}{2\lambda_R}} \frac{d\mu}{\mu^2}\right]\right.$$

$$\left. + \Theta(-\lambda_R)\left[\Theta(-y)\left(\int_{-\infty}^0 \frac{d\mu}{\mu^2} + \int_0^{-\frac{|\lambda|^2}{2\lambda_R}} \frac{d\mu}{\mu^2}\right) - \Theta(y)\int_{-\frac{|\lambda|^2}{2\lambda_R}}^{+\infty} \frac{d\mu}{\mu^2}\right]\right\} e^{\frac{ix}{\mu} + \frac{y}{\mu^2} + \frac{2y}{\lambda\mu}}. \quad (4.4.4)$$

This Green function is analytic nowhere:

$$\frac{\partial G(x,y;\lambda)}{\partial \bar\lambda} = \frac{1}{2\pi}\frac{\lambda^2}{|\lambda|^4}\text{sgn}(\lambda_R)e^{(ipx+iqy)} \quad (4.4.5)$$

where

$$p = \frac{\lambda + \bar\lambda}{|\lambda|^2}, \quad q = i\frac{\bar\lambda^2 - \lambda^2}{|\lambda|^4}. \quad (4.4.6)$$

Hence, the solutions of equation (4.4.3) are also analytic nowhere. Following the standard $\bar\partial$-method, one now has to calculate $\partial\mu/\partial\bar\lambda$. Using (4.4.5), one gets

$$\frac{\partial\mu(x,y;\lambda,\bar\lambda)}{\partial\bar\lambda} = F(\lambda,\bar\lambda)e^{ipx+iqy} + \left[G(\cdot;\lambda,\bar\lambda)u(\cdot)\left(\partial + \frac{i}{\lambda}\right)\frac{\partial\mu}{\partial\bar\lambda}\right](x,y) \quad (4.4.7)$$

where

$$F(\lambda,\bar\lambda) = \frac{\lambda^2}{2\pi|\lambda|^4}\text{sgn}(\lambda_R)$$

$$\times \iint_{-\infty}^{+\infty} d\xi d\eta \, e^{-ip\xi-iq\eta}u(\xi,\eta)\left(\partial_\xi + \frac{i}{\lambda}\right)\mu(\xi,\eta;\lambda,\bar\lambda). \quad (4.4.8)$$

Introducing the function $N(x,y;\lambda,\bar\lambda)$ which obeys the integral equation

$$N(x,y;\lambda,\bar\lambda) = e^{ipx+iqy} + \left[G(\cdot;\lambda,\bar\lambda)u(\cdot)\left(\partial + \frac{i}{\lambda}\right)N(\cdot;\lambda,\bar\lambda)\right](x,y) \quad (4.4.9)$$

and assuming that the homogeneous equation (4.4.3) has no nontrivial solutions, one gets

$$\frac{\partial\mu}{\partial\bar\lambda} = F(\lambda,\bar\lambda)N(x,y;\lambda,\bar\lambda). \quad (4.4.10)$$

The interrelation between $N(x,y;\lambda,\bar\lambda)$ and $\mu(x,y;\lambda,\bar\lambda)$ follows from (4.4.9), (4.4.3) and the symmetry property of the Green function (4.4.4)

$$G(x,y;\lambda,\bar\lambda) = G(x,y;-\bar\lambda,-\lambda)e^{ipx+iqy} \quad (4.4.11)$$

and the identity

$$\left(\partial_x + \frac{i}{\lambda}\right)(e^{ipx+iqy}\phi(x,y)) = e^{ipx+iqy}\left(\partial_x - \frac{i}{\lambda}\right)\phi(x,y) \quad (4.4.12)$$

where $\phi(x,y)$ is an arbitrary function. Indeed, multiplying (4.4.9) by $e^{-ipx-iqy}$, comparing the obtained equation with equation (4.4.3) with the change of variable $\lambda \to -\bar{\lambda}$ and taking into account (4.4.11) and (4.4.12), one finds

$$N(x,y;\lambda,\bar{\lambda}) = e^{ipx+iqy}\mu(x,y;-\bar{\lambda},-\lambda). \tag{4.4.13}$$

So, we have the following $\bar{\partial}$-problem [201]

$$\frac{\partial \mu(x,y;\lambda,\bar{\lambda})}{\partial \bar{\lambda}} = F(\lambda,\bar{\lambda})e^{ipx+iqy}\mu(x,y;-\bar{\lambda},-\lambda) \tag{4.4.14}$$

where the function $F(\lambda,\bar{\lambda})$ is given by (4.4.8). The $\bar{\partial}$-equation (4.4.14) generates the integral equation

$$\mu(x,y;\lambda,\bar{\lambda}) = 1 + \frac{1}{2\pi i}\iint_C \frac{d\mu \wedge d\bar{\mu}}{\mu - \lambda} F(\mu,\bar{\mu})e^{i\hat{p}x+i\hat{q}y}\mu(x,y;-\bar{\mu},-\mu) \tag{4.4.15}$$

where

$$\hat{p} = -\left(\frac{1}{\mu} + \frac{1}{\bar{\mu}}\right), \quad \hat{q} = i\left(\frac{1}{\bar{\mu}^2} - \frac{1}{\mu^2}\right)$$

via the generalized Cauchy formula.

Equation (4.4.15) is the inverse problem equation for the problem (4.4.1) and the function $F(\lambda,\bar{\lambda})$ is the inverse problem data. Equation (4.4.15) is uniquely solvable at least for small $F(\lambda,\bar{\lambda})$.

The time dependence of the function $F(\lambda,\bar{\lambda})$ is defined similar to the mKP-I case. Thus one has

$$F(\lambda,\bar{\lambda};t) = F(\lambda,\bar{\lambda};0)e^{\left(-4i\left(\frac{1}{\lambda^3}+\frac{1}{\bar{\lambda}^3}\right)t\right)} \tag{4.4.16}$$

where $F(\lambda,\bar{\lambda};0)$ is an arbitrary function.

The formulae (4.4.15), (4.4.16) and (4.4.2) linearize the initial value problem for the mKP-II equation via the standard IST scheme.

For the real-valued potential $u(x,y,t)$ equation (4.4.3) with the use of (4.4.11) and (4.4.12) gives

$$\overline{\mu(x,y;\lambda,\bar{\lambda})} = \mu(x,y;-\bar{\lambda},-\lambda). \tag{4.4.17}$$

So, for the real-valued $u(x,y,t)$ the $\bar{\partial}$-problem (4.4.14) is of the form

$$\frac{\partial \mu(x,y;\lambda,\bar{\lambda})}{\partial \bar{\lambda}} = F(\lambda,\bar{\lambda})e^{ipx+iqy}\overline{\mu(x,y;\lambda,\bar{\lambda})}. \tag{4.4.18}$$

In the terms of $\psi(x,y,t)$ (4.3.1) one has

$$\frac{\partial \psi(x,y;\lambda,\bar{\lambda})}{\partial \bar{\lambda}} = F(\lambda,\bar{\lambda})\overline{\psi(x,y;\lambda,\bar{\lambda})}. \tag{4.4.19}$$

Thus, for the real-valued $u(x,y,t)$ the bounded solution of the problem (4.4.1) is nothing but the pseudo(generalized)-analytical function. The properties of the pseudo-analytic functions guarantee the solvability of equation (4.4.19) and, hence, the inverse problem equations for the mKP-II equation with real $u(x,y,t)$ for the smooth data $F(\lambda, \bar{\lambda})$.

Emphasize that the inverse problem for (4.4.1) can be solved in a manner when one starts from the very beginning by looking for its solution within the class of pseudo (generalized)-analytic functions.

## 4.5. The Miura map and the 2+1-dimensional Gardner equation

Similar to the KdV and mKdV equations their 2+1-dimensional counterparts, the KP and mKP equations are closely connected.

In fact, it is not difficult to show that if $u_{\text{mKP}}$ is the solution of the mKP equation (4.0.3) then the function

$$u_{\text{KP}} = -\frac{1}{2}\sigma^2 \partial_x^{-1} u_{\text{mKP}y} - \frac{1}{2}\sigma u_{\text{mKP}} - \frac{1}{4}\sigma^2 u_{\text{mKP}}^2 \qquad (4.5.1)$$

solves the KP equation (3.0.1). This map is the direct 2+1-dimensional generalization of the celebrated Miura transformation from mKdV to KdV equation (see e. g. [39]). The transformation (4.5.1) has been established in [84] and [85] within the completely different approaches.

Both the mKP and KP equations have the wide classes of exact solutions. An interesting problem is to analyze what type of correspondence between the solutions of the mKP and KP equation is provided by the Miura map (4.5.1).

It is convenient to use, for this purpose, the formula (4.1.9), i. e.

$$u_{\text{mKP}} = -\frac{2}{\sigma}(\ln \chi_{0\text{mKP}})_x. \qquad (4.5.2)$$

Substituting the expression (4.5.2) into (4.5.1) one first finds

$$u_{\text{KP}} = -\frac{\sigma(\chi_{0\text{mKP}}^{-1})_y + (\chi_{0\text{mKP}}^{-1})_{xx}}{\chi_{0\text{mKP}}^{-1}} \qquad (4.5.3)$$

which clearly demonstrates that $u_{\text{KP}}$ solves the KP equation. Further, using the relation (4.1.16), one obtains

$$u_{\text{KP}} = -2i\left(\frac{\chi_{1\text{mKP}}}{\chi_{0\text{mKP}}}\right)_x. \qquad (4.5.4)$$

This formula gives us a simple way for calculating the solutions of the KP equation using the known wave function $\chi$ for the mKP equation.

Using the formula (4.5.4), one can establish the relation between the classes of solutions of the mKP and KP equations presented in the previous sections [203]. In particular, one can show that the solutions of the mKP equation with functional parameters given by (4.2.19) are converted into the solutions of the KP equation with functional parameters given by (3.2.12). The correspondence between the functions $\xi$ and $\eta$ is of the form [203]

$$\xi_k^{KP} = \xi_k^{mKP}, \quad \eta_k^{KP} = \eta_{kx}^{mKP}. \tag{4.5.5}$$

Then the line solitons of the mKP equations are converted into the line solitons of the KP equation under the transformation (4.5.1). Similarly for the rational solutions and, in particular, for lumps. The detailed analysis of this correspondence has been given in [203]. Note that only in the case $\sigma = 1$ the Miura map interrelates the real-valued solutions of the mKP equation with the real-valued solutions of the KP equation.

In fact, the $\bar{\partial}$-dressing method provides even more deep and strict connection between KP and mKP equations. Indeed, the KP equation can be imbedded into the $\bar{\partial}$-dressing method using the $\bar{\partial}$-problem with the different normalizations of $\chi$ and the choices of $I_i(\lambda)$. Namely, one can show [145] that the scalar $\bar{\partial}$-problem

$$\frac{\partial \chi^{KP}(\lambda, \bar{\lambda})}{\partial \bar{\lambda}} = \pi \delta(\lambda) + \iint_C d\mu \wedge d\bar{\mu}\, \chi^{KP}(\mu, \bar{\mu}) R^{KP}(\mu, \bar{\mu}; \lambda, \bar{\lambda}) \tag{4.5.6}$$

with the normalization

$$\chi^{KP}(\lambda) \to \frac{1}{\lambda} \quad \text{as} \quad \lambda \to 0 \tag{4.5.7}$$

and the choice

$$I_1 = \frac{i}{\lambda}, \quad I_2(\lambda) = \frac{1}{\sigma \lambda^2}, \quad I_3(\lambda) = \frac{4i}{\lambda^3} \tag{4.5.8}$$

gives rise to the KP equation too.

Note that

$$R^{KP}(\mu, \bar{\mu}; \lambda, \bar{\lambda})$$
$$= \exp\left(i\left(\frac{1}{\mu} - \frac{1}{\lambda}\right)x + \frac{1}{\sigma}\left(\frac{1}{\mu^2} - \frac{1}{\lambda^2}\right)y + 4i\left(\frac{1}{\mu^3} - \frac{1}{\lambda^3}\right)t\right) R_0^{KP}(\mu, \bar{\mu}; \lambda, \bar{\lambda}). \tag{4.5.9}$$

The function $\chi^{mKP}(\lambda)$ is the solution of the $\bar{\partial}$-problem

$$\frac{\partial \chi^{mKP}(\lambda, \bar{\lambda})}{\partial \bar{\lambda}} = 1 + \iint_C d\mu \wedge d\bar{\mu}\, \chi^{mKP}(\mu, \bar{\mu}) R^{mKP}(\mu, \bar{\mu}; \lambda, \bar{\lambda}) \tag{4.5.10}$$

where $R^{mKP}$ has the same form (4.5.9) as $R^{KP}$.

It is easy to see that the function $\tilde{\chi} = \frac{1}{\lambda} \chi^{mKP}(\lambda, \bar{\lambda})$ solves the $\bar{\partial}$-problem

$$\frac{\partial \hat{\chi}(\lambda, \bar{\lambda})}{\partial \bar{\lambda}} = \pi \delta(\lambda) \chi_0^{mKP} + \iint_C d\mu \wedge d\bar{\mu}\, \hat{\chi}(\mu, \bar{\mu}) \frac{\mu}{\lambda} R^{mKP}(\mu, \bar{\mu}; \lambda, \bar{\lambda}) \tag{4.5.11}$$

where $\chi_0^{\mathrm{mKP}} = \chi^{\mathrm{mKP}}(\lambda = 0)$. Comparing (4.5.11) and (4.5.6), one concludes that the function

$$\frac{1}{\lambda} \frac{\chi^{\mathrm{mKP}}(\lambda, \bar{\lambda})}{\chi_0^{\mathrm{mKP}}}$$

solves the same $\bar{\partial}$-problem as the function $\chi^{\mathrm{KP}}(\lambda, \bar{\lambda})$ but with the kernel

$$\frac{\mu}{\lambda} R^{\mathrm{mKP}}(\mu, \bar{\mu}; \lambda, \bar{\lambda}).$$

So one can identify

$$\chi^{\mathrm{KP}}(\lambda, \bar{\lambda}) = \frac{1}{\lambda} \frac{\chi^{\mathrm{mKP}}(\lambda, \bar{\lambda})}{\chi_0^{\mathrm{mKP}}} \tag{4.5.12}$$

and

$$R^{\mathrm{KP}}(\mu, \bar{\mu}; \lambda, \bar{\lambda}) = \frac{\mu}{\lambda} R^{\mathrm{mKP}}(\mu, \bar{\mu}; \lambda, \bar{\lambda}). \tag{4.5.13}$$

The relation (4.5.12) is, in fact, the gauge transformation between the mKP and KP equation. It is not difficult to show that the relation (4.5.12) provides exactly the Miura transformation (4.5.1) in terms of $u^{\mathrm{mKP}}$ and $u^{\mathrm{KP}}$. This formula establishes the relation between the kernels of the $\bar{\partial}$-problem under the Miura transformation. The author is grateful to S. Manakov for the discussion in which this result has been obtained.

Now let us consider briefly the 2+1-dimensional integrable equation which is closely connected both with the KP and mKP equation. It is the 2+1-dimensional Gardner equation [199]

$$u_t + u_{xxx} + 6\beta u u_x - \frac{3}{2}\alpha^2 u^2 u_x + 3\sigma^2 \partial_x^{-1} u_{yy} - 3\alpha\sigma u_x \partial_x^{-1} u_y = 0 \tag{4.5.14}$$

where $\sigma^2 = \pm 1$ and $\alpha, \beta$ are arbitrary constants. Equation (4.5.14) is equivalent to the compatibility condition for the linear system [199]

$$\sigma \psi_y + \psi_{xx} + \alpha u \psi_x + \beta u \psi = 0,$$

$$\psi_x + 4\psi_{xxx} + \alpha u \psi_{xx} + \left(3\alpha u_x + \frac{3}{2}\alpha^2 u^2 + 6\beta u - 3\alpha\sigma \partial_x^{-1} u_y\right)\psi_x$$

$$+ \left(3\beta u_x + \frac{3}{2}\alpha u^2 + 3\beta\sigma \partial_x^{-1} u_y\right)\psi = 0. \tag{4.5.15}$$

At $\alpha = 0$ equation (4.5.14) coincides with the KP equation while at $\beta = 0$ it is converted into the mKP equation. Equation (4.5.14) has the properties which are similar to that of the 1+1-dimensional Gardner equation [39].

Within the $\bar{\partial}$-dressing method equation (4.5.14) is associated with the scalar $\bar{\partial}$-problem with the canonical normalization and [204]

$$I_1(\lambda) = i\left(i\frac{\beta}{\alpha} + \frac{1}{\lambda}\right), \quad I_2(\lambda) = \frac{1}{\sigma}\left(i\frac{\beta}{\alpha} + \frac{1}{\alpha}\right)^2, \quad I_3(\lambda) = 4i\left(i\frac{\beta}{\alpha} + \frac{1}{\lambda}\right)^3. \tag{4.5.16}$$

The application of the $\bar{\partial}$-dressing to the Gardner equation is similar to that for the mKP equation. One can construct the explicit exact solutions of equation (4.5.14) with functional parameter, line solitons and breathers and decaying and line lumps. The detailed analysis was given in [204].

# Chapter 5

# The Davey–Stewartson equation

The Davey—Stewartson (DS) equation

$$iq_t + \frac{1}{2}(q_{xx} + \sigma^2 q_{yy}) + \kappa |q|^2 q + q\varphi = 0,$$

$$\varphi_{yy} - \sigma^2 \varphi_{xx} - 2\kappa(|q|^2)_{xx} = 0 \qquad (5.0.1)$$

is the straightforward 2+1-dimensional integrable generalization of the NLS equation (1.1.7).

Equation (5.0.1) at $\sigma^2 = -1$ was first derived in the fluid dynamics [88] as the shallow water limit of the Benney–Roskes equation [205]. It describes the evolution of waves of slowly varying amplitude on a two-dimensional water surface under gravity. In this case $q$ is the amplitude of a surface wave packet and $\varphi$ is the velocity potential. If the effect of surface tension dominates gravity then one has equation (5.0.1) but with $\sigma^2 = 1$ [88, 205]. Equation (5.0.1) is referred to as the DS-I equation in the case $\sigma^2 = 1$ and as the DS-II equation at $\sigma^2 = -1$. The cases $\kappa = \pm 1$ also have the different properties. We will consider the case $\kappa = -1$. The DS equation can also be applied in plasma physics and nonlinear optics.

The DS equation is a universal equation since it arises as the multiscale limit of the wide class of 2+1-dimensional nonlinear equations [206, 207].

Note that equation (5.0.1) has the solution $q = 0$, $\varphi = u_1(x+\sigma y) + u_2(x-\sigma y)$ where $u_1$ and $u_2$ are arbitrary functions. Such solutions of the DS-I equation are the backgrounds for the construction of the exponentially localized solitons of the DS-I equation.

## 5.1. $\bar{\partial}$-dressing

The DS equation (5.0.1) is, in fact, the system of equations for $q$ and $\bar{q}$. So in this case we start with the matrix $\bar{\partial}$-problem

$$\frac{\partial \chi(\lambda, \bar{\lambda})}{\partial \bar{\lambda}} = \iint_C d\mu \wedge d\bar{\mu}\, \chi(\mu, \bar{\mu}) R(\mu, \bar{\mu}; \lambda, \bar{\lambda}) \qquad (5.1.1)$$

where $\chi$ and $R$ are the $2 \times 2$ matrices. We assume that the problem (5.1.1) is uniquely solvable with the canonical normalization $\chi \to 1$ as $\lambda \to \infty$, i. e.

$$\chi = 1 + \frac{1}{\lambda}\chi_{-1} + \frac{1}{\lambda^2}\chi_{-2} + \ldots \quad (\lambda \to \infty). \tag{5.1.2}$$

The choice of the operators $\hat{B}_i$ ($i = 1, 2, 3$) in the case of trivial background is the following

$$\hat{B}_1 = i\lambda, \quad \hat{B}_2 = -\frac{1}{\sigma}i\lambda\sigma_3, \quad \hat{B}_3 = -i\lambda^2\sigma_3 \tag{5.1.3}$$

where $\sigma_3 = \begin{pmatrix} 1 & 0 \\ 0 & -1 \end{pmatrix}$. The operators $D_i$ ($i = 1, 2, 3$) in this case act as

$$D_1 f = \partial_x f + i\lambda f, \quad D_2 f = \partial_y f - \frac{i}{\sigma}\lambda f \sigma_3, \quad D_3 f = \partial_t f - i\lambda^2 f \sigma_3. \tag{5.1.4}$$

The $x$, $y$ and $t$ dependence of the kernel $R$ is of the form

$$R(\mu, \bar{\mu}; \lambda, \bar{\lambda}; x, y, t) = e^{F(\mu)} R_0(\mu, \bar{\mu}; \lambda, \bar{\lambda}) e^{-F(\lambda)} \tag{5.1.5}$$

where $R_0$ is an arbitrary $2 \times 2$ matrix valued function and

$$F(\lambda) = i\lambda x - \frac{i}{\sigma}\lambda\sigma_3 y - i\lambda^2 \sigma_3 t. \tag{5.1.6}$$

Let us apply now the general scheme of the $\bar{\partial}$-dressing. First we note that both $D_1\chi$ and $D_2\chi$ have the first order singularities as $\lambda \to \infty$. So their linear combination may not have such singularity. Indeed, it is easy to show that the superposition

$$\sigma D_2 \chi + \sigma_3 D_1 \chi = \sigma \partial_y \chi + \sigma_3 \partial_x \chi + i\lambda[\sigma_3, \chi] \tag{5.1.7}$$

obey this condition. Then

$$\sigma D_2 \chi + \sigma_3 D_1 \chi \to i[\sigma_3, \chi_{-1}] \tag{5.1.8}$$

as $\lambda \to \infty$. Using (5.1.8), one concludes that the quantity

$$\sigma D_2 \chi + \sigma_3 D_1 \chi + P\chi \tag{5.1.9}$$

with

$$P \equiv \begin{pmatrix} 0 & q \\ r & 0 \end{pmatrix} = -i[\sigma_3, \chi_{-1}] \tag{5.1.10}$$

obeys both conditions (2.3.15) and (2.3.16).

Thus, the first linear problem is

$$\sigma D_2 \chi + \sigma_3 D_1 \chi + P\chi = 0 \tag{5.1.11}$$

where
$$P = \begin{pmatrix} 0 & q \\ r & 0 \end{pmatrix} = -i[\sigma_3, \chi_{-1}] = -2i\sigma_3(\chi_{-1})_{\text{off}}. \tag{5.1.12}$$

Here $\chi_{\text{off}} \doteq \chi - \chi_{\text{diag}}$ and $\chi_{\text{diag}}$ is the diagonal part of the matrix $\chi$.

Equation (5.1.11) also implies

$$\sigma\chi_{-1y} + \sigma_3\chi_{-1x} + i[\sigma_3, \chi_{-2}] + P\chi_{-1} = 0, \tag{5.1.13}$$

$$\sigma\chi_{-2y} + \sigma_3\chi_{-2x} + i[\sigma_3, \chi_{-3}] + P\chi_{-2} = 0. \tag{5.1.14}$$

Further we pass to the second linear problem. The second order singularity in the quantity $D_3\chi$ can be compensated by the term of the type $\sim D_1^2\chi$. Indeed

$$D_3\chi - i\sigma_3 D_1^2\chi = \partial_t\chi - i\sigma_3\partial_x^2\chi + 2\lambda\sigma_3\partial_x\chi + i\lambda^2[\sigma_3, \chi]. \tag{5.1.15}$$

However, due to (5.1.2) the term $\lambda^2[\sigma_3, \chi]$ contributes only to the first order singularity. To annihilate the first order singularity in (5.1.15) we add the term $\tilde{P}D_1\chi$. The requirement that the quantity

$$\begin{aligned}
D_3\chi - &i\sigma_3 D_1^2\chi + \tilde{P}D_1\chi \\
&= \partial_t\chi - i\sigma_3\partial_x^2\chi + 2\lambda\sigma_3\partial_x\chi + i\lambda^2[\sigma_3, \chi] + \tilde{P}\partial_x\chi + i\lambda\tilde{P}\chi
\end{aligned} \tag{5.1.16}$$

has no first order singularity as $\lambda \to \infty$ gives

$$\tilde{P} = -[\sigma_3, \chi_{-1}] = -iP \tag{5.1.17}$$

where $P$ is given by (5.1.12)

At last, to satisfy the condition (2.3.16) one should add to (5.1.16) the term $Q\chi$. Namely

$$D_3\chi - i\sigma_3 D_1^2\chi - iPD_1\chi + Q\chi \to 0 \tag{5.1.18}$$

as $\lambda \to \infty$ if

$$Q = -2\sigma_3\chi_{-1x} - i[\sigma_3, \chi_{-2}] - P\chi_{-1}. \tag{5.1.19}$$

Using (5.1.13), one gets

$$Q = (\sigma\partial_y - \sigma_3\partial_x)\chi_{-1}. \tag{5.1.20}$$

Denoting $(\chi_{-1})_{\text{diag}} = iC$, one has

$$Q = i(\sigma\partial_y - \sigma_3\partial_x)\left(C + \frac{1}{2}\sigma_3 P\right). \tag{5.1.21}$$

The diagonal matrix $C$ is, in fact, related to $P$. Indeed, considering the diagonal part of equation (5.1.13) and using (5.1.12), one gets

$$\sigma C_y + \sigma_3 C_x = \frac{1}{2}\sigma_3 P^2. \tag{5.1.22}$$

Thus, the second desired linear problem is of the form

$$D_3\chi - i\sigma_3 D_1^2\chi - iPD_1\chi + i\left(\sigma C_y - \sigma_3 C_x + \frac{1}{2}\sigma_3\sigma P_y - \frac{1}{2}P_x\right)\chi = 0 \qquad (5.1.23)$$

where the diagonal matrix $C$ is related to $P$ by equation (5.1.22)

The linear problems (5.1.11) and (5.1.23) form the basis of the linear problems in the case under consideration. It is easy to show that one can construct another second linear problem using the operator $D_2$ instead of $D_1$. But these two problems are equivalent modulo the first problem (5.1.11).

The linear problems (5.1.11) and (5.1.23) are compatible by construction and directly provide the corresponding nonlinear system. Substituting the asymptotic expansion (5.1.2) into equation (5.1.23), considering the terms of order $\lambda^{-1}$ and eliminating $\chi_{-3}$ with the use of (5.1.14), one gets

$$\chi_{-1t} - i\sigma_3\chi_{-1xx} - (\sigma\partial_y - \sigma_3\partial_x)\chi_{-2} - iP\chi_{-1x}$$

$$+ i(\sigma\partial_y - \sigma_3\partial_x)C + \frac{i}{2}\sigma_3(\sigma\partial_y - \sigma_3\partial_x)P\chi_{-1} = 0. \qquad (5.1.24)$$

Then taking the off-diagonal part of (5.1.24) and using (5.1.13), one finally obtains

$$iP_t + \frac{1}{2}\sigma_3(\sigma^2 P_{yy} + P_{xx}) - [\sigma C_y - \sigma_3 C_x, P] = 0 \qquad (5.1.25)$$

together with equation (5.1.22), i. e.

$$\sigma C_y + \sigma_3 C_x = \frac{1}{2}\sigma_3 P^2. \qquad (5.1.26)$$

This is the nonlinear system we are looking for. Denoting $\tilde{C} \doteq \sigma C_y - \sigma_3 C_x$, one represents the system (5.1.25), (5.1.26) in the equivalent form

$$iP_t + \frac{1}{2}\sigma_3(\sigma^2 P_{yy} + P_{xx}) - [\tilde{C}, P] = 0,$$

$$\sigma\tilde{C}_y + \sigma_3\tilde{C}_x = \frac{1}{2}\sigma_3(\sigma\partial_y - \sigma_3\partial_x)P^2. \qquad (5.1.27)$$

At last, in terms of components $P = \begin{pmatrix} 0 & q \\ r & 0 \end{pmatrix}$ and $\Phi \doteq \tilde{C}_2 - \tilde{C}_1$ the system (5.1.27) looks like

$$iq_t + \frac{1}{2}(\sigma^2 q_{yy} + q_{xx}) + \Phi q = 0,$$

$$ir_t - \frac{1}{2}(\sigma^2 r_{yy} + r_{xx}) - \Phi r = 0, \qquad (5.1.28)$$

$$\sigma^2\Phi_{yy} - \Phi_{xx} = \sigma^2(qr)_{yy} + (qr)_{xx}.$$

Chapter 5

Introducing $\varphi = \Phi - qr$, one arrives at the system

$$iq_t + \frac{1}{2}(\sigma^2 q_{yy} + q_{xx}) + q^2 r + \varphi q = 0,$$

$$ir_t - \frac{1}{2}(\sigma^2 r_{yy} + r_{xx}) - r^2 q - \varphi r = 0, \qquad (5.1.29)$$

$$\sigma^2 \varphi_{yy} - \varphi_{xx} = 2(qr)_{xx}.$$

Finally, under the reduction $r = \kappa \bar{q}$ the system (5.1.29) is converted into the DS equation (5.0.1). The reduction $r = \kappa \bar{q}$, due to (5.1.26) and (5.1.27), implies also that at $\sigma = 1$ $\bar{C} = C$ and $\bar{\tilde{C}} = \tilde{C}$, and $C_2 = \bar{C}_1$ and $\tilde{C}_2 = \bar{\tilde{C}}_1$ at $\sigma = i$.

Emphasize that since

$$P = -2i(\chi_{-1})_{\text{off}}, \quad C = (\chi_{-1})_{\text{diag}} \qquad (5.1.30)$$

the system (5.1.25), (5.1.26) is, in fact, the system of equations for the entire matrix $\chi_{-1}(x,y,t)$.

According to the general scheme (section (2.3), formulae (2.3.23), (2.3.24)) the linear problems (5.1.11), (5.1.23) can be reformulated in the usual form by transiting to the function $\psi$ defined by

$$\psi = \chi \exp\left(i\lambda x - \frac{i\lambda}{\sigma}\sigma_3 y - i\lambda^2 \sigma_3 t\right). \qquad (5.1.31)$$

They are

$$\sigma \psi_y + \sigma_3 \psi_x + P\psi = 0,$$

$$\psi_t - i\sigma_3 \psi_{xx} - iP\psi_x + i\left(\sigma C_y - \sigma_3 C_x - \frac{\sigma\sigma_3}{2}P_y - \frac{1}{2}P_x\right)\psi = 0 \qquad (5.1.32)$$

that is nothing but (1.2.6). The compatibility condition for the system (5.1.32) is, of course, equivalent to the DS system (5.0.1).

Now let us consider the $\bar{\partial}$-dressing on nontrivial background. Following the general approach of section (2.5) and our previous experience we choose the operator $\hat{B}_i$ ($i = 1, 2, 3$) as the convolutive operators such that

$$\hat{B}_1 \to i\lambda, \quad \hat{B}_2 \to -\frac{i\lambda}{\sigma}\sigma_3, \quad \hat{B}_3 \to -i\lambda^2 \sigma_3 \qquad (5.1.33)$$

as $\lambda \to \infty$. In dual space $\{s\}$ defined by the transformation

$$X(s,x,y,t) \doteq \frac{1}{2\pi} \iint_C d\lambda \wedge d\bar{\lambda}\, e^{i(\lambda s + \bar{\lambda}\bar{s})} f(\lambda, \bar{\lambda}; x, y, t) \qquad (5.1.34)$$

the operators $\hat{B}_i$ ($i = 1, 2, 3$) are the differential operators of the first and second orders.

Choosing in (2.5.14) ($x_1 \equiv x$, $x_2 \equiv y$, $x_3 \equiv t$) $n = 1$, $m = 2$ and

$$\tilde{u}_1 = -\frac{\sigma_3}{\sigma}, \quad \tilde{u}_0 = -P_0(z,y,t),$$

$$\tilde{v}_2 = i\sigma_3, \quad \tilde{v}_1 = iP(z,y,t), \quad \tilde{v}_0 = -i\left(\sigma C_{0y} - \sigma_3 C_{0z} - \frac{\sigma\sigma_3}{2}P_{0y} - \frac{1}{2}P_{0z}\right), \quad (5.1.35)$$

one has in the dual space the linear system

$$(\sigma\sigma_y + \sigma_3\sigma_s + P_0(s+x,y,t))X = 0,$$

$$\partial_t - i\sigma_3\partial_s^2 - iP_0(s+x,y,t)\partial_s + i\left(\sigma C_{0y} - \sigma_3 C_{0s} + \frac{\sigma\sigma_3}{2}P_{0y} - \frac{1}{2}P_{0s}\right)X = 0 \quad (5.1.36)$$

where

$$\sigma C_{0y} + \sigma_3 C_{0s} = \frac{1}{2}\sigma_3 P_0^2. \quad (5.1.37)$$

This system is nothing but the linear system for the DS equation for potential $u_0$. So we have chosen the solutions of the DS equation as the background.

In the original spectral space $\{\lambda\}$ the operators $\hat{B}_i$ act as follows

$$\hat{B}_1 = i\lambda,$$

$$(\hat{B}_2 f)(\lambda) = -\frac{i\lambda}{\sigma}\sigma_3 f(\lambda) - \frac{1}{\sigma}\iint_C d\mu \wedge d\bar{\mu}\, \hat{P}_0(\lambda - \mu, x, y, t) f(\mu), \quad (5.1.38)$$

$$(\hat{B}_3 f)(\lambda) = -i\lambda^2\sigma^3 f(\lambda) + i\iint_C d\mu \wedge d\bar{\mu}\, \hat{P}_0(\lambda - \mu, x, y, t) i\mu f(\mu)$$

$$- \iint_C d\mu \wedge d\bar{\mu}\, \hat{Q}_0(\lambda - \mu, x, y, t) f(\mu)$$

where

$$Q_0 = i(\sigma\partial_y - \sigma_3\partial_s)\left(C_0 + \frac{1}{2}\sigma_3 P_0\right) \quad (5.1.39)$$

and

$$\hat{P}_0(\lambda, x, y, t) \doteq \frac{1}{2\pi}\iint_C ds \wedge d\bar{s}\, e^{-i(\lambda s + \bar{\lambda}\bar{s})} P_0(s+x,y,t). \quad (5.1.40)$$

Thus, the operators $D_i = \partial_{x_i} + \hat{B}_i^*$ with nontrivial background act according to

$$(D_1\chi)(\lambda) = (\partial_x + i\lambda)\chi(\lambda),$$

$$(D_2\chi)(\lambda) = \partial_y\chi(\lambda) - \frac{i\lambda}{\sigma}\chi(\lambda)\sigma_3 - \frac{1}{\sigma}\iint_C d\mu \wedge d\bar{\mu}\, \chi(\lambda+\mu)\hat{P}_0(\mu, x, y, t), \quad (5.1.41)$$

$$(D_3\chi)(\lambda) = \partial_t\chi(\lambda) - i\lambda^2\chi(\lambda)\sigma_3 - \iint_C d\mu \wedge d\bar{\mu}\, (\lambda+\mu)\chi(\lambda+\mu)\hat{P}_0(\mu, x, y, t)$$

$$- \iint_C d\mu \wedge d\bar{\mu}\, \chi(\lambda+\mu)\hat{Q}_0(\mu, x, y, t).$$

Using these operators $D_i$ ($i = 1, 2, 3$), we will construct the operators $L_1$ and $L_2$. Analogously to the case of trivial background we start with the quantity

$$\sigma D_2\chi + \sigma_3 D_1\chi = \sigma\chi_y + \sigma_3\chi_x + i\lambda[\sigma_3, \chi(\lambda)]$$
$$- \iint_C d\mu \wedge d\bar{\mu}\, \chi(\lambda + \mu)\hat{P}_0(\mu, x, y, t). \tag{5.1.42}$$

In virtue of (5.1.2) it again has no singularity as $\lambda \to \infty$. To satisfy the condition (2.3.16) let us add to (5.1.42) the term $P(x, y, t)\chi$. Calculating the $\lambda \to \infty$ limit of $\sigma D_2\chi + \sigma_3 D_1\chi + P\chi$ and using (5.1.40), one finds that

$$P(x, y, t) = P_0(x, y, t) - i[\sigma_3, \chi_{-1}]. \tag{5.1.43}$$

Thus, the first linear problem is given by the equation

$$\sigma D_2\chi + \sigma_3 D_1\chi + P\chi = 0 \tag{5.1.44}$$

where $D_1$, $D_2$ are defined by (5.1.42) and $P$ is related to $P_0$ via (5.1.43).

In construction of the second problem we again start with the operators $D_3 - i\sigma_3 D_1^2$. One has

$$D_3\chi - i\sigma_3 D_1^2\chi = \partial_t\chi - i\sigma_3\partial_x^2\chi + 2\lambda\sigma_3\partial_x\chi + i\lambda^2[\sigma_3, \chi]$$
$$- \iint_C d\mu \wedge d\bar{\mu}\, \chi(\lambda + \mu)\hat{P}_0(\mu, x, y, t) - \iint_C d\mu \wedge d\bar{\mu}\, \chi(\lambda + \mu)\hat{Q}_0(\mu, x, y, t). \tag{5.1.45}$$

It has no second order singularity as $\lambda \to \infty$. To compensate the first order singularity we add the term $\hat{P}D_1\chi$. The requirement of the absence of singularity of the first in the quantity

$$D_3\chi - i\sigma_3 D_1^2\chi + \hat{P}D_1\chi \tag{5.1.46}$$

gives

$$i\hat{P} + i[\sigma_3, \chi_{-1}] - P_0(x, y, t) = 0 \tag{5.1.47}$$

or

$$\tilde{P} = iP \tag{5.1.48}$$

where $P$ is given by (5.1.43).

Further, in order to satisfy the condition (2.3.16) we add the term $Q\chi$ to (5.1.46). Demanding that

$$D_3\chi - i\sigma_3 D_1^2\chi - iPD_1\chi + Q\chi \to 0 \tag{5.1.49}$$

as $\lambda \to \infty$ and eliminating $\chi_{-2}$ with the use of the following equation

$$\sigma\chi_{-1y} + \sigma_3\chi_{-1x} + i[\sigma_3, \chi_{-2}] - \chi_{-1}P_0 + P\chi_{-1} = 0 \tag{5.1.50}$$

which is a consequence of (5.1.44), one finds

$$Q = (\sigma\partial_y - \sigma_3\partial_x)\chi_{-1} + Q_0. \qquad (5.1.51)$$

Introducing the diagonal matrix $C$ via

$$C = C_0 - i(\chi_{-1})_{\text{diag}} \qquad (5.1.52)$$

where $C_0 = -i(\chi_{0_{-1}})_{\text{diag}}$, one gets

$$Q = i(\sigma\partial_y - \sigma_3\partial_x)\left(C + \frac{1}{2}\sigma_3 P\right). \qquad (5.1.53)$$

In virtue of (5.1.50), one also has

$$\sigma C_y + \sigma_3 C_x = \frac{1}{2}\sigma_3 P^2. \qquad (5.1.54)$$

So the second desired problem is of the form

$$D_3\chi - i\sigma_3 D_1^2\chi - iPD_1\chi + i\left(\sigma C_y - \sigma_3 C_x + \frac{1}{2}\sigma\sigma_3 P_y - \frac{1}{2}P_x\right)\chi = 0 \qquad (5.1.55)$$

where $P$ is given by (5.1.43) and $C$ by (5.1.52).

Thus, the linear system associated with the nontrivial background $P_0$ is given by the equations

$$\sigma D_2\chi + \sigma_3 D_1\chi + P\chi = 0,$$

$$D_3\chi - i\sigma_3 D_1^2\chi - iPD_1\chi + i\left(\sigma C_y - \sigma_3 C_x + \frac{1}{2}\sigma\sigma_3 P_y - \frac{1}{2}P_x\right)\chi = 0 \qquad (5.1.56)$$

where the operators $D_i$ ($i = 1, 2, 3$) are given by (5.1.41), $C$ is related to $P$ via the equation

$$\sigma C_y + \sigma_3 C_x = \frac{1}{2}\sigma_3 P^2 \qquad (5.1.57)$$

and

$$P = P_0 - i[\sigma_3, \chi_{-1}],$$

$$C = C_0 - i(\chi_{-1})_{\text{diag}}. \qquad (5.1.58)$$

Emphasize that the system (5.1.56), (5.1.57) is exactly the same as that for the trivial background. The only difference is in the form of the commuting operators $D_i$ ($i = 1, 2, 3$). This fact readily implies that the matrix function $P(x, y, t)$ obeys system (5.1.25) or (5.1.27) with

$$\tilde{C} = \sigma C_{0y} - \sigma_3 C_{0x} - i(\sigma\partial_y - \sigma_3\partial_x)(\chi_{-1})_{\text{diag}}. \qquad (5.1.59)$$

One can also derive the nonlinear system (5.1.25) by substituting the expansion (5.1.2) into equation (5.1.55) and considering the terms of order $\lambda^{-1}$. Under the reduction $r = \kappa \bar{q}$ we again have the DS equation (5.1.29)

Thus, the linear system (5.1.56) and the formulae (5.1.58) provide us the $\bar{\partial}$-dressing for the DS equation on the nontrivial background $P_0$, $C_0$. For the first time such dressing for the DS equation on nontrivial background has been done in [148] within the framework of the nonlocal Riemann—Hilbert problem.

The formulae (5.1.58) allow us to construct the solutions of the DS equation starting with an arbitrary background. One simple and almost trivial background is of a special interest. It corresponds to the solution $P_0 \equiv 0$ of the DS system. In this case the matrix $\tilde{C}_0$, due to (5.1.26), is of the form

$$\tilde{C}_0 = \begin{pmatrix} u_1(y - \sigma x, t) & 0 \\ 0 & u_2(y + \sigma x, t) \end{pmatrix} \tag{5.1.60}$$

where $u_1$ and $u_2$ are arbitrary functions.

All formulae are simplified essentially for such background. The operators $\hat{\tilde{B}}_i$ are of the form

$$\hat{\tilde{B}}_1 = \partial_s, \quad \hat{\tilde{B}}_2 = -\frac{\sigma_3}{\sigma}\partial_s, \quad \hat{\tilde{B}}_3 = i\sigma_3\partial_s^2 - i\tilde{C}_0(x + s, y, t) \tag{5.1.61}$$

where the matrix $\tilde{C}_0$ is given by (5.1.60).

The Fourier transform $\tilde{R}(s, \tilde{s}, x, y, t)$ of the kernel $R$ defined by (2.5.15) obeys the system of equation (2.5.16) ($x_1 \equiv x$, $x_2 \equiv y$, $x_3 \equiv t$), i. e.

$$\tilde{R}_x(s, \tilde{s}; x, y, t) - \tilde{R}_s - \tilde{R}_{\tilde{s}} = 0,$$

$$\sigma \tilde{R}_y(s, \tilde{s}; x, y, t) + \sigma_3 \tilde{R}_s + \tilde{R}_{\tilde{s}}\sigma_3 = 0, \tag{5.1.62}$$

$$i\tilde{R}_t(s, \tilde{s}, x, y, t) + \sigma_3 \tilde{R}_{ss} - \tilde{R}_{\tilde{s}\tilde{s}}\sigma_3 - \tilde{C}_0(s + x, y, t)\tilde{R} + \tilde{R}\tilde{C}_0(\tilde{s} + x, y, t) = 0.$$

The system (5.1.62) can be solved by separation of variables $s$ and $\tilde{s}$

$$\tilde{R}(s, \tilde{s}; x, y, t) = \sum_{i,k=1}^{N} \rho_{ik} X_{0i}(s, x, y, t) Y_{0k}^*(\tilde{s}, x, y, t)$$

where the matrix functions $X_0 i$ and $Y_0 k$ obey the system

$$X_{0x} - X_{0s} = 0,$$

$$\sigma X_{0y} + \sigma_3 X_{0s} = 0, \tag{5.1.63}$$

$$iX_{0t} + \sigma_3 X_{0ss} - \tilde{C}_0(s + x, y, t)X_0 = 0$$

and adjoint equations for $Y_0^*(s, x, y, t)$.

The first and second equations (5.1.63) are readily solvable. The third equation (5.1.63) is, in fact, the first linear problem (3.0.2) associated with the KP-I equation. So, using the known results for the KP-I equation, we can find explicit exact solutions of the system (5.1.63). This gives us the exact kernels $\tilde{R}$. Then solving the $\bar{\partial}$-problem, we obtain the solutions of the DS equation via

$$P = -2i[\sigma_3, \chi_{-1}],$$

$$\tilde{C} = \tilde{C}_0 - i(\chi_{-1})_{\text{diag}} \tag{5.1.64}$$

for given background $\tilde{C}_0$.

Note that for the off-diagonal components of $\tilde{R}$, for instance, for $\tilde{R}_{12} = S$, the third equation (5.1.62) looks like

$$iS_t + S_{ss} + S_{\tilde{s}\tilde{s}} - (\tilde{C}_{01}(x+s,x,y,t) - \tilde{C}_{02}(x+\tilde{s},x,y,t))S = 0. \tag{5.1.65}$$

So after the separation of variables one has the scalar equation for $X_0$ in the variable $z = x + s$

$$iX_{0t} + X_{0zz} + \tilde{u}_1(z,y,t)X_0 = 0 \tag{5.1.66}$$

that is exactly the first linear problem for the KP-I equation.

Note that $\tilde{C}_0$ is, in fact, the asymptotic boundary value of the matrix $\tilde{C}$. Thus, the dressing presented above provides the exact solutions of the DS equation with nontrivial boundary values of $\tilde{C}$ or $\varphi$.

Emphasize that in the $\bar{\partial}$-dressing we did not assume the boundedness of $\tilde{C}_0$. The procedure effectively works in both cases $\sigma^2 = 1$ and $\sigma^2 = -1$. The situation changes essentially if one requires the boundedness of all functions $q$, $r$ and $\tilde{C}$.

In the case $\sigma = 1$ the functions $u_1$ and $u_2$ from (5.1.60) are still arbitrary but bounded functions of the real variables $y - x$ and $y + x$ respectively. The case $\sigma^2 = -1$ is crucially different. In virtue of the well-known properties of analytic and anti-analytic functions the only bounded functions $u_1$ and $u_2$ are constants.

In the case $P_0 = 0$ and $\tilde{C}_0$ given by (5.1.60) the linear problems (5.1.56) are simplified too. In terms of the function

$$\tilde{\psi}(x,y,t,\lambda) = \chi \exp\left(i\lambda x - \frac{i\lambda}{\sigma}\sigma_3 y\right) \tag{5.1.67}$$

the problems (5.1.56) look like

$$\sigma\tilde{\psi}_y + \sigma_3\psi_x + P(x,y,t)\tilde{\psi} = 0, \tag{5.1.68}$$

$$\tilde{\psi}_t(\lambda) - i\sigma_3\tilde{\psi}_{x\lambda}(\lambda) - iP\tilde{\psi}_x(\lambda) + i\left(\tilde{C} + \frac{1}{2}\sigma\sigma_3 P_y - \frac{1}{2}P_x\right)\tilde{\psi}(\lambda)$$

$$- i\iint_C d\mu \wedge d\bar{\mu}\,\tilde{\psi}(x,y,t;\lambda + \mu)\tilde{C}_0(x,y,t;\mu) = 0. \tag{5.1.69}$$

The problems (5.1.68), (5.1.69) are the same as for the trivial background. All the difference is in the last term in the second problem. The fact that consideration of the nontrivial boundaries demands such a modification of the second (temporal) linear problem has been established for the first time in [116] within a different approach.

The background $P_0 = 0$ and $\tilde{C}_0$ of the form (5.1.60) within the framework of dressing method based on the nonlocal Riemann—Hilbert method was discussed in [148].

The $\bar{\partial}$-dressing discussed in this section is adequate for construction of solutions of the DS equation such that $q$ and $r$ do not tend to some nonzero constants as $x^2 + y^2 \to \infty$. Indeed, in the case

$$P \xrightarrow[x^2+y^2\to\infty]{} \begin{pmatrix} 0 & q_\infty \\ r_\infty & 0 \end{pmatrix} \tag{5.1.70}$$

the function $\exp\left(i\lambda x - \frac{i\lambda}{\sigma}\sigma_3 y\right)$ is unappropriate solution of the linear problem (5.1.32) as $x^2 + y^2 \to \infty$. This means, in fact, that the operators $\hat{B}_i(\lambda)$ in the case (5.1.70) should be different from (5.1.3). We will consider the $\bar{\partial}$-dressing for the DS equation with the constant asymptotics (5.1.70) in chapter VIII.

## 5.2. Exact solutions

Using the results of the previous section we are in a position to construct wide classes of the exact explicit solutions of the DS equation. The solutions $P$ and $\tilde{C}$ are given by the formulae

$$P = P_0 - i[\sigma_3, \chi_{-1}],$$

$$C = C_0 - i(\chi_{-1})_{\text{diag}}. \tag{5.2.1}$$

Using the integral equation (2.2.3) for $\chi$, one gets

$$q(x,y,t) = q_{0t} + \frac{1}{\pi}\iint_C d\lambda \wedge d\bar{\lambda} \iint_C d\mu \wedge d\bar{\mu}\,(\chi(\mu,\bar{\mu})R(\mu,\bar{\mu};\lambda,\bar{\lambda}))_{12},$$

$$r(x,y,t) = r_0 - \frac{1}{\pi}\iint_C d\lambda \wedge d\bar{\lambda} \iint_C d\mu \wedge d\bar{\mu}\,(\chi(\mu,\bar{\mu})R(\mu,\bar{\mu};\lambda,\bar{\lambda}))_{21}, \tag{5.2.2}$$

$$C(x,y,t) = C_0 + \frac{1}{2\pi}\iint_C d\lambda \wedge d\bar{\lambda} \iint_C d\mu \wedge d\bar{\mu}\,(\chi(\mu,\bar{\mu})R(\mu,\bar{\mu};\lambda,\bar{\lambda}))_{\text{diag}}$$

where for $q_0 = r_0 = C_0 = 0$

$$R(\mu,\bar{\mu};\lambda,\bar{\lambda};x,y,t) = e^{F(\mu)}R_0(\mu,\bar{\mu};\lambda,\bar{\lambda})e^{-F(\lambda)} \tag{5.2.3}$$

and
$$F(\lambda) = i\lambda x - \frac{i\lambda}{\sigma}\sigma_3 y - i\lambda^2 \sigma_3 t. \tag{5.2.4}$$

We will consider first the case $q_0 = r_0 = C_0 = 0$. For small data $R$ one has $\chi \sim 1$ and, consequently,

$$q = \frac{1}{\pi} \iint_C d\lambda \wedge d\bar{\lambda} \iint_C d\mu \wedge d\bar{\mu}\, R_{012}(\mu,\bar{\mu};\lambda,\bar{\lambda}) e^{i(\mu-\lambda)x - \frac{i}{\sigma}(\mu+\lambda)y - i(\mu^2+\lambda^2)t},$$

$$r = -\frac{1}{\pi} \iint_C d\lambda \wedge d\bar{\lambda} \iint_C d\mu \wedge d\bar{\mu}\, R_{021}(\mu,\bar{\mu};\lambda,\bar{\lambda}) e^{i(\mu-\lambda)x + \frac{i}{\sigma}(\mu+\lambda)y + i(\mu^2+\lambda^2)t} \tag{5.2.5}$$

and

$$C_1 = \frac{1}{2\pi} \iint_C d\lambda \wedge d\bar{\lambda} \iint_C d\mu \wedge d\bar{\mu}\, R_{011}(\mu,\bar{\mu};\lambda,\bar{\lambda}) e^{i(\mu-\lambda)x - \frac{i}{\sigma}(\mu-\lambda)y - i(\mu^2-\lambda^2)t},$$

$$C_2 = \frac{1}{2\pi} \iint_C d\lambda \wedge d\bar{\lambda} \iint_C d\mu \wedge d\bar{\mu}\, R_{022}(\mu,\bar{\mu};\lambda,\bar{\lambda}) e^{i(\mu-\lambda)x + \frac{i}{\sigma}(\mu-\lambda)y + i(\mu^2-\lambda^2)t}. \tag{5.2.6}$$

These expressions provide necessary conditions for the reduction $r = -\bar{q}$ in terms of $R$. Taking into account that in the case $\sigma = 1$ also $\bar{C} = C$ and at $\sigma^2 = -1$, $C_2 = \bar{C}_1$ one gets, for the DS-I equation ($\sigma = 1$)

$$R^+(\mu,\bar{\mu};\lambda,\bar{\lambda}) = R(\bar{\lambda},\lambda;\bar{\mu},\mu) \tag{5.2.7}$$

where $+$ denotes the Hermitian conjugation of matrix and for the DS-II equation ($\sigma^2 = -1$)

$$\bar{R}(\mu,\bar{\mu};\lambda,\bar{\lambda}) = \sigma_2 R(-\bar{\mu},-\mu;-\bar{\lambda},-\lambda)\sigma_2^{-1}. \tag{5.2.8}$$

At first, we will consider the **solutions with functional parameters**. They, as usual, correspond to the degenerate kernel $R_0$

$$R_0(\mu,\bar{\mu};\lambda,\bar{\lambda}) = \sum_{k=1}^{N} f_{0k}(\mu,\bar{\mu}) g_{0k}(\lambda,\bar{\lambda}) \tag{5.2.9}$$

where $f_{0k}$ and $g_{0k}$ are arbitrary $2 \times 2$ matrix-valued functions and $N$ is an arbitrary integer. Using the formulae (2.2.20) – (2.2.21) and (5.2.1), one gets

$$q(x,y,t) = \frac{1}{\pi} \sum_{k,l=1}^{N} (\xi_k (1+A)_{kl}^{-1} \eta_l)_{12},$$

$$r(x,y,t) = -\frac{1}{\pi} \sum_{k,l=1}^{N} (\xi_k (1+A)_{kl}^{-1} \eta_l)_{21}, \tag{5.2.10}$$

$$C(x,y,t) = \frac{1}{2\pi} \sum_{k,l=1}^{N} (\xi_k (1+A)_{kl}^{-1} \eta_l)_{\text{diag}},$$

where
$$\xi_k(x,y,t) = \iint_C d\lambda \wedge d\bar{\lambda}\, e^{i\lambda x - \frac{i\lambda \sigma_3}{\sigma} y - i\lambda^2 \sigma_3 t} f_{0k}(\lambda, \bar{\lambda}),$$

$$\eta_l(x,y,t) = \iint_C d\lambda \wedge d\bar{\lambda}\, g_{0k}(\lambda, \bar{\lambda}) e^{-i\lambda x + \frac{i\lambda \sigma_3}{\sigma} y + i\lambda^2 \sigma_3 t} \quad (5.2.11)$$

and
$$A_{lk}(x,y,t) = \frac{1}{2\pi i} \iint_C d\mu \wedge d\bar{\mu} \iint_C \frac{d\lambda \wedge d\bar{\lambda}}{\lambda - \mu} g_{0l}(\mu, \bar{\mu})$$

$$\times \exp\left(i(\mu - \lambda)x - \frac{i}{\sigma}(\lambda - \mu)\sigma_3 y - i(\lambda^2 - \mu^2)\sigma_3 t\right) f_{0k}(\lambda, \bar{\lambda}). \quad (5.2.12)$$

Using the formal identity $\frac{1}{i\lambda} e^{i\lambda x} = \partial_x^{-1}(e^{i\lambda x})$, one represents the matrix $A$ in the form

$$A_{lk}(x,y,t) = \frac{1}{2\pi} \partial_x^{-1}(\eta_l \cdot \xi_k) \quad (5.2.13)$$

where $\xi_k$ and $\eta_l$ are given by (5.2.11).

The solutions (5.2.10)–(5.2.13) are parametrized by $8N$ arbitrary scalar functions of two variables. The functions $\xi_k$ and $\eta_l$ satisfy the systems of equations

$$\sigma \xi_y + \sigma_3 \xi_x = 0,$$

$$\xi_t - i\sigma_3 \xi_{xx} = 0 \quad (5.2.14)$$

and

$$\sigma \eta_y + \eta_x \sigma_3 = 0,$$

$$\eta_t - i\eta_{xx}\sigma_3 = 0. \quad (5.2.15)$$

Both $\xi$ and $\eta$ are also the solutions of the special type (5.2.11) of the linear part of the DS system

$$i\tilde{P}_t + \frac{1}{2}\sigma_3(\sigma^2 \tilde{P}_{yy} + \tilde{P}_{xx}) = 0. \quad (5.2.16)$$

Under the reduction $q = -\bar{r}$ for the DS-I equation one has for arbitrary $N$

$$g_{0k}(\lambda, \bar{\lambda}) = f_{0k}^+(\bar{\lambda}, \lambda) \quad (5.2.17a)$$

and, consequently,

$$\eta_k(x,y,t) = \xi_k^+(x,y,t). \quad (5.2.17b)$$

For $N = 2M$ where $M$ is arbitrary integer in addition to (5.2.17) the condition $q = -\bar{r}$ is satisfied also if

$$f_{0k+M}(\lambda, \bar{\lambda}) = g_{0k}^+(\bar{\lambda}, \lambda) \quad g_{0k+M}(\lambda, \bar{\lambda}) = f_{0k}^+(\bar{\lambda}, \lambda) \quad (k = 1, \ldots, N), \quad (5.2.18a)$$

and, hence,

$$\xi_{k+M}(x,y,t) = -\eta_k^+(x,y,t), \quad \eta_{k+M}(x,y,t) = -\xi_k^+(x,y,t), \quad (k=1,\ldots,M). \quad (5.2.18b)$$

Hence, the solutions with the functional parameters of the DS-I equation, for instance, in the case (5.2.17), are of the form

$$q = -\frac{1}{\pi} \sum_{k,l=1}^{N} (\xi_k(1+A)^{-1}_{kl}\xi_l^+)_{12},$$

$$C = -\frac{1}{2\pi} \sum_{k,l=1}^{N} (\xi_k(1+A)^{-1}_{kl}\xi_l^+)_{\text{diag}}, \quad (5.2.19)$$

where

$$A_{lk}(x,y,t) = \frac{1}{2\pi}\partial_x^{-1}(\xi_l^+\xi_k). \quad (5.2.20)$$

For the DS-II equation the reduction $r = \bar{q}$ implies

$$\overline{f_{0k}(\lambda,\bar{\lambda})} = \sigma_2 f_{0k}(-\bar{\lambda},-\lambda)\sigma_2^{-1},$$

$$\overline{g_{0k}(\lambda,\bar{\lambda})} = \sigma_2 g_{0k}(-\bar{\lambda},-\lambda)\sigma_2^{-1} \quad (5.2.21)$$

where $\sigma_2 = \begin{pmatrix} 0 & 1 \\ -1 & 0 \end{pmatrix}$ and, consequently,

$$\bar{\xi}_k(x,y,t) = \sigma_2 \xi_k(x,y,t)\sigma_2^{-1},$$

$$\bar{\eta}_k(x,y,t) = \sigma_2 \eta_k(x,y,t)\sigma_2^{-1} \quad (5.2.22)$$

Now we proceed to **line solitons**. They correspond to the choice of $f_{0k}$ and $g_{0k}$ as the delta-functions. Taking into account the condition (5.2.17), one has for the DS-I equation

$$R_0(\mu,\bar{\mu};\lambda,\bar{\lambda}) = \sum_{k=1}^{N} \tilde{f}_{0k}\tilde{f}_{0k}^+\delta(\mu-\lambda_k)\delta(\lambda-\bar{\lambda}_k) \quad (5.2.23)$$

where $\tilde{f}_{0k}$ are constant $2\times 2$ matrices and $\lambda_k$ are arbitrary distinct complex parameters. For the DS-II equation the conditions (5.2.22) imply

$$R_0(\mu,\bar{\mu};\lambda,\bar{\lambda}) = \sum_{k=1}^{N} \tilde{f}_{0k}\tilde{g}_{0k}\delta(\mu-i\mu_k)\delta(\lambda-i\lambda_k) \quad (5.2.24)$$

where $\mu_k, \lambda_k$ are arbitrary real constants and $f_{0k}, g_{0k}$ are arbitrary $2 \times 2$ matrices which obey the conditions

$$\overline{\tilde{f}_{0k}} = \sigma_2 f_{0k}\sigma_2^{-1}, \quad \tilde{g}_{0k} = \sigma_2 g_{0k}\sigma_2^{-1}. \quad (5.2.25)$$

For the DS-I equation one has

$$\xi_k(x,y,t) = -2i\exp(i\lambda_k x - i\lambda_k\sigma_3 y - i\lambda_k^2\sigma_3 t)\tilde{f}_{0k} \qquad (5.2.26)$$

and

$$A_{lk} = \frac{2i}{\pi} \frac{\tilde{f}_{0l}^+ \exp(i(\lambda_k - \bar{\lambda}_l)x - i(\lambda_k - \bar{\lambda}_l)\sigma_3 y - i(\lambda_k^2 - \bar{\lambda}_l^2)\sigma_3 t)\tilde{f}_{0k}}{\lambda_k - \bar{\lambda}_l}. \qquad (5.2.27)$$

So the line solitons of the DS-I equation is given by the formula (5.2.19) with $\xi_k$ and $A$ of the form (5.2.26) and (5.2.27).

Respectively, for the DS-II equation the line solitons are given by the formula (5.2.10) with

$$\xi_k(x,y,t) = -2i\exp(-\mu_k x - i\mu_k\sigma_3 y + i\mu_k^2\sigma_3 t)\tilde{f}_{0k},$$

$$\eta_l(x,y,t) = -2i\tilde{g}_{0k}\exp(\lambda_l x + i\lambda_l\sigma_3 y - i\lambda_l^2\sigma_3 t) \qquad (5.2.28)$$

and

$$A_{lk}(x,y,t) = \frac{2i}{\pi} \frac{\tilde{g}_{0l}^+ e^{(\lambda_l - \mu_k)x - i(\lambda_l - \mu_k)\sigma_3 y + i(\lambda_l^2 - \mu_k^2)\sigma_3 t}\tilde{f}_{0k}}{\lambda_l - \mu_k} \quad (l,k=1,\ldots N) \qquad (5.2.29)$$

where the matrices $\tilde{f}_{0k}$ and $\tilde{g}_{0k}$ obey the constraints (5.2.25).

The line solitons of the DS equation are apparently the two-dimensional generalizations of the solitons of the NLS equation. Similarly to the NLS case the line solitons collide elastically. The corresponding phase shift can be calculated explicitly.

For the first time the line solitons of the DS equation have been calculated by the Zakharov—Shabat dressing method in [208] and by the Hirota method in [179, 209].

The third standard class of exact solutions for the integrable equations are rational solutions. For the matrix systems like the DS equation associated with the matrix $\bar{\partial}$-problem the situation with such type of solutions is more complicated in comparison with the scalar case (the KP and mKP equations). In general, the solutions which correspond to the kernel $R$ of the $\bar{\partial}$-problem of the form (2.2.31) contain both rational and exponential dependence on $x$, $y$ and $t$. Different choices of more general kernels (2.2.36) lead to the solutions of different type. The most strong requirement for such a solutions consists in their boundedness. It occurs that for the DS equation the proper choice correspond to the case when each column of the matrix $\chi$ has its own set of simple poles, i. e.

$$R_{0\alpha\beta}(\mu,\bar{\mu};\lambda,\bar{\lambda}) = \sum_{k=1}^{N}\sum_{\gamma=1}^{2}(f_{0k}(\mu))_{\alpha\gamma}\delta(\mu - \lambda_\gamma^{(k)})(g_{0k}(\lambda))_{\gamma\beta}\delta(\lambda - \lambda_\beta^{(k)}) \qquad (5.2.30)$$

$$\alpha,\beta = 1,2$$

where $\lambda_\gamma^{(k)}$ are arbitrary distinct complex numbers. Using (2.2.37), one finds that in this case the function $\chi$ is of the form

$$\chi_{\alpha\beta}(\lambda,\bar{\lambda}) = \delta_{\alpha\beta} + \frac{2}{\pi i}\sum_{k=1}^{N}\sum_{\delta,\gamma=1}^{2}\frac{\chi_{\alpha\delta}\left(\lambda_\gamma^{(k)}\exp\left(i\lambda_\delta^{(k)}x - \frac{i}{\sigma}\lambda_\delta^{(k)}\sigma_3 y - i\lambda_\delta^{(k)^2}\sigma_3 t\right)(f_{0k})_{\delta\gamma}(\lambda_\gamma^{(k)})\right)}{\lambda - \lambda_\beta^{(k)}}$$

$$\times(g_{0k\gamma\beta})(\lambda_\beta^{(k)})e^{-i\lambda_\beta^{(k)} + \frac{i}{\sigma}\lambda_\beta^{(k)}\sigma_3 y + i\lambda_\beta^{(k)^2}\sigma_3 t} \qquad (5.2.31)$$

and the quantities $\chi_{\alpha\delta}(\lambda_\gamma^{(k)})$ are determined from the system (2.2.38), (2.2.39) in the particular case $\lambda_{\alpha\beta}^{(k)} = \lambda_\beta^{(k)}$ and $\alpha, \beta = 1,2$. One can also show that the reduction $r = -\bar{q}$ implies $\lambda_2 = \bar{\lambda}_1$. Then the functions $q = -2i(\chi_{-1})_{12}$ give us the rational-exponential solutions of the DS equation. Just for a change we will construct the bounded rational-exponential solutions of the DS-II equation within the solution of the initial-value problem in section (5.4).

Next we will consider **a novel class of solutions** of the 2+1-dimensional integrable equations which were not found for the KP and mKP equations. These are the **solutions exponentially localized in all directions** on the plane $x, y$. Such solutions are associated with the specific background $q_0 = r_0 = 0$ and the matrix $\tilde{C}_0$ given by (5.1.61).

We confine ourselves by the DS-I equation. So the functions $u_1(y-x)$ and $u_2(y+x)$ are arbitrary bounded functions. Further the boundedness of $q$ and $r$ due to (5.2.5) implies that ($\sigma = 1$)

$$R(\mu,\bar{\mu};\lambda,\bar{\lambda}) = R(\mu,\lambda)\delta(\mu-\bar{\mu})\delta(\lambda-\bar{\lambda}). \qquad (5.2.32)$$

Under the constraint (5.2.32) the $\bar{\partial}$-problem is reduced to the nonlocal Riemann—Hilbert problem with the jump across the real axis. Consequently, all the integrals over the plane $\lambda$ which have appeared above in the $\bar{\partial}$-dressing on nontrivial background convert into the integrals along the real axis. The variable $s$ also become real one. Then, the condition $r = -\bar{q}$ implies that the constraints (5.2.7) on the kernel $R$. At last, we will consider the particular solutions which corresponds to the off-diagonal kernel $R$. So we choose the kernel $R$ as

$$R(\mu,\lambda) = \begin{pmatrix} 0 & S(\mu,\lambda) \\ \overline{S(\lambda,\mu)} & 0 \end{pmatrix}. \qquad (5.2.33)$$

The Fourier transform of $S(\mu,\lambda)$

$$\tilde{S}(s,\tilde{s}) \doteq \frac{1}{2\pi}\int_R\int_R d\mu d\lambda\, e^{2i\mu s - 2i\lambda\tilde{s}}S(\mu,\lambda) \qquad (5.2.34)$$

obeys equation (5.1.65), i. e.

$$i\tilde{S}_t + \frac{1}{4}\tilde{S}_{zz} + \frac{1}{4}\tilde{S}_{\tilde{z}\tilde{z}} + (u_1(y-z,t) + u_2(y+\tilde{z},t))\tilde{S} = 0 \qquad (5.2.35)$$

where $z = s + x, \tilde{z} = \tilde{s} + x$. Introducing the variables $\xi = \frac{1}{2}(y - z)$ and $\eta = \frac{1}{2}(y + \tilde{z})$, one obtains

$$i\tilde{S}_t + \tilde{S}_{\xi\xi} + \tilde{S}_{\eta\eta} + (u_1(\xi,t) + u_2(\eta,t))\tilde{S} = 0. \tag{5.2.36}$$

Equation (5.2.36) is solvable by the separation of variables

$$\tilde{S}(\xi,\eta,t) = \sum_{i,k=1}^{N} \rho_{ik} X_i(\xi,t) Y_k(\eta,t) \tag{5.2.37}$$

and the functions $X$ and $Y$ obey the equations

$$iX_t + X_{\xi\xi} + u_1(\xi,t)X = 0, \tag{5.2.38}$$

$$iY_t + Y_{\eta\eta} + u_2(\eta,t)Y = 0. \tag{5.2.39}$$

These equations are exactly the first linear problems for the KP-I equation with the change $y \to t$. So we have a wide class of the exact solutions for equations (5.2.38) and (5.2.39).

In order to construct now the exact solutions of the DS-I equation for the given $u_1(\xi,t)$ and $u_2(\eta,t)$ one should combine the proper generalization of the formulae for the solutions with functional parameters of the DS-I equation and exact solutions of equations (5.2.38) and (5.2.39).

To follow this line we first note the degenerate kernel $R$ of the type (5.2.33), suitable to our problem looks like

$$R(\mu,\lambda,x,y,t)$$
$$= \sum_{i,k=1}^{N} \rho_{0ik} \left\{ \begin{pmatrix} f_i(\mu) \\ 0 \end{pmatrix} \otimes (0, g_k(\lambda)) + \bar{\rho}_{0ik} \begin{pmatrix} 0 \\ \bar{g}_i(\mu) \end{pmatrix} \otimes (\bar{f}_k(\lambda), 0) \right\}, \tag{5.2.40}$$

where $\otimes$ denotes a tensor product and $f_i(\mu)$, $g_i(\lambda)$ are arbitrary scalar functions the dependence of which on $x$, $y$, $t$ is defined by equations (5.1.62). Note that now for nontrivial background $f_k(\mu)$ and $g_k(\lambda)$ are not of the form $f_k(\mu) = \exp(F(\mu))f_{0k}(\mu)$, $g_k(\lambda) = g_{0k}(\lambda)\exp(-F(\lambda))$ where $F(\lambda)$ is given by (5.2.4).

Further, the Fourier transform of the kernel $R$ (5.2.40) is

$$R(s,\tilde{s}) = \sum_{i,k=1}^{N} \left\{ \rho_{0ik} \begin{pmatrix} X_i(s) \\ 0 \end{pmatrix} \otimes (0, Y_k(\tilde{s})) \right.$$
$$\left. + \bar{\rho}_{0ik} \begin{pmatrix} 0 \\ \bar{Y}_i(s) \end{pmatrix} \otimes (\bar{X}_k(\tilde{s}), 0) \right\} = \sum_{i,k=1}^{N} \begin{pmatrix} 0 & \rho_{ik} X_i(s) Y_k(\tilde{s}) \\ \bar{\rho}_{ik} \bar{Y}_i(s) \bar{X}_k(\tilde{s}) & 0 \end{pmatrix}. \tag{5.2.41}$$

The quantity $\tilde{S} = R_{12} = \sum_{i,k=1}^{N} \rho_{0ik} X_i(s) Y_k(\tilde{s})$ obeys equation (5.2.35) and the scalar functions $X_i, Y_k$ satisfy the linear equations (5.2.38) and (5.2.39).

The solutions of the DS-I equation which correspond to the kernel $R$ of the type (5.2.40) can be constructed with the use of the results of section (2.2). The reconstruction formula (5.2.1) and the formula (2.2.22) for $f$ and $g$ of the special form defined in (5.2.40) give

$$q(x,y,t) = \frac{1}{\pi} \sum_{k,l=1}^{N} \rho_{0kl} \xi_k \eta_l (((1+A)^{-1})_{kl})_{11} \qquad (5.2.42)$$

where

$$\xi_k = \int_R d\lambda\, f_k(\lambda), \quad \eta_k = \int_R d\lambda\, g_k(\lambda) \qquad (5.2.43)$$

and

$$A_{kl} = \int_R d\lambda \int_R d\mu \, \frac{1}{\lambda - \mu - i0} \begin{pmatrix} 0 & \sum_{i,l=1}^{N} \rho_{0il} f_i(\mu) g_l(\lambda) \\ \sum_{i,l=1}^{N} \bar{\rho}_{0il} \bar{g}_i(\mu) \bar{f}_l(\lambda) & 0 \end{pmatrix}. \qquad (5.2.44)$$

Then, in our case instead of (2.5.12) one has

$$f_k(\mu) = \int_R ds\, e^{-2i\mu s} X_k(s+x,y,t),$$

$$g_k(\lambda) = \int_R d\tilde{s}\, e^{-2i\lambda \tilde{s}} Y_k(\tilde{s}+x,y,t). \qquad (5.2.45)$$

Hence

$$\xi_k = X_k(x,y,t),$$

$$\eta_k = Y_k(x,y,t). \qquad (5.2.46)$$

In virtue of (5.1.63)

$$X_k = X_k(y-x,t),$$

$$Y_k = Y_k(y+x,t). \qquad (5.2.47)$$

Substituting (5.2.45) and (5.2.46) into (5.2.42), after straightforward but cumbersome calculations one finally obtains

$$q = \frac{2}{\pi} \sum_{k,l=1}^{N} X_k \rho_{kl} Y_l \qquad (5.2.48)$$

where the matrix $\rho$ is defined via

$$(1+A)\rho = \rho_0 \qquad (5.2.49)$$

where

$$A = \rho_0 \beta \rho_0^\dagger \bar{\alpha}. \qquad (5.2.50)$$

The matrices $\alpha$ and $\beta$ in (5.2.49) are of the form

$$\alpha_{ij} = \int_{-\infty}^{\frac{1}{2}(y-x)} d\xi' \, X_i(\xi',t)\bar{X}_j(\xi',t),$$

$$\beta_{ij} = \int_{-\infty}^{\frac{1}{2}(y+x)} d\eta' \, Y_i(\eta',t)\bar{Y}_j(\eta',t) \qquad (5.2.51)$$

where $\xi = \frac{1}{2}(y-x), \eta = \frac{1}{2}(y+x)$.

The formula (5.2.48) gives us the solution of the DS-I equation for arbitrary background $q_0 = r_0 = 0$, $C_0$ or, equivalently boundary functions $u_1(\xi, t)$ and $u_2(\eta, t)$. For the first time the formula (5.2.48) has been derived by another approach in [210, 117].

In order to present the explicit solutions (5.2.48) one should choose the functions $u_1$ and $u_2$ such that one can calculate the solutions of equations (5.2.38) and (5.2.39) explicitly.

The multiline soliton solutions of the KP-I equation are the proper candidates. Dropping the time dependence in these solutions (3.2.12), (3.2.24) and changing $y \to t$, we obtain the explicit solutions of the problems (5.2.38) and (5.2.39). Thus, one has

$$u_1(\xi, t) = 2\partial_\xi^2 \ln \det(1 + A^{(\xi)}) \qquad (5.2.52)$$

where

$$A_{lk}^{(\xi)} = \frac{\gamma_k \bar{\gamma}_l}{i(\lambda_k - \bar{\lambda}_l)} e^{i(\lambda_k - \bar{\lambda}_l)(\xi - (\lambda_k + \bar{\lambda}_l)t)} \quad (k,l = 1, \ldots, N) \qquad (5.2.53)$$

and $\lambda_k, \gamma_k$ are arbitrary complex constants and $N$ is an arbitrary integer. Analogously, for $u_2(\eta, t)$

$$u_2(\eta, t) = 2\partial_\eta^2 \ln \det(1 + A^{(\eta)}) \qquad (5.2.54)$$

where

$$A_{lk}^{(\eta)} = \frac{\tilde{\gamma}_k \bar{\tilde{\gamma}}_l}{i(\mu_k - \bar{\mu}_l)} e^{i(\mu_k - \bar{\mu}_l)(\eta - (\mu_k + \bar{\mu}_l)t)} \quad (k,l = 1, \ldots, M) \qquad (5.2.55)$$

where $\mu_k$, $\gamma_k$ are arbitrary complex constants and $M$ is an arbitrary integer.

Then the solutions $X_k$ and $Y_k$ of the linear problems (5.2.38) and (5.2.39) are

$$X_k = \chi_k^{(\xi)} e^{i\lambda_k \xi - i\lambda_k^2 t},$$

$$Y_k = \chi_k^{(\eta)} e^{i\mu_k \eta - i\mu_k^2 t}, \qquad (5.2.56)$$

where the functions $\chi_k$ are the solutions of the problems of the type (3.3.5), i. e.

$$i\chi_{kt}^{(\xi)} + \chi_{k\xi\xi}^{(\xi)} + 2i\lambda_k \chi_{k\xi}^{(\xi)} + u_1(\xi,t)\chi_k^{(\xi)} = 0 \qquad (5.2.57)$$

and

$$i\chi_{kt}^{(\eta)} + \chi_{k\eta\eta}^{(\eta)} + 2i\mu_k \chi_{k\eta}^{(\eta)} + u_2(\eta,t)\chi_k^{(\eta)} = 0. \qquad (5.2.58)$$

Using the results of section (2.2), one readily shows that the functions $\chi_k$ obey the system of equations

$$\chi_l = \gamma_l + \sum_{k=1}^{N} \chi_k A_{kl} \tag{5.2.59}$$

where the matrix $A$ is of the form (5.2.53). Therefore

$$X_k = \sum_{l=1}^{N} \gamma_l e^{i\lambda_l \xi - i\lambda_l^2 t}(1 + A^{(\xi)})^{-1}_{lk}, \tag{5.2.60}$$

$$Y_k = \sum_{l=1}^{M} \tilde{\gamma}_l e^{i\mu_l \eta - i\mu_l^2 t}(1 + A^{(\eta)})^{-1}_{lk}. \tag{5.2.61}$$

For such a choice of the functions $X_k$ and $Y_k$ one can calculate the matrices $\alpha$ and $\beta$ (5.2.51) in the following manner [210, 117]. First note that

$$X_i \bar{X}_j = \sum_{k,m}(1+A^T)^{-1}_{ik}\gamma_k\bar{\gamma}_m e^{i(\lambda_k-\bar{\lambda}_m)\xi - i(\lambda_k^2-\bar{\lambda}_m^2)t}(1+A^T)^{-1}_{mj} \tag{5.2.62}$$

where we use the equality $A^+ = A$. Then

$$\gamma_k \bar{\gamma}_m e^{i(\lambda_k-\bar{\lambda}_m)\xi - i(\lambda_k^2-\bar{\lambda}_m^2)t} = \partial_\xi (1 + A^T)_{km}. \tag{5.2.63}$$

As a result

$$X_i(\xi,t)\bar{X}_j(\xi,t) = -\partial_\xi((1+A^{(\xi)T})^{-1})_{ij}. \tag{5.2.64}$$

Analogously

$$Y_i(\eta,t)\bar{Y}_j(\eta,t) = -\partial_\eta((1+A^{(\eta)T})^{-1})_{ij}. \tag{5.2.65}$$

Using (5.2.64) and (5.2.65), one gets

$$\alpha(\xi,t) = -(1+A^{(\xi)T})^{-1} \tag{5.2.66}$$

and

$$\beta(\eta,t) = -(1+A^{(\eta)T})^{-1}. \tag{5.2.67}$$

Thus, the matrix $A$ (5.2.50) is

$$A = \rho_0(1+A^{(\eta)T})^{-1}\rho_0^+(1+A^{(\xi)})^{-1}. \tag{5.2.68}$$

So, the exact solutions of the DS-I equation associated with the multisoliton boundaries $u_1$ and $u_2$ are of the form

$$q = \frac{1}{\pi}\sum_{k,l=1}^{N} X_k((1+A)\rho_0)_{kl} Y_l \tag{5.2.69}$$

Chapter 5

where $(\rho_0)_{kl}$ are arbitrary constants, the matrix $A$ is given by (5.2.68) and the functions $X_k$, $Y_k$ by (5.2.60), (5.2.61). For the first time the formula (5.2.69) has been derived in [210, 117] by another approach.

The simplest solution of the type (5.2.69) is associated with the choice of $u_1$ and $u_2$ as one-soliton solutions (3.2.25) ($N = M = 1$)

$$u_1(\xi, t) = \frac{2\rho^2}{\cosh^2(\rho(\xi + 2\nu t) - \xi_0)}, \qquad (5.2.70)$$

$$u_2(\eta, t) = \frac{2\tilde{\rho}^2}{\cosh^2(\tilde{\rho}(\eta + 2\tilde{\nu}t) - \eta_0)} \qquad (5.2.71)$$

and has a form [115, 116]

$$q(x, y, t) = \frac{\rho_{011}\sqrt{\rho\tilde{\rho}}}{\pi}$$

$$\times \frac{\exp(-[\rho(\hat{\xi} - \xi_0) + \tilde{\rho}(\hat{\eta} - \eta_0)] + i[-(\nu\hat{\xi} + \tilde{\nu}\hat{\eta}) + (\rho^2 + \nu^2 + \tilde{\rho}^2 + \tilde{\nu}^2)t + \arg(\gamma, \tilde{\gamma}_1)])}{(1 + e^{-2\rho(\hat{\xi}-\xi_0)})(1 + e^{-2\tilde{\rho}(\hat{\eta}-\eta_0)}) + |\rho_{011}|^2} \qquad (5.2.72)$$

where

$$\hat{\xi} = \frac{1}{2}(y - x) + 2\nu t, \quad \hat{\eta} = \frac{1}{2}(y + x) + 2\tilde{\nu}t. \qquad (5.2.73)$$

It is not difficult to see that the solution (5.2.72) is localized exponentially in all directions on $x$, $y$ plane. It moves with the velocity $(-2\nu, -2\tilde{\nu})$ in the $(\xi, \eta)$ coordinates. This velocity is determined by the velocities $-2\nu$ and $-2\tilde{\nu}$ with which the boundaries $u_1(\xi, t)$ and $u_2(\eta, t)$ move.

The solution (5.2.72) exemplifies a novel class of exact solutions of the 2+1-dimensional integrable equations. The existence of such solutions of the DS-I equation has been discovered in the paper [115] with the use of elementary Bäcklund transformations. The connection of such solutions with the initial-boundary value problem has been established in [116] soon after. Changing the velocity of the boundaries $u_1$ and $u_2$ one can derive the solution (5.2.72) [117]. By this reason the solutions of the type (5.2.72) has been referred in [117] as dromions.

General solution (5.2.69) represents the localized solution of the type $(N, M)$. In the case $N = M$ it describes the scattering of $N$ simple localized objects (5.2.72).

Localized solutions of the DS-I equation of the type (5.2.69) have a number of interesting properties. They have studied in great detail mainly by the two groups of authors (Boiti, Léon, Martina and Pempinelly) [115, 211–223] and (Fokas, Santini [116, 117, 210, 224]) by essentially different approaches. For detail we address the readers to their papers. The exponentially localized solutions of the DS equation has also been obtained by the

Hirota method [225, 226], direct linearization method [227], by the method of Darboux transformations [73] and by other approaches [228, 229].

We emphasize one more time that from the point of view of the $\bar{\partial}$-dressing method the exponentially localized solutions of the DS-I equations arise as the particular case of the dressing on the almost trivial background $q_0 = 0$ and $C_0$ given by (5.1.61). This fact has been established, in essence, for the first time in [148].

The general formula (5.2.48) is valid, of course, not only for bounded $u_1(\xi,t)$ and $u_2(\eta,t)$. Equations (5.2.38) and (5.2.39) are the one-dimensional Schrödinger equations where $-u_1(\xi,t)$ and $-u_2(\eta,t)$ play the role of potentials. So to construct the exact solutions of the DS-I equation, one can use in addition to (5.2.71), (5.2.72), the other exactly solvable cases (see e. g. [230]). The harmonic oscillator is the most well known among them. Choosing the moving harmonic oscillator potential as $-u_1$ and $u_2$, namely,

$$u_1 = -\frac{1}{4}(\xi - \sqrt{2}\,t)^2,$$

$$u_2 = -\frac{1}{4}(\eta - \sqrt{2}\,t)^2 \tag{5.2.74}$$

and considering the ground states, one has [230]

$$X(\xi,t) = \frac{1}{(2\pi)^{1/4}} \exp\left(-\frac{(\xi-\sqrt{2}\,t)^2}{4} + \frac{i}{\sqrt{2}}(\xi - \sqrt{2}\,t)\right),$$

$$Y(\eta,t) = \frac{1}{(2\pi)^{1/4}} \exp\left(-\frac{(\eta-\sqrt{2}\,t)^2}{4} + \frac{i}{\sqrt{2}}(\eta - \sqrt{2}\,t)\right). \tag{5.2.75}$$

The corresponding solution of the DS-I equation is of the form

$$q(x,y,t)$$

$$= \frac{\sqrt{2}\,\rho_0}{\pi^{3/2}} \frac{\exp\left(-\left[\left(\frac{1}{2}(y-x) - \sqrt{2}\,t\right)^2 + \left(\frac{1}{2}(y+x) - \sqrt{2}\,t\right)^2\right] + i\left(\frac{y}{\sqrt{2}} - t\right)\right)}{1 + \frac{|\rho_0|^2}{2\pi} \int_{+\infty}^{\frac{1}{2}(y-x)} d\xi' \exp\left(-\frac{(\xi'-\sqrt{2}\,t)^2}{2}\right) \cdot \int_{+\infty}^{\frac{1}{2}(y+x)} d\eta' \exp\left(-\frac{(\eta'-\sqrt{2}\,t)^2}{2}\right)}. \tag{5.2.76}$$

This solution has a Gaussian localization while the boundaries $u_1$ and $u_2$ are unbounded.

The formula (5.2.48) also provides the convenient form different from (5.2.19) for the solutions of the DS-I equation with functional parameters for the trivial background. In the form (5.2.48) these functional parameters are arbitrary solutions of the free nonstationary Schrödinger equations

$$iX_t + X_{\xi\xi} = 0,$$

$$iY_t + Y_{\eta\eta} = 0. \tag{5.2.77}$$

The well known Gaussian packet

$$X(t,x) = \frac{1}{\sqrt{(a+it)}} \exp\left(-\frac{x^2}{4(a+it)}\right) \quad (5.2.78)$$

where $a$ is an arbitrary real constant is the particular solution of the Schrödinger equation. Choosing $X$ and $Y$ in the form (5.2.78), one gets the following exact solution of the DS-I equation

$$q(x,y,t) = \frac{2\rho_0}{\pi} \frac{1}{\sqrt{(a+it)(b+it)}} \exp\left(-\frac{1}{16}\left[\frac{(y-x)^2}{a+it} + \frac{(y+x)^2}{b+it}\right]\right)$$

$$\times \frac{1}{1 + |\rho_0|^2 \frac{1}{\sqrt{(a^2+t^2)(b^2+t^2)}} \int_\infty^{\frac{1}{2}(y-x)} d\xi \, e^{-\frac{a\xi^2}{2(a^2+t^2)}} \int_\infty^{\frac{1}{2}(y+x)} d\eta \, e^{-\frac{b\eta^2}{2(b^2+t^2)}}}. \quad (5.2.79)$$

The solution (5.2.79) has a Gaussian localization but vanish as $t \to \pm\infty$. For the first time the solution (5.2.79) has been derived by different approach in [231, 232]. Such Gaussian solutions of different types for the DS-I equation have been considered in [229].

## 5.3. Initial value problem for the DS-I equation

Similar to the KP and mKP equations the study of the initial value problem for the DS equation requires the analysis of the forward and inverse problems for the linear system (5.1.33). We will assume that $q$ and $r$ decay sufficiently rapidly as $x^2 + y^2 \to \infty$.

The spectral parameter is introduced into the linear system (5.1.33) by considering the special class of solutions of the type

$$\psi = \mu(x,y,t,\lambda) \exp\left(i\lambda x - \frac{i\lambda}{\sigma}\sigma_3 y\right) \quad (5.3.1)$$

where $\lambda$ is an arbitrary complex parameter and $\mu \to 1$ as $\lambda \to \infty$. The function $\mu$ obeys the following linear system of equations

$$\sigma\mu_y + \sigma_3\mu_x + i\lambda[\sigma_3,\mu] + P\mu = 0, \quad (5.3.2)$$

$$\mu_t - i\sigma_3(\mu_{xx} + 2i\lambda\mu_x) - iP(\mu_x + i\lambda\mu) + i\left(\sigma C_y - \sigma_3 C_x - \frac{\sigma\sigma_3}{2}P_y - \frac{1}{2}P_x\right)\mu = 0. \quad (5.3.3)$$

In this section we will discuss the DS-I equation. The initial-value problem for the DS-I equation has been studied in [233, 234, 129].

It is convenient to use the characteristic variables

$$\xi = \frac{1}{2}(y-x), \quad \eta = \frac{1}{2}(y+x).$$

The linear problem (5.3.2) is now of the form

$$\begin{pmatrix} \partial_\eta & 0 \\ 0 & \partial_\xi \end{pmatrix} \mu + i\lambda[\sigma_3, \mu] + P\mu = 0. \tag{5.3.4}$$

Note that the general solution of equation (5.3.4) with $P \equiv 0$ is

$$\mu_0(\xi, \eta, \lambda, \mu) = \begin{pmatrix} \alpha_1(\xi) & \alpha_2(\xi)e^{-2i\lambda\eta} \\ \alpha_3(\eta)e^{2i\lambda\xi} & \alpha_4(\eta) \end{pmatrix} \tag{5.3.5}$$

where $\alpha_1$, $\alpha_2$, $\alpha_3$, $\alpha_4$ are arbitrary functions.

To specify the solutions of equation (5.3.4) by certain integral equation one should first define the Green function. It can be calculated by different approaches. We will use here the trick analogous to that proposed for the DS-II equation in [124].

Namely, one can verify that operator $L_0 = \begin{pmatrix} \partial_\eta & 0 \\ 0 & \partial_\xi \end{pmatrix} + i\lambda[\sigma_3, \cdot]$ is representable in the factorized form

$$L_0 = E_\lambda^{-1} D E_\lambda \tag{5.3.6}$$

where $D = \begin{pmatrix} \partial_\eta & 0 \\ 0 & \partial_\xi \end{pmatrix}$ and the operator $E_\lambda$ acts by the rule

$$E_\lambda B = \begin{pmatrix} B_{11} & e^{2i\lambda\eta} B_{12} \\ e^{-2i\lambda\xi} & B_{22} \end{pmatrix}. \tag{5.3.7}$$

The operator $L_0$ (5.3.6) is not bounded as the function on $\lambda$. However, this disadvantage is more or less compensated by the simple structure of the operator $D^{-1} = \begin{pmatrix} \partial_\eta^{-1} & 0 \\ 0 & \partial_\xi^{-1} \end{pmatrix}$. As a result, the inverse operator $\hat{G} = L_0^{-1}$ is given by the simple formula

$$\hat{G} = E_\lambda^{-1} D^{-1} E_\lambda, \tag{5.3.8}$$

i. e.

$$(\hat{G}B)(\xi, \eta) = \begin{pmatrix} \partial_\eta^{-1} B_{11}(\xi', \eta') & \partial_\eta^{-1}(e^{-2i\lambda(\eta-\eta')} B_{12}(\xi', \eta')) \\ \partial_\xi^{-1}(e^{2i\lambda(\xi-\xi')} B_{21}(\xi', \eta')) & \partial_\xi^{-1} B_{22}(\xi', \eta') \end{pmatrix}. \tag{5.3.9}$$

The kernel $G(\xi, \eta, \lambda)$ of the operator $\hat{G}$ is the usual Green function for the operator $L_0$.

The main feature of the Green function (5.3.9) is that it is analytic function on the entire complex plane of $\lambda$.

Another feature of the Green function (5.3.9) is that it is defined nonuniquely. The freedom in the definition of the Green function is connected with the possibility to choose

Chapter 5

the different concrete realizations of the formal operators $\partial_\eta^{-1}$ and $\partial_\xi^{-1}$. Let us use this freedom in order to construct the bounded Green functions. Choosing $\partial_\eta^{-1} f = \int_{+\infty}^\eta d\eta' f(\xi, \eta')$ and $\partial_\xi^{-1} f = \int_{-\infty}^\xi d\xi' f(\xi', \eta)$ we define the Green function $G^+$:

$$(\hat{G}^+(\cdot, \lambda) B(\cdot))(\xi, \eta)$$
$$\doteq \begin{pmatrix} \int_{+\infty}^\eta d\eta' B_{11}(\xi, \eta') & \int_{+\infty}^\eta d\eta' e^{-2i\lambda(\eta-\eta')} B_{12}(\xi, \eta') \\ \int_{-\infty}^\xi d\xi' e^{2i\lambda(\xi-\xi')} B_{21}(\xi', \eta) & \int_{-\infty}^\xi d\xi' B_{22}(\xi', \eta) \end{pmatrix}. \quad (5.3.10a)$$

The choice $\partial_\xi^{-1} f = \int_{+\infty}^\xi d\xi' f(\xi', \eta)$ and $\partial_\eta^{-1} f = \int_{-\infty}^\eta d\eta' f(\xi, \eta')$ gives the Green function $G^-$:

$$(\hat{G}^-(\cdot, \lambda) B(\cdot))(\xi, \eta)$$
$$\doteq \begin{pmatrix} \int_{-\infty}^\eta d\eta' B_{11}(\xi, \eta') & \int_{-\infty}^\eta d\eta' e^{-2i\lambda(\eta-\eta')} B_{12}(\xi, \eta') \\ \int_{+\infty}^\xi d\xi' e^{2i\lambda(\xi-\xi')} B_{21}(\xi', \eta) & \int_{+\infty}^\xi d\xi' B_{22}(\xi', \eta) \end{pmatrix}. \quad (5.3.10b)$$

It is easy to see that the Green function $G^+(x, y, \lambda)$ is bounded in the upper half-plane $\text{Im}\lambda > 0$ while the Green function $G^-(x, y, \lambda)$ is bounded in the lower half-plane $\text{Im}\lambda < 0$.

Now let us consider the solutions $\mu^+$ and $\mu^-$ of the problem (5.3.4) which simultaneously are the solutions of the integral equations

$$\mu^+(\xi, \eta, \lambda) = 1 - [G^+(\cdot, \lambda) P(\cdot) \mu^+(\cdot, \lambda)](\xi, \eta) \quad (5.3.11)$$

and

$$\mu^-(\xi, \eta, \lambda) = 1 - [G^-(\cdot, \lambda) P(\cdot) \mu^-(\cdot, \lambda)](\xi, \eta) \quad (5.3.12)$$

where $G^+$ and $G^-$ are given by the formulae (5.3.10) and $P = \begin{pmatrix} 0 & p \\ r & 0 \end{pmatrix}$. The solutions $\mu^+$ and $\mu^-$ are analytic and bounded functions in the upper and lower half-plane respectively. Further, since $G^+ - G^- \neq 0$ at $\text{Im}\lambda = 0$ then $\mu^+ - \mu^- \neq 0$ at $\text{Im}\lambda = 0$ too. So one can define the function $\mu \doteq \begin{cases} \mu^+, & \text{Im}\lambda > 0 \\ \mu^-, & \text{Im}\lambda < 0 \end{cases}$ which is analytic on the entire complex plane and has a jump across the real axis. We will assume that the homogeneous equations (5.3.11) and (5.3.12) have no nontrivial solutions.

At this stage our purpose is to find the relation between the functions $\mu^+$ and $\mu^-$ on the real axis. We denote $K(\xi, \eta, \lambda) \doteq \mu^+(\xi, \eta, \lambda) - \mu^-(\xi, \eta, \lambda)$ for $\text{Im}\lambda = 0$. Subtracting (5.3.12) from (5.3.11) and taking into account (5.3.10), we obtain

$$K(\xi, \eta, \lambda) = T(\xi, \eta, \lambda) - [g^-(\cdot, \lambda) P(\cdot) K(\cdot, \lambda)](\xi, \eta) \quad (5.3.13)$$

where

$$T(\xi, \eta, \lambda) = \begin{pmatrix} -\int_{-\infty}^{+\infty} d\eta' (q\mu_{21}^+)(\xi, \eta') & -\int_{-\infty}^{+\infty} d\eta' e^{-2i\lambda(\eta-\eta')} (q\mu_{22}^+)(\xi, \eta') \\ \int_{-\infty}^{+\infty} d\xi' e^{2i\lambda(\xi-\xi')} (r\mu_{11}^+)(\xi', \eta) & \int_{-\infty}^{+\infty} d\xi' (r\mu_{12}^+)(\xi', \eta) \end{pmatrix}.$$

Further, let us introduce the quantity

$$\Delta(\xi,\eta,\lambda) = \int_{-\infty}^{+\infty} d\lambda' \mu^-(\xi,\eta,\lambda') \Sigma_{\lambda'}(\xi,\eta) f(\lambda',\lambda) \Sigma_\lambda^{-1}(\xi,\eta) \quad (5.3.14)$$

where $\Sigma_\lambda(\xi,\eta) = \begin{pmatrix} e^{2i\lambda\eta} & 0 \\ 0 & e^{-2i\lambda\xi} \end{pmatrix}$ and $f(\lambda',\lambda) = \begin{pmatrix} 0 & f_{12} \\ f_{21} & f_{22} \end{pmatrix}$ where

$$f_{12}(\lambda',\lambda) = \frac{2}{\pi} \int_{-\infty}^{+\infty} \int_{-\infty}^{+\infty} d\xi d\eta \, r(\xi,\eta,t) \mu_{11}^+(\xi,\eta,\lambda) e^{2i\lambda\eta + 2i\lambda'\xi}, \quad (5.3.15)$$

$$f_{21}(\lambda',\lambda) = -\frac{2}{\pi} \int_{-\infty}^{+\infty} \int_{-\infty}^{+\infty} d\xi d\eta \, q(\xi,\eta,t) \mu_{22}^-(\xi,\eta,\lambda) e^{-2i\lambda'\eta - 2i\lambda\xi} \quad (5.3.16)$$

and

$$f_{22}(\lambda',\lambda) = \int_{-\infty}^{+\infty} d\mu \, f_{21}(\lambda',\mu) f_{12}(\mu,\lambda). \quad (5.3.17)$$

One can show that the quantity $\Delta$ obeys the same equation (5.3.13) as the jump $K(\xi,\eta,\lambda)$. As a result, in virtue of the absence of the nontrivial solutions for the homogeneous equation (5.3.13), one has $\Delta = K$, i. e.

$$\mu^+(\xi,\eta,\lambda) = \mu^-(\xi,\eta,\lambda) + \int_{-\infty}^{+\infty} d\lambda' \mu^-(\xi,\eta,\lambda') \Sigma_\lambda f(\lambda',\lambda) \Sigma_\lambda. \quad (5.3.18)$$

Thus, we have constructed the function $\mu = \begin{cases} \mu^+, & \text{Im}\lambda > 0 \\ \mu^-, & \text{Im}\lambda < 0 \end{cases}$ which is analytic on the entire complex plane $\lambda$ and has the jump across the real axis which is given by (5.3.18). So we have arrived at the standard regular nonlocal Riemann—Hilbert problem. The solution of this problem is given by the formula (1.1.28). Passing in this formula to the limit $\lambda_I \to i0$ (see equation (1.1.29)), we obtain

$$\mu^-(\xi,\eta,\lambda) = 1 + \frac{1}{2\pi i} \int_{-\infty}^{+\infty} \int_{-\infty}^{+\infty} d\mu d\lambda' \frac{\mu^-(\xi,\eta,\lambda') \Sigma_{\lambda'} f(\lambda',\mu) \Sigma_\mu}{\mu - \lambda + i0}. \quad (5.3.19)$$

The formula for the reconstruction of the potential $(q,r)$ via $\mu^-$ and $f(\lambda',\lambda)$ is of the form

$$\begin{pmatrix} 0 & q \\ r & 0 \end{pmatrix} = -\frac{1}{4\pi} \left[ \sigma_3, \int_{-\infty}^{+\infty} \int_{-\infty}^{+\infty} d\mu d\lambda \mu^-(\xi,\eta,\lambda) e^{2i\lambda y\sigma_3 - 2i\lambda x} f(\lambda,\mu) e^{-2i\mu y\sigma_3 + 2i\mu x} \right]. \quad (5.3.20)$$

The integral equation (5.3.19) and the reconstruction formula (5.3.20) give the complete solution of the inverse problem for equation (5.3.4) [129]. The functions $f_{12}(\lambda',\lambda)$ and $f_{21}(\lambda',\lambda)$ are the inverse problem data.

The time-dependence of the inverse problem data can be found in the standard manner. One has

$$f_{12}(\lambda',\lambda,t) = e^{-\frac{i}{4}(\lambda'^2 + \lambda^2)t} f_{12}(\lambda',\lambda,0),$$

$$f_{21}(\lambda',\lambda,t) = e^{\frac{i}{4}(\lambda'^2+\lambda^2)t} f_{21}(\lambda',\lambda,0). \tag{5.3.21}$$

The forward and inverse problem equations for the problem (5.3.4) and formulae (5.3.20) give the solution of the initial value problem for the DS-I equation within the class of decreasing solutions.

The classes of the exact solutions of the DS-I equation with the functional parameters can be calculated by the standard procedure in the case of the factorised functions $f_{12}(\lambda',\lambda)$ and $f_{21}(\lambda',\lambda)$. These solutions are the particular case of the solutions with functional parameters described in the previous section.

Note that the reduction $r = -q$ implies $\overline{f(\mu,\lambda)} = -f(\lambda,\mu)$.

The mathematically rigorous analysis of the forward and inverse problems for equation (5.3.4) has been given in [235–240, 70] for the first time within the method of the operators of transformations. Another approach has been proposed in [232] and [241]. Within the modern formalism the problem (5.3.4) has been treated rigorously in [242, 196].

The IST method allows us also to linearise a special initial-boundary value problem for the DS-I equation in addition to the initial-value problem described above [116, 117].

Indeed, the DS-I equation (5.0.1) contains the two fields $q$ and $\varphi$. The second equation implies

$$\varphi_\xi = u_1(\xi,t) + \frac{1}{2} \int_{-\infty}^{\eta} d\xi' |q|^2,$$

$$\varphi_\eta = u_2(\eta,t) + \frac{1}{2} \int_{-\infty}^{\xi} d\eta' |q|^2 \tag{5.3.22}$$

where $u_1$ and $u_2$ are arbitrary functions. They are the boundary values of $\varphi_\xi$ and $\varphi_\eta$:

$$u_1(\xi,t) = \lim_{\eta \to -\infty} \varphi_\xi,$$

$$u_2(\eta,t) = \lim_{\xi \to -\infty} \varphi_\eta. \tag{5.3.23}$$

Let us consider the initial-boundary value problem for constructing solution of the DS-I equation for given $q(x,y,t)$ and $u_1(\xi,t)$, $u_2(\eta,t)$ [116, 117, 224].

The solution of the forward spectral problem for equation (5.3.4) gives us the inverse problem data $f(\mu,\lambda,0)$ and its expression (5.3.15) via the initial data $q(x,y,0)$. To find the time evolution of the data $f(\mu,\lambda,t)$ one should use the time part of the auxiliary linear problems. For nonzero $u_1$ and $u_2$ the second equation (5.1.32) requires a modification since $e^{i\lambda x - i\lambda^2 y}$ is not its solution now. Using the first linear problem one can find this modification [116]. It is the problem (5.1.69). Using equation (5.1.69), one finds the linear integral equation which cover the time evolution of the data $f(\mu,\lambda,t)$. In terms of the Fourier transform

$$S(\xi,\eta,t) = \int_R d\lambda \int_R d\mu\, e^{i(\lambda\xi+\mu\eta)} f(\mu,\lambda,t) \tag{5.3.24}$$

this equation becomes [116]

$$iS_t + S_{\xi\xi} + S_{\eta\eta} + (u_1(\xi,t) + u_2(\eta,t))S = 0. \tag{5.3.25}$$

Further, this equation can be solved by separation of variables $S(\xi,\eta,t) = X(\xi,t)Y(\eta,t)$ and the functions $X$ and $Y$ obey equations (5.2.38) and (5.2.39). Then, using the solution of the inverse problem equation (6.3.14) with degenerate data $f(\mu,\lambda)$, one finds the expression for $q$ via $X$ and $Y$. It is, obviously, given by (5.2.48). So, in principle, we have the solution of the initial-boundary value problem [116, 117]

$$\{q(x,y,0),\ u_1(y-x,t),\ u_2(y+x,t)\} \to f(\mu,\lambda,0)$$

$$\to \{q(x,y,t),\ u_1(y-x,t),\ u_2(y+x,t)\}. \tag{5.3.26}$$

The analysis of this problem basically uses the formula (5.2.48). It allows, for instance, to calculate the asymptotics essentially of $q$ as $t \to \infty$. This asymptotics essentially is determined by the particular exponentially localized solutions (5.2.69). The detailed study of the initial-boundary value problem for the DS-I equation has been given in [117, 224].

## 5.4. Initial value problem for the DS-II equation

We will describe the solution of the initial value problem for the DS-II equation following the papers [233, 234, 124, 243].

For the DS-II equation it is convenient to use the complex coordinates $z = \frac{1}{2}(y - ix)$ and $\bar{z} = \frac{1}{2}(y + ix)$. In this variables the problem (5.3.2) looks like

$$\begin{pmatrix} \partial_{\bar{z}} & 0 \\ 0 & \partial_z \end{pmatrix} \mu + \lambda[\sigma_3,\mu] - i\begin{pmatrix} 0 & q \\ r & 0 \end{pmatrix}\mu = 0. \tag{5.4.1}$$

The problem (5.4.1) has an important difference in comparison with the case $\sigma = 1$. Namely, under the reduction $r = -\bar{q}$ it possesses the involution

$$\sigma_2 \mu(\lambda,\bar{\lambda})\sigma_2^{-1} = \overline{\mu(-\bar{\lambda},-\lambda)} \tag{5.4.2}$$

where $\sigma_2 = \begin{pmatrix} 0 & i \\ -i & 0 \end{pmatrix}$. The involution (5.4.2) means that the matrix $\mu$ is of the form

$$\mu = \begin{pmatrix} \mu_{11} & -\overline{\mu_{21}} \\ \mu_{21} & \overline{\mu_{11}} \end{pmatrix}.$$

So, in fact, only one column of $\mu$ is independent. We will use this fact in further constructions.

Chapter 5

First of all note that the general solution $\mu_0$ of this equation (5.4.1) at $q = r = 0$ bounded for all $\lambda$ is of the form

$$\mu_0(z,\bar{z},\lambda) = \begin{pmatrix} \alpha_1(z) & \alpha_2(z)e^{-2\lambda\bar{z}+2\bar{\lambda}z} \\ \alpha_3(\bar{z})e^{2\lambda z-2\bar{\lambda}\bar{z}} & \alpha_4(\bar{z}) \end{pmatrix}$$

where $\alpha_1(z)$, $\alpha_2(z)$, $\alpha_3(\bar{z})$, $\alpha_4(\bar{z})$ are arbitrary functions. Then, the operator $L_0 \doteq L_{q=r=0}$ $= \begin{pmatrix} \partial_{\bar{z}} & 0 \\ 0 & \partial_z \end{pmatrix} + \lambda[\sigma_3, \cdot]$ can be represented in the factorized form [124]

$$L_0 = E_\lambda^{-1} D E_\lambda \tag{5.4.3}$$

where $D = \begin{pmatrix} \partial_{\bar{z}} & 0 \\ 0 & \partial_z \end{pmatrix}$ and the operator $E_\lambda$ acts on the any $2 \times 2$ matrix $B$ as follows

$$E_\lambda B(z) = \begin{pmatrix} B_{11}(z) & e^{2\bar{z}\lambda - 2z\bar{\lambda}} B_{12}(z) \\ e^{-2z\lambda + 2\bar{z}\bar{\lambda}} B_{21}(z) & B_{22}(z) \end{pmatrix}. \tag{5.4.4}$$

It is important that the operator $E_\lambda$ is bounded for all $\lambda$.

Now let us consider the solutions $\mu(z, \bar{z}, \lambda)$ of the problem (5.4.1) bounded in all $\lambda$ and canonically normalized ($\mu \to 1$ as $\lambda \to \infty$). Such solutions of equation (5.4.1) are defined by the integral equation

$$\mu(z, \bar{z}, \lambda) = 1 - (G(\cdot, \lambda) P(\cdot) \mu(\cdot, \lambda))(z, \bar{z}) \tag{5.4.5}$$

where $P = -i \begin{pmatrix} 0 & q \\ r & 0 \end{pmatrix}$ and $G(z, \bar{z}, \lambda)$ is the Green function of the operator $L_0$. The explicit form of the operator $\hat{G}$ with the kernel $G(z, \bar{z}, \lambda)$ obviously follows from (5.4.3):

$$\hat{G} = E_\lambda^{-1} D^{-1} E_\lambda. \tag{5.4.6}$$

The operator $D^{-1}$, i. e. $\begin{pmatrix} \partial_{\bar{z}}^{-1} & 0 \\ 0 & \partial_z^{-1} \end{pmatrix}$, acts as

$$(D^{-1} f)(z, \bar{z}) = \frac{1}{2\pi i} \iint_C dz' \wedge d\bar{z}' \begin{pmatrix} (z'-z)^{-1} & 0 \\ 0 & (\bar{z}'-\bar{z})^{-1} \end{pmatrix} f(z', \bar{z}'). \tag{5.4.7}$$

Taking into account (5.4.4) and (5.4.7), we obtain from (5.4.6) the explicit formula which determines the action of the operator $\hat{G}$ on any $2 \times 2$ matrix $B$ [124]:

$$(\hat{G}B)(z, \bar{z})$$
$$= \frac{1}{2\pi i} \iint_C dz' \wedge d\bar{z}' \begin{pmatrix} \frac{B_{11}(z',\bar{z}')}{z'-z} & \frac{\exp(-2\lambda(\bar{z}-\bar{z}')+2\bar{\lambda}(z-z'))B_{12}(z',\bar{z}')}{z'-z} \\ \frac{\exp(2\lambda(z-z')-2\bar{\lambda}(\bar{z}-\bar{z}'))B_{21}(z',\bar{z}')}{\bar{z}'-\bar{z}} & \frac{B_{22}(z',\bar{z}')}{\bar{z}'-\bar{z}} \end{pmatrix}.$$
$$\tag{5.4.8}$$

The formula (5.4.8) manifestly demonstrates that the Green function $G$ is founded for all $\lambda$ and has no jump anywhere. But, the most important fact is that the Green function $G$ is evidently analytic nowhere ($\partial G/\partial\bar{\lambda} \neq 0$ for $\forall \lambda \in C$). In virtue of the nonanalyticity of $G(z,\bar{z};\lambda,\bar{\lambda})$, the solutions $\mu$ of equation (5.4.5) are analytic nowhere on the entire complex plane too. In addition, the solutions $\mu$ of equation (5.4.5) are bounded with the exception of the points in which the Fredholm determinant $\Delta(\lambda)$ vanishes. We will assume that $\Delta(\lambda)$ has a finite number of simple zeros $\lambda_1, \ldots, \lambda_n$, i. e. the homogeneous equation (5.4.5) has nontrivial solutions in the finite number of points $\lambda_1, \ldots, \lambda_n$. In these points the solution $\mu$ has the poles. So, the solutions of equation (5.4.5) are of the form

$$\mu(z,\bar{z};\lambda,\bar{\lambda}) = \sum_{i=1}^{n} \frac{\mu_i(z,\bar{z})}{\lambda - \lambda_i} + \tilde{\mu}(z,\bar{z};\lambda,\bar{\lambda}) \qquad (5.4.9)$$

where $\tilde{\mu}$ is the function founded for all $\lambda$.

Firstly let us consider the solutions $\mu(\lambda)$ of equation (5.4.5) for the values of $\lambda$ which are different from $\lambda_1, \ldots, \lambda_n$. According to the general rule we must calculate $\partial\mu/\partial\bar{\lambda}$. Differentiation of (5.4.5) over $\bar{\lambda}$ gives

$$\frac{\partial\mu(z,\bar{z};\lambda,\bar{\lambda})}{\partial\bar{\lambda}} = -\left(\frac{\partial G}{\partial\bar{\lambda}}(\cdot;\lambda,\bar{\lambda})P(\cdot)\mu(\cdot;\lambda,\bar{\lambda})\right)(z,\bar{z})$$

$$-\left(G(\cdot;\lambda,\bar{\lambda})P(\cdot)\frac{\partial\mu}{\partial\bar{\lambda}}(\cdot;\lambda,\bar{\lambda})\right)(z,\bar{z}). \qquad (5.4.10)$$

In virtue of (5.4.8) one has

$$\frac{\partial\hat{G}}{\partial\bar{\lambda}} = E_\lambda^{-1}\Gamma E_\lambda \qquad (5.4.11)$$

where the operator $\Gamma$ acts by the rule

$$\Gamma f \doteq \begin{pmatrix} 0 & -\hat{f}_{12} \\ \hat{f}_{21} & 0 \end{pmatrix} \qquad (5.4.12)$$

and

$$\hat{f} = \frac{1}{2\pi}\iint_C dz \wedge d\bar{z}\, f(z,\bar{z}). \qquad (5.4.13)$$

Note that the form of the r.h.s. of (5.4.11) is in agreement with the obvious equation $L_0(\partial G/\partial\bar{\lambda}) = 0$.

Then, due to (5.4.11)–(5.4.13), equation (5.4.10) is equivalent to the following

$$\frac{\partial\mu(z,\bar{z};\lambda,\bar{\lambda})}{\partial\bar{\lambda}} = \begin{pmatrix} 0 & F_1(\lambda)e^{-2\lambda\bar{z}+2\bar{\lambda}z} \\ F_2(\lambda)e^{2\lambda z - 2\bar{\lambda}\bar{z}} & 0 \end{pmatrix}$$

$$-\left(G(\cdot;\lambda,\bar{\lambda})P(\cdot)\frac{\partial\mu(\cdot;\lambda,\bar{\lambda})}{\partial\bar{\lambda}}\right)(z,\bar{z}) \qquad (5.4.14)$$

where
$$F(\lambda) \doteq \begin{pmatrix} 0 & F_1 \\ F_2 & 0 \end{pmatrix} = -\Gamma E_\lambda P(\cdot)\mu(\cdot;\lambda,\bar\lambda),$$
i. e.
$$F_1(\lambda,\bar\lambda) = -\frac{i}{2\pi}\iint_C dz \wedge d\bar z\, e^{2\bar z\lambda - 2z\bar\lambda} q(z,\bar z,t)\mu_{22}(z,\bar z;\lambda,\bar\lambda), \qquad (5.4.15)$$

$$F_2(\lambda,\bar\lambda) = \frac{i}{2\pi}\iint_C dz \wedge d\bar z\, e^{-2z\lambda + 2\bar z\bar\lambda} r(z,\bar z,t)\mu_{11}(z,\bar z;\lambda,\bar\lambda). \qquad (5.4.16)$$

In the case $r = -\bar q$ one has $F_2(\lambda,\bar\lambda) = -\overline{F_1(\bar\lambda,\lambda)}$.

It follows from (5.4.14) that it is convenient to consider the function $N(z,\bar z;\lambda,\bar\lambda)$ which is the solution of the integral equation

$$N(z,\bar z;\lambda,\bar\lambda) = \begin{pmatrix} 0 & e^{-2\bar z\lambda + 2z\bar\lambda} \\ e^{2z\lambda - 2\bar z\bar\lambda} & 0 \end{pmatrix} - (\hat G P(\cdot) N(\cdot;\lambda,\bar\lambda))(z,\bar z). \qquad (5.4.17)$$

Multiplying (5.4.17) by the matrix $\begin{pmatrix} F_2 & 0 \\ 0 & F_1 \end{pmatrix}$ from the right, subtracting the obtained equation from (5.4.14) and taking into account that for the values of $\lambda$ under consideration the homogeneous equation (5.4.5) has no nontrivial solutions, we obtain

$$\frac{\partial \mu(z,\bar z;\lambda,\bar\lambda)}{\partial \bar\lambda} = N(z,\bar z;\lambda,\bar\lambda)\begin{pmatrix} F_2(\lambda,\bar\lambda) & 0 \\ 0 & F_1(\lambda,\bar\lambda) \end{pmatrix}. \qquad (5.4.18)$$

Now one must find out the relation between the functions $N$ and $\mu$. To do this note that the Green function $G(z,\bar z;\lambda,\bar\lambda)$ has the following symmetry property [124]

$$[G(\cdot;\lambda,\bar\lambda)(f(\cdot)\Sigma_\lambda(\cdot))](z,\bar z) = [G(\cdot;-\lambda,-\bar\lambda)f(\cdot)](z,\bar z)\Sigma_\lambda(z,\bar z) \qquad (5.4.19)$$

where
$$\Sigma_\lambda(z,\bar z) = \begin{pmatrix} 0 & e^{-2\bar z\lambda + 2z\bar\lambda} \\ e^{2z\lambda - 2\bar z\bar\lambda} & 0 \end{pmatrix}.$$

Comparing (5.4.5) with (5.4.19) and using (5.4.19), one gets

$$N(z,\bar z;\lambda,\bar\lambda) = \mu(z,\bar z;-\bar\lambda,\lambda)\Sigma_\lambda(z,\bar z). \qquad (5.4.20)$$

Substituting this expression for $N$ into equation (5.4.18), we obtain the linear $\bar\partial$-problem [124]

$$\frac{\partial \mu(z,\bar z;\lambda,\bar\lambda)}{\partial \bar\lambda} = \mu(z,\bar z;-\bar\lambda,-\lambda) F(z,\bar z;\lambda,\bar\lambda) \qquad (5.4.21)$$

where
$$F(z,\bar z;\lambda,\bar\lambda) \doteq \Sigma_\lambda(z,\bar z)\begin{pmatrix} F_2 & 0 \\ 0 & F_1 \end{pmatrix} = \begin{pmatrix} 0 & F_1(\lambda,\bar\lambda)e^{-2\bar z\lambda + 2z\bar\lambda} \\ F_2(\lambda,\bar\lambda)e^{2z\lambda - 2\bar z\bar\lambda} & 0 \end{pmatrix} \qquad (5.4.22)$$

and the functions $F_1(\lambda,\bar\lambda)$, $F_2(\lambda,\bar\lambda)$ are given by the formulae (5.4.15), (5.4.16).

The linear $\bar{\partial}$-problem (5.4.21) is the main object of our construction. However, equation (5.4.21) is not sufficient for the complete solution of the inverse problem since (5.4.21) takes place only for the values of $\lambda$ which do not coincide with the zeros of the Fredholm determinant. Now one should take into account the existence of the nontrivial solutions for the homogeneous equation (5.4.5).

These solutions have rather special structure. Indeed, let the homogeneous equation (5.4.5) has the nontrivial solution $\mu \doteq \chi(z, \bar{z}; \lambda_i, \bar{\lambda}_i)$ at the point $\lambda_i$, i. e.

$$\mu_i(z, \bar{z}) = -[G(\cdot; \lambda_i, \bar{\lambda}_i) P(\cdot) \mu_i(\cdot)](z, \bar{z}). \tag{5.4.23}$$

Using (5.4.19), one can show that together with (5.4.23) one has

$$\mu_i(z, \bar{z}) \Sigma_{-\bar{\lambda}_i}(z, \bar{z}) = -[G(\cdot; -\bar{\lambda}_i, -\lambda_i) P(\cdot) \mu_i(\cdot) \Sigma_{-\bar{\lambda}_i}(\cdot)](z, \bar{z}). \tag{5.4.24}$$

Hence, the function $a\mu_i(z, \bar{z}) \Sigma_{\bar{\lambda}_i}(z, \bar{z})$ where $a$ is arbitrary constant is the solution of the homogeneous equation (5.4.5) too, but at the point $-\bar{\lambda}_i$. For example, if equation (5.4.23) has solution of the form $\mu_i = \begin{pmatrix} \mu_{(i)11} & 0 \\ \mu_{(i)21} & 0 \end{pmatrix}$ at the point $\lambda_i$, then it also has the solution of the form $\mu_i = \begin{pmatrix} 0 & e^{\bar{z}\bar{\lambda}_i - z\lambda_i} \mu_{(i)11} \\ 0 & e^{-2\bar{z}\bar{\lambda}_i + z\lambda_i} \mu_{(i)21} \end{pmatrix}$ at the point $-\bar{\lambda}_i$.

Further one should take into account the involution (6.4.2) which implies

$$\mu_i^s = \begin{pmatrix} \mu_{i11}^{(s)} & -\overline{\mu_{i21}^{(s)}} \\ \mu_{i21}^{(s)} & \overline{\mu_{i11}^{(s)}} \end{pmatrix}. \tag{5.4.25}$$

It is not difficult to show that the nontrivial $\mu^{(s)}$ which obeys the involution (5.4.25) correspond to the choice $a = 0$. So we have

$$\mu^{(s)} = \sum_{i=1}^{N} \frac{C_i}{\lambda - \lambda_i} \begin{pmatrix} \mu_{11i} & 0 \\ \mu_{21i} & 0 \end{pmatrix} + \sum_{i=1}^{N} \frac{\bar{C}_i}{\lambda + \bar{\lambda}_i} \begin{pmatrix} 0 & -\overline{\mu_{21i}} \\ 0 & \overline{\mu_{11i}} \end{pmatrix}. \tag{5.4.26}$$

For the first time such a structure of the pole terms has been proposed in [243].

So the full $\bar{\partial}$-problem for the problem (5.4.1) is of the form

$$\frac{\partial \mu}{\partial \bar{\lambda}} = \mu \begin{pmatrix} 0 & \overline{F(\lambda, \bar{\lambda})} e^{-2\lambda z + 2\bar{\lambda}\bar{z}} \\ F(\lambda, \bar{\lambda}) e^{2\lambda z - 2\bar{\lambda}\bar{z}} & 0 \end{pmatrix}$$

$$+ \sum_{i=1}^{N} C_i \pi \delta(\lambda - \lambda_i) \begin{pmatrix} \mu_{11i} & 0 \\ \mu_{21i} & 0 \end{pmatrix} + \sum_{i=1}^{N} \bar{C}_i \pi \delta(\lambda + \bar{\lambda}_i) \begin{pmatrix} 0 & -\overline{\mu_{21i}} \\ 0 & \overline{\mu_{11i}} \end{pmatrix}. \tag{5.4.27}$$

*Chapter 5*

In virtue of (5.4.25) one can be restricted to consider only of the equation for the first column $\mu^{(1)} = \begin{pmatrix} \mu_{11} \\ \mu_{21} \end{pmatrix}$. It is

$$\frac{\partial \mu^{(1)}(\lambda, \bar{\lambda})}{\partial \bar{\lambda}} = \pi \sum_{i=1}^{N} \delta(\lambda - \lambda_i) \mu_i^{(1)} - iF(\lambda, \bar{\lambda})e^{2\lambda z - 2\bar{\lambda}\bar{z}} \sigma_2 \overline{\mu^{(1)}(\lambda, \bar{\lambda})}, \quad (5.4.28)$$

where we normalize $\mu_i^{(1)}$ as $\mu_i^{(1)} \xrightarrow[|z|\to\infty]{} \frac{i}{z}\begin{pmatrix} 1 \\ 0 \end{pmatrix}$, i. e. $C_i = i$. The generalized Cauchy formula gives

$$\mu^{(1)}(\lambda, \bar{\lambda}) = \begin{pmatrix} 1 \\ 0 \end{pmatrix} + \sum_{i=1}^{N} \frac{i\mu_i^{(1)}(z, \bar{z})}{\lambda - \lambda_i}$$

$$+ \frac{1}{2\pi} \iint_C \frac{d\lambda' \wedge d\bar{\lambda}'}{\lambda' - \lambda} F(\lambda', \bar{\lambda}')e^{2\lambda'z - 2\bar{\lambda}'\bar{z}} \sigma_2 \overline{\mu^{(1)}(\lambda', \bar{\lambda}')}. \quad (5.4.29)$$

Equation (5.4.29) is the part of the inverse problem equations. To derive the rest of such equations one should proceed in equation (5.4.29) to the limit $\lambda \to \lambda_k$ to obtain the system of equations for $\mu_i^{(1)}$. To do this we must calculate the limit

$$\lim_{\lambda \to \lambda_i} \left( \mu^{(1)} - \frac{i\mu_i^{(1)}}{\lambda - \lambda_i} \right). \quad (5.4.30)$$

Completely similar to the KP equation one can show that the quantity

$$\lim_{\lambda \to \lambda_i} \left( \mu^{(1)} - \frac{i\mu_i^{(1)}}{\lambda - \lambda_i} \right) - z\mu_i^{(1)} \quad (5.4.31)$$

is also the solution of the homogeneous equation (5.4.5) provided the certain constraints on $\theta$ and $\mu_i^{(1)}$. Therefore the expression (5.4.31) is the linear superposition of the two independent solutions of the homogeneous equation (5.4.5), i. e.

$$\lim_{\lambda \to \lambda_i} \left( \mu^{(1)} - \frac{i\mu_i^{(1)}}{\lambda - \lambda_i} \right) - z\mu_i^{(1)} = \gamma_i \mu_i^{(1)} - i\tilde{\gamma}_i i\sigma_2 \overline{\mu_i^{(1)}}. \quad (5.4.32)$$

where $\gamma_i$ and $\tilde{\gamma}_i$ are some constants. Note that the fact that the l.h.s. of the relation (5.4.32) is the superposition of the two independent solutions is very important. This has been realized for the first time in [243].

Now, proceeding to the limit $\lambda \to \lambda_k$ in equation (5.4.29) and taking into account the identity (5.4.31), we obtain

$$(z + \gamma_k)\mu_k^{(1)} + \tilde{\gamma}_k \sigma_2 \bar{\mu}_k^{(1)} e^{2\lambda_k z - 2\bar{\lambda}_k \bar{z}}$$

$$= \begin{pmatrix} 1 \\ 0 \end{pmatrix} + \sum_{j \neq k}^{N} \frac{i\mu_j^{(1)}}{\lambda_j - \lambda_k} + \frac{1}{2\pi} \iint_C \frac{d\mu \wedge d\bar{\mu}}{\mu - \lambda_k} F(\mu, \bar{\mu}) e^{2\mu z - 2\bar{\mu}\bar{z}} \sigma_2 \overline{\mu^{(1)}(\mu, \bar{\mu})}, \quad (5.4.33)$$

$$k = 1, \ldots, N.$$

At last, the reconstruction formula for the potential $q$ is

$$q = 2i \sum_{k=1}^{N} \bar{\mu}_{21k} - \frac{1}{\pi i} \iint_C d\mu \wedge d\bar{\mu}\, F(\mu, \bar{\mu}) e^{2\mu z - 2\bar{\mu} \bar{z}} \bar{\mu}_{11}(\mu, \bar{\mu}). \tag{5.4.34}$$

The formulae (5.4.29), (5.4.33) and (5.4.34) form the complete set of the inverse problem equations for the linear problem (5.4.1). The quantities $\{F(\lambda, \bar{\lambda}),\ \lambda_i,\ \gamma_i,\ \mu_i,\ (i = 1, \ldots, N)\}$ are the inverse problem data $\mathcal{F}$.

Thus we have the solution of the inverse problem for the spectral problem (6.4.1). In order to use this solution for the integration of the DS-II equation one should find the time evolution of the inverse problem data $\mathcal{F}(\lambda, \bar{\lambda}, t)$.

One can do this by the same method as the KP equation, namely, noting that $\partial \mu / \partial \bar{\lambda}$ obeys the same equations as $\mu$ and then considering these equations at the limit $|z| \to \infty$. There is also another way which is based on the use of the formulae (5.4.34) and directly DS equation. For the small $q$ when one can neglect by the nonlinear terms in the system (5.4.1) and when $\chi_{11} \sim 1, \chi_{22} \sim 1$ one has

$$q(x, y, t) = \frac{i}{\pi} \iint d\lambda_R d\lambda_I\, F(\lambda_R, \lambda_I) e^{-2\lambda_I x + 2\lambda_R y}. \tag{5.4.35}$$

If one substitutes this expression for $q$ into the linearized DS equation, one gets

$$i \frac{\partial F_1}{\partial t} + 2(\lambda_I^2 - \lambda_R^2) F_1 = 0. \tag{5.4.36}$$

Hence

$$F(\lambda, \bar{\lambda}, t) = e^{(\lambda^2 - \bar{\lambda}^2) t} F(\lambda, \bar{\lambda}, 0). \tag{5.4.37}$$

Using the second linear problem, one can also show that

$$\frac{\partial \lambda_k}{\partial t} = 0, \quad \frac{\partial \gamma_k}{\partial t} = 4\lambda_k, \quad \frac{\partial \tilde{\gamma}_k}{\partial t} = (\bar{\lambda}_k^2 - \lambda_k^2) \tilde{\gamma}_k.$$

Hence

$$\lambda_k(t) = \lambda_k(0),$$

$$\gamma_k(t) = 4\lambda_k t + \gamma_k(0), \tag{5.4.38}$$

$$\tilde{\gamma}_k(t) = \tilde{\gamma}_k(0) e^{(\bar{\lambda}_k^2 - \lambda_k^2) t}.$$

The formulae (5.4.37) and (5.4.38) allow us to solve the initial value problem for the DS-II equation according to the standard IST scheme:

$$\{q(x, y, 0)\} \xrightarrow{I} \{\mathcal{F}(\lambda, \bar{\lambda}, 0)\} \xrightarrow{II} \{\mathcal{F}(\lambda, \bar{\lambda}, t)\} \xrightarrow{III} \{q(x, y, t)\}.$$

Similar to the case of the KP and mKP equations the formulae (5.4.29), (5.4.33), (5.4.34), (5.4.37) and (5.4.38) can be used for the construction of the infinite families of the exact solutions of the DS-II equation.

In the case of the pure discrete data $F(\lambda, \bar{\lambda}) \equiv 0$ the solutions of the DS-II equation are given by the formula

$$q = 2i \sum_{k=1}^{N} \overline{\mu_{k21}} \tag{5.4.39}$$

where the quantities $\overline{\mu_{k21}}$ are determined from the algebraic system

$$(z + 4\lambda_k t + \gamma_{k0})\mu_k^{(1)} - \sum_{j \neq k}^{N} \frac{i\mu_j^{(1)}}{\lambda_j - \lambda_k}$$

$$+ \tilde{\gamma}_{k0} \exp(-(\lambda_k^2 - \bar{\lambda}_k^2)t + 2\lambda_k z - 2\bar{\lambda}_k \bar{z}) i\sigma_2 \overline{\mu_k^{(1)}} = \begin{pmatrix} 1 \\ 0 \end{pmatrix} \tag{5.4.40}$$

$$(k = 1, \ldots, N).$$

Solving the system (5.4.40) together with the complex conjugate system, one can find $\overline{\mu_{21k}}$. This gives us the rational-exponential solutions of the DS-II equation. The simplest of them is of the form ($N = 1$) [243]

$$q(x, y, t) = \frac{2i\tilde{\gamma}_{10} e^{-i[2(\lambda_1 z - \bar{\lambda}_1 \bar{z}) + (\lambda_1^2 - \bar{\lambda}_1^2)t]}}{|z + 4\lambda_1 t + \gamma_{01}|^2 + |\tilde{\gamma}_{01}|^2} \tag{5.4.41}$$

where $z = \frac{1}{2}(y - ix)$. This solution is apparently nonsingular. For such solution $|q|^2$ is a rational function. At $\lambda_1 = 0$ the solution (5.4.41) is itself rational. So it is close to the lumps of the KP and mKP equations.

The general solution (5.4.39), (5.4.40) describes the elastic scattering of $N$ simple objects (5.4.41).

Note that such solutions, in particular, the solution (5.4.41) are obviously absent ($q \equiv 0$) for $\tilde{\gamma}_{0k} = 0$. This fact emphasizes the importance of the second term in the r.h.s. of (5.4.32).

In the case $\tilde{\gamma}_{0k} = 0$ for all $k$ one can also construct the rational-exponential solutions of the DS-II equation. They correspond to the choice of solutions defined by (5.4.24) with $a \neq 0$ [234]. But these solutions are singular.

# Chapter 6

# The Ishimori equation

The Heisenberg ferromagnet model equation (1.1.19) is the important 1+1-dimensional integrable equation. Its 2+1-dimensional integrable generalization

$$\vec{S}_t + \frac{1}{2}\vec{S} \times (\sigma^2 \vec{S}_{yy} + \vec{S}_{xx}) + \Phi_y \vec{S}_x + \Phi_x \vec{S}_y = 0,$$

$$\sigma^2 \Phi_{yy} - \Phi_{xx} = \sigma^2 \vec{S} \cdot (\vec{S}_x \times \vec{S}_y) \tag{6.0.1}$$

where $\vec{S}\vec{S} = S_1^2 + S_2^2 + S_3^2 = 1$ and $\sigma^2 = \pm 1$ has been derived by Ishimori in [92] for the description of the time evolution of the system of static spin vortices on the plane $(x, y)$. The corresponding auxiliary linear problem is [92]

$$\sigma \Psi_y + P(x, y, t)\Psi_x = 0,$$

$$\Psi_t - iP\Psi_{xx} - \frac{1}{2}\left(iP_x + i\sigma P_y P + \frac{1}{\sigma}P\Phi_x - \Phi_y\right)\Psi_x = 0 \tag{6.0.2}$$

where $P \doteq \sum_{i=1}^{3} \sigma_i \vec{S}_i(x, y, t)$ and $\sigma_1, \sigma_2, \sigma_3$ are Pauli matrices.

The Ishimori equation (6.0.1) may describe the classical spin system on the plane. It is of great interest also because it possesses the topological invariant [92]

$$Q = \frac{1}{4\pi} \iint dx dy\, \vec{S} \cdot (\vec{S}_x \times \vec{S}_y) \tag{6.0.3}$$

and its solutions are classified by the integer value $N = 0, \pm 1, \pm 2, \ldots$ of $Q$.

We will refer equation (6.0.1) at $\sigma = 1$ as the Ishimori-I equation and at $\sigma^2 = -1$ as the Ishimori-II equation. Note the change of notations in comparison with [76].

The Ishimori equation (6.0.1) has another interesting form. In terms of the stereographic variable $q = \frac{S_1 + iS_2}{1 + S_3}$ it looks like [244]

$$iq_t + \frac{1}{2}(\sigma^2 q_{yy} + q_{xx}) + \frac{\bar{q}(q_x^2 - \sigma^2 q_y^2)}{1 + |q|^2} - i\sigma(\Phi_x q_y + \Phi_y q_x) = 0,$$

$$\sigma^2 \Phi_{yy} - \Phi_{xx} = 2i\sigma^2 \frac{q_x \bar{q}_y - \bar{q}_x q_y}{(1 + |q|^2)^2}. \tag{6.0.4}$$

So it belongs, in fact, to the same family of nonlinear equations as the DS equation.

## 6.1. $\bar{\partial}$-dressing

We will start with the $2 \times 2$ matrix $\bar{\partial}$-problem (5.1.1) which is assumed to be uniquely solvable and canonically normalized

$$\chi \underset{\lambda \to \infty}{\longrightarrow} 1 + \frac{1}{\lambda}\chi_{-1} + \frac{1}{\lambda^2}\chi_{-2} + \cdots . \qquad (6.1.1)$$

First we consider the case of trivial background. We choose

$$\hat{B}_1 = i\frac{1}{\lambda}, \quad \hat{B}_2 = -\frac{i}{\sigma\lambda}\sigma_3, \quad \hat{B}_3 = -\frac{i}{\lambda^2}\sigma_3. \qquad (6.1.2)$$

The operators $D_i$ are given by

$$D_1 f = \partial_x f + \frac{i}{\lambda}f, \quad D_2 f = \partial_y f - \frac{i}{\sigma\lambda}f\sigma_3, \quad D_3 f = \partial_t f - \frac{i}{\lambda^2}f\sigma_3. \qquad (6.1.3)$$

Correspondingly the kernel $R$ is

$$R(\mu, \bar{\mu}; \lambda, \bar{\lambda}; x, y, t) = e^{F(\mu)} R_0(\mu, \bar{\mu}; \lambda, \bar{\lambda}) e^{-F(\lambda)} \qquad (6.1.4)$$

where

$$F(\lambda) = \frac{i}{\lambda}x - \frac{i\sigma_3}{\sigma\lambda}y - \frac{i\sigma_3}{\lambda^2}t. \qquad (6.1.5)$$

The case of the Ishimori equation is similar in some respects to the mKP case. The quantity $\sigma D_2 \chi$ has the first order pole at $\lambda = 0$. The first order singularity at the origin is absent in the combination

$$\sigma D_2 \chi + P D_1 \chi = \sigma\chi_y - \frac{i}{\lambda}\chi\sigma_3 + P\chi_x + \frac{i}{\lambda}P\chi \qquad (6.1.6)$$

with

$$P(x, y, t) = \chi_0 \sigma_3 \chi_0^{-1}. \qquad (6.1.7)$$

The function $\chi_0(x, y, t)$ is defined by the expansion

$$\chi(x, y, t; \lambda, \bar{\lambda}) = \chi_0 + \lambda\chi_1 + \lambda^2\chi_2 + \ldots \qquad (6.1.8)$$

near the point $\lambda = 0$.

Then one readily concludes that

$$\sigma D_2 \chi + P D_1 \chi \to 0 \qquad (6.1.9)$$

as $\lambda \to \infty$. Hence, the quantity (6.1.6) obeys both conditions (2.3.15) and (2.3.16). Thus, one has the first auxiliary linear problem

$$\sigma D_2 \chi + P D_1 \chi = 0 \qquad (6.1.10)$$

where $P$ is given by (6.1.7). Since $P^2 = 1$ and $\text{tr}P = 0$ then the $2 \times 2$ matrix $P$ can be parametrized by the three-component unit vector $\vec{S}$:

$$P(x,y,t) = \sum_{i=1}^{3} \sigma_i S_i(x,y,t) \qquad (6.1.11)$$

where $\vec{S}\vec{S} = S_1^2 + S_2^2 + S_3^2 = 1$. Representation (6.1.7) implies that the trivial $\vec{S}$ is the vector $(0,0,1)$.

Equation (6.1.10), in addition to (6.1.7), gives

$$i[\sigma_3, \chi_0^{-1}\chi_1] = -\chi_0^{-1}(\sigma\chi_{0y} + P\chi_{0x}) = -\sigma\chi_0^{-1}\chi_{0y} - \sigma_3\chi_0^{-1}\chi_{0x}, \qquad (6.1.12)$$

$$i[\sigma_3, \chi_0^{-1}\chi_2] = -\chi_0^{-1}(\sigma\chi_{1y} + P\chi_{1x}). \qquad (6.1.13)$$

To construct the second linear problem we should use, as usual, $D_3\chi$. It has the second order pole at $\lambda = 0$. To cancel this singularity one can add to $D_3\chi$ the term $VD_1^2\chi$ similar to the previous cases. Just for a change let us consider another proper combination

$$D_3\chi + VD_1D_2\chi. \qquad (6.1.14)$$

It is easy to check that this quantity has no second order pole if $V = 2i\sigma$. In order to compensate the first order singularity in $D_3\chi + 2i\sigma D_1D_2\chi$ let us add to it both the terms with $D_1\chi$ and $D_2\chi$. The requirement that the residue of

$$D_3\chi + i\sigma D_1D_2\chi + V_1 D_1\chi + V_2 D_2\chi \qquad (6.1.15)$$

at $\lambda = 0$ vanishes is satisfied if

$$V_1 = -i\sigma\chi_{0y}\chi_0^{-1},$$

$$V_2 = -i\sigma\chi_{0x}\chi_0^{-1}. \qquad (6.1.16)$$

Further it is easy to check that

$$D_3\chi + i\sigma D_1D_2\chi - i\sigma\chi_{0y}\chi_0^{-1}D_1\chi - i\sigma\chi_{0x}\chi_0^{-1}D_2\chi \to 0 \qquad (6.1.17)$$

as $\lambda \to \infty$. Hence

$$D_3\chi + i\sigma D_1D_2\chi - i\sigma\chi_{0y}\chi_0^{-1}D_1\chi - i\sigma\chi_{0x}\chi_0^{-1}D_2\chi = 0. \qquad (6.1.18)$$

That is the desired second linear problem. The matrices $V_1$ and $V_2$ can be expressed via $P$. Indeed, equation (6.1.12) gives

$$(\sigma\chi_0^{-1}\chi_{0y} + \sigma_3\chi_0^{-1}\chi_{0x})_{\text{diag}} = 0. \qquad (6.1.19)$$

Then (6.1.16) implies

$$\mathrm{tr}\, V_1 = i\sigma\, \mathrm{tr}(\chi_{0y}\chi_0^{-1}) = i\sigma\partial_y \ln\det\chi_0,$$

$$\mathrm{tr}\, V_2 = i\sigma\, \mathrm{tr}(\chi_{0x}\chi_0^{-1}) = i\sigma\partial_y \ln\det\chi_0. \qquad (6.1.20)$$

Using (6.1.7), (6.1.19), (6.1.20), one gets

$$V_1 = \frac{1}{2}(\Phi_y - iP_x), \quad V_2 = \frac{1}{2}(\Phi_x + i\sigma^2 P_y) \qquad (6.1.21)$$

where

$$\Phi = -i\sigma\ln\det\chi_0. \qquad (6.1.22)$$

So one has

$$D_3\chi + i\sigma D_1 D_2\chi + \frac{1}{2}(\Phi_y - iP_x)D_1\chi + \frac{1}{2}(\Phi_x + i\sigma^2 P_y)D_2\chi = 0. \qquad (6.1.23)$$

So the linear problems associated with the case (6.1.2) are given by (6.1.10) and (6.1.23). Instead of (6.1.23) one can construct another second linear problem

$$D_3\chi - iPD_1^2\chi - \frac{1}{2}\left(iP_x + i\sigma P_y P + \frac{1}{\sigma}P\Phi_x - \Phi_y\right)D_1\chi = 0 \qquad (6.1.24)$$

which is equivalent to (6.1.23) modulo (6.1.10).

Transiting to the function $\Psi$ defined by

$$\Psi = \chi e^{\frac{i}{\lambda}x - \frac{i}{\lambda}\sigma_3 y - \frac{i}{\lambda^2}\sigma_3 t}, \qquad (6.1.25)$$

one converts the problems (6.1.10) and (6.1.23) into the following

$$\sigma\Psi_y + P\Psi_x = 0,$$

$$\Psi_t + i\sigma\Psi_{xy} + \frac{1}{2}(\Phi_y - iP_x)\Psi_x + \frac{1}{2}(\Phi_x + i\sigma^2 P_y)\Psi_y = 0. \qquad (6.1.26)$$

The pair of equations (6.1.10) and (6.1.24) is converted into the original linear problem (6.0.2).

The system of equations (6.1.10), (6.1.23) is automatically compatible and provides us the corresponding nonlinear integrable equation. Evaluating equation (6.1.18) at $\lambda = 0$, one obtains

$$i\chi_{0t} + \sigma\chi_{0xy} + \sigma\chi_0((\chi_0^{-1})_x\chi_{0y} + (\chi_0^{-1})_y\chi_{0x}) - i\chi_0(\sigma(\chi_0^{-1}\chi_1)_y - (\chi_0^{-1}\chi_1)_x\sigma_3) = 0. \qquad (6.1.27)$$

Then equation (6.1.12) implies

$$2i(\chi_0^{-1}\chi_1)_{\mathrm{off}} = -\sigma\chi_0^{-1}\chi_{0y} - \sigma_3\chi_0^{-1}\chi_{0x}. \qquad (6.1.28)$$

Further, equation (6.1.13) after some transformations gives

$$2i(\sigma\partial_y + \sigma_3\partial_x)(\chi_0^{-1}\chi_1)_{\text{diag}} = \sigma_3(\sigma\chi_0^{-1}\chi_{0y} + \sigma_3\chi_0^{-1}\chi_{0x})^2. \tag{6.1.29}$$

Denoting

$$C = 2i(\chi_0^{-1}\chi_1)_{\text{diag}}, \tag{6.1.30}$$

one finds that equation (6.1.27) is equivalent to the system

$$2i\chi_{0t} + \sigma^2\chi_{0yy} - \sigma_3\chi_{0xx}\sigma_3 + 2\sigma\chi_{0xy} + \sigma[\sigma_3,\chi_{0xy}] - \sigma\chi_{0x}\chi_0^{-1}\chi_{0y} - 2\sigma\chi_{0y}\chi_0^{-1}\chi_{0x}$$
$$-\sigma^2\chi_{0y}\chi_0^{-1}\chi_{0y} - \sigma\sigma_3\chi_{0y}\chi_0^{-1}\chi_{0x} + \sigma_3\chi_{0x}\chi_0^{-1}\chi_{0x}\sigma_3 - \chi_0(\sigma C_y - C_x\sigma_3) = 0, \tag{6.1.31}$$

$$\sigma C_y + \sigma_3 C_x = -\sigma_3(\sigma\chi_0^{-1}\chi_{0y} + \sigma_3\chi_0^{-1}\chi_{0x})^2.$$

Finally, in terms of $P = \chi_0\sigma_3\chi_0^{-1} = \sum_i \sigma_i S_i$ and $\Phi = -i\sigma\ln\det\chi_0$, after cumbersome transformations, one gets the Ishimori equation (6.0.1). The Ishimori equation arises also as the compatibility condition for the system (6.1.26).

Note that within the described approach equation (6.1.31) for $\chi_0$ is the primary one. One can also express the topological charge $Q$ (6.0.3) via $\chi_0$ [245]. Note first that

$$\frac{1}{4\pi}\vec{S}\cdot(\vec{S}_x\times\vec{S}_y) = \frac{i}{8\pi}\text{tr}(PP_xP_y). \tag{6.1.32}$$

Using (6.1.7), one gets

$$\text{tr}(PP_xP_y) = 2\text{tr}\{\sigma_3((\chi_0^{-1})_y\chi_{0x} - (\chi_0^{-1})_x\chi_{0y})\}. \tag{6.1.33}$$

Differentiating (6.1.12) with respect to $x$ and $y$, one obtains

$$i\sigma_3[\sigma_3,\chi_0^{-1}\chi_1]_x = -\sigma\sigma_3(\chi_0^{-1})_x\chi_{0y} - \sigma\sigma_3\chi_0^{-1}\chi_{0xy} - (\chi_0^{-1}\chi_{0x})_x, \tag{6.1.34}$$

$$i\sigma[\sigma_3,\chi_0^{-1}\chi_1]_y = -\sigma^2(\chi_0^{-1}\chi_{0y})_y - \sigma\sigma_3(\chi_0^{-1})_y\chi_{0x} - \sigma\sigma_3\chi_0^{-1}\chi_{0xy}. \tag{6.1.35}$$

The subtraction of (6.1.35) from (6.1.34) gives

$$\sigma\sigma_3\{(\chi_0^{-1})_y\chi_{0x} - (\chi_0^{-1})_x\chi_{0y}\}$$
$$= -i(\sigma\partial_y - \sigma_3\partial_x)[\sigma_3,\chi_0^{-1}\chi_1] - \sigma^2(\chi_0^{-1}\chi_{0y})_y + (\chi_0^{-1}\chi_{0x})_x. \tag{6.1.36}$$

Then we note that

$$\text{tr}\{(\sigma\partial_y - \sigma_3\partial_x)[\sigma_3,\chi_0^{-1}\chi_1]\} = 0. \tag{6.1.37}$$

Substituting now (6.1.36) into (6.1.33) and taking into account (6.1.37), one gets

$$\text{tr}(PP_xP_y) = \frac{2}{\sigma}\text{tr}((\chi_0^{-1}\chi_{0x})_x - \sigma^2(\chi_0^{-1}\chi_{0y})_y). \tag{6.1.38}$$

*Chapter 6* 169

Finally using the well known formula $\text{tr}(\chi_0^{-1}\chi_{0x}) = \partial_x \ln \det \chi_0$, we obtain

$$\text{tr}(PP_xP_y) = \frac{2}{\sigma}(\partial_x^2 - \sigma^2 \partial_y^2) \ln \det \chi_0. \tag{6.1.39}$$

So the topological charge $Q$ is [245]

$$Q = \frac{i}{\sigma 4\pi} \iint dxdy \, (\partial_x^2 - \sigma^2 \partial_y^2) \ln \det \chi_0. \tag{6.1.40}$$

Further we proceed to the case of nontrivial background. We choose $\hat{B}_i$ $(i = 1, 2, 3)$ as the integral operators such that

$$\hat{B}_1 \to \frac{i}{\lambda}, \quad \hat{B}_2 \to -\frac{i}{\sigma\lambda}\sigma_3, \quad \hat{B}_3 \to -\frac{i}{\lambda^2}\sigma_3 \tag{6.1.41}$$

as $\lambda \to 0$. In the dual space defined by the Fourier transformation

$$X(s,\bar{s};x,y,t) = \frac{1}{2\pi} \iint_C \frac{d\lambda \wedge d\bar{\lambda}}{|\lambda|^4} \exp\left(i\left(\frac{s}{\lambda} + \frac{\bar{s}}{\bar{\lambda}}\right)\right) f(\lambda,\bar{\lambda};x,y,t) \tag{6.1.42}$$

the operators $\hat{B}_i$ $(i = 1, 2, 3)$ are the first and second order differential operators.

Our choice of the system (2.5.8) is the following

$$(\partial_x - \partial_s)X_0 = 0,$$

$$(\sigma\partial_y + P_0\partial_s)X_0 = 0, \tag{6.1.43}$$

$$\left(\partial_t - iP\partial_s^2 + \frac{1}{2}\left(\Phi_{0y} - iP_{0s} - \frac{1}{\sigma}\Phi_{0s}P - i\sigma P_{0y}P_0\right)\partial_s\right)X_0 = 0$$

where $P_0^2 = 1$. So, $P_0(x+s,y,t)$ and $\Phi_0(x+s,y,t)$ are the solutions of the background Ishimori equation.

Thus, the operators $\hat{\tilde{B}}_i$ $(i = 1, 2, 3)$ are chosen to be

$$\hat{\tilde{B}}_1 = \partial_s,$$

$$\hat{\tilde{B}}_2 = -\frac{1}{\sigma}P_0\partial_s, \tag{6.1.44}$$

$$\hat{\tilde{B}}_3 = iP\partial_s^2 - \frac{1}{2}(\Phi_{0y} - iP_{0s} - \frac{1}{\sigma}\Phi_{0s}P_0 - i\sigma\Phi_{0y}P_0)\partial_s.$$

In the spectral space $\lambda$ these operators, in virtue of (4.1.30), act as follows

$$(\hat{B}_1 f)(\lambda) = \frac{i}{\lambda}f(\lambda),$$

$$(\hat{B}_2 f)(\lambda) = -\frac{i}{\sigma\lambda}\sigma_3 f(\lambda) - \frac{1}{\sigma}\iint_C \frac{d\mu \wedge d\bar{\mu}}{|\mu|^4} \hat{P}_0^a\left(\frac{\lambda\mu}{\mu - \lambda}\right)\frac{i}{\mu}f(\mu), \tag{6.1.45}$$

$$(\hat{B}_3 f)(\lambda) = -\frac{i}{\lambda^2}\sigma_3 f(\lambda) - i \iint_C \frac{d\mu \wedge d\bar{\mu}}{|\mu|^4} \hat{P}_0^a\left(\frac{\lambda\mu}{\mu-\lambda}\right) \frac{1}{\mu^2} f(\mu)$$

$$-\frac{i}{2}\iint_C \frac{d\mu \wedge d\bar{\mu}}{|\mu|^4} \left\{ \hat{\Phi}_{0y}\left(\frac{\lambda\mu}{\mu-\lambda}\right) + \left(\frac{1}{\lambda} - \frac{1}{\mu}\right)\hat{P}_0\left(\frac{\lambda\mu}{\mu-\lambda}\right) - \frac{1}{\sigma}(\widehat{\Phi_{0s}P})\left(\frac{\lambda\mu}{\mu-\lambda}\right) \right.$$

$$\left. - i\sigma(\widehat{P_{0y}P})\left(\frac{\lambda\mu}{\mu-\lambda}\right) \right\} \frac{1}{\mu} f(\mu)$$

where $P_0^a \doteq P_0 - \sigma_3$ and

$$\hat{P}_0(\lambda, x, y, t) = \frac{1}{2\pi}\iint_C ds \wedge d\bar{s}\, \exp\left(-i\left(\frac{s}{\lambda} + \frac{\bar{s}}{\bar{\lambda}}\right)\right) P_0(s+x,y,t),$$

$$\hat{\Phi}_0(\lambda, x, y, t) = \frac{1}{2\pi}\iint_C ds \wedge d\bar{s}\, \exp\left(-i\left(\frac{s}{\lambda} + \frac{\bar{s}}{\bar{\lambda}}\right)\right) \Phi_0(s+x,y,t). \tag{6.1.46}$$

Therefore, the operators $D_i = \partial_{x_i} + B_i^*$ in the case of nontrivial background $P_0, \Phi_0$ act according to

$$(\hat{D}_1 \chi)(\lambda) = \left(\partial_x + \frac{i}{\lambda}\right)\chi(\lambda),$$

$$(\hat{D}_2 \chi)(\lambda) = \partial_y \chi(\lambda) - \frac{i}{\sigma\lambda}\chi(\lambda)\sigma_3 - \frac{i}{\sigma\lambda}\iint_C \frac{d\mu \wedge d\bar{\mu}}{|\mu|^4} \chi\left(\frac{\lambda\mu}{\lambda+\mu}\right)\hat{P}_0^a(\mu), \tag{6.1.47}$$

$$(\hat{D}_3 \chi)(\lambda) = \partial_t \chi(\lambda) - \frac{i}{\lambda^2}\chi(\lambda)\sigma_3 - \frac{i}{\lambda^2}\iint_C \frac{d\mu \wedge d\bar{\mu}}{|\mu|^4} \chi\left(\frac{\lambda\mu}{\lambda+\mu}\right)\hat{P}_0^a(\mu)$$

$$-\frac{i}{2\lambda}\iint_C \frac{d\mu \wedge d\bar{\mu}}{|\mu|^4} \chi\left(\frac{\lambda\mu}{\lambda+\mu}\right)\left\{\hat{\Phi}_{0y}(\mu) + \frac{1}{\mu}\hat{P}_0(\mu) - \frac{1}{\sigma}(\widehat{\Phi_{0s}P})(\mu) - i\sigma(\widehat{P_{0y}P})(\mu)\right\}.$$

Emphasize that similar to the mKP equation the operators $D_i$ (6.1.47) are not the convolutive integral operators in contrast to the KP and DS cases.

Using these operators $D_i$ we will construct the corresponding linear problems. Similarly for the trivial background case ($P_0 = \sigma_3$, $\Phi_0 = 0$) we start with the combination

$$(\sigma D_2 + \chi D_1)\chi(\lambda) = \partial_y \chi(\lambda) + P(x,y,t)\partial_x \chi(\lambda)$$

$$+\frac{i}{\lambda}\left(P\chi(\lambda) - \iint_C \frac{d\mu \wedge d\bar{\mu}}{|\mu|^4} \chi\left(\frac{\lambda\mu}{\lambda+\mu}\right)\hat{P}_0^a(\mu,x,y,t)\right) - \chi(\lambda)\sigma_3. \tag{6.1.48}$$

The requirement of the absence of singularity in (6.1.48) at $\lambda = 0$ gives

$$P(x,y,t)\chi(0) - \iint_C \frac{d\mu \wedge d\bar{\mu}}{|\mu|^4} \chi(0)\hat{P}_0^a(\mu,x,y,t) - \chi(0)\sigma_3 = 0 \tag{6.1.49}$$

where $\chi(0) \doteq \chi(\lambda = 0)$. Using (6.1.46), one gets

$$P(x,y,t) = \chi(0)P_0(x,y,t)\chi^{-1}(0). \tag{6.1.50}$$

Then, it is easy to show that

$$\sigma D_2\chi(\lambda) + P(D_1\chi)(\lambda) \to 0 \qquad (6.1.51)$$

as $\lambda \to \infty$.

Thus, the first linear problem in the case of nontrivial background is

$$\sigma D_2\chi + PD_1\chi = 0 \qquad (6.1.52)$$

where $P$ is given by (6.1.50).

To construct the second linear problem we start again with the quantity

$$D_3\chi + i\sigma D_1 D_2\chi = \chi_t - \frac{i}{2\lambda}\iint_C \frac{d\mu \wedge d\bar{\mu}}{|\mu|^4}\chi\left(\frac{\lambda\mu}{\lambda+\mu}\right)\left(\hat{\Phi}_{0y}(\mu) + \frac{1}{\mu}\hat{P}_0(\mu)\right)$$

$$-\frac{1}{\sigma}(\widehat{\Phi_{0s}P})(\mu) - i\sigma(\widehat{P_{0y}P})(\mu)\Big) + i\sigma\chi_{xy}(\lambda) + \frac{1}{\lambda}\chi_x(\lambda)\sigma_3 \qquad (6.1.53)$$

$$+\frac{1}{\lambda}\iint_C \frac{d\mu \wedge d\bar{\mu}}{|\mu|^4}\left(\chi\left(\frac{\lambda\mu}{\lambda+\mu}\right)\hat{P}_0^a(\mu)\right)_x - \frac{\sigma}{\lambda}\chi_y$$

which apparently has no second order singularity at $\lambda = 0$. To compensate the first order singularity we add to (6.1.53) the term $V_1 D_1\chi + V_2 D_2\chi$. The requirement, that the residue in the first order pole at $\lambda = 0$ of the quantity

$$D_3\chi + i\sigma D_1 D_2\chi + V_1 D_1\chi + V_2 D_2\chi \qquad (6.1.54)$$

vanishes, gives

$$-\frac{i}{2}\iint \frac{d\mu \wedge d\bar{\mu}}{|\mu|^4}\chi(0)\left(\hat{\Phi}_{0y}(\mu) + \frac{1}{\mu}\hat{P}_0(\mu) + \frac{1}{\sigma}(\widehat{\Phi_{0s}P_0})(\mu) + i\sigma(\widehat{P_{0y}P_0})(\mu)\right) + \chi_x(0)\sigma_3$$

$$+ \iint \frac{d\mu \wedge d\bar{\mu}}{|\mu|^4}(\chi(0)\hat{P}_0^a(\mu))_x - \sigma\chi_y(0) + iV_1\chi(0) - \frac{i}{\sigma}V_2\chi(0)\sigma_3 \qquad (6.1.55)$$

$$+ \frac{i}{\sigma}\iint_C \frac{d\mu \wedge d\bar{\mu}}{|\mu|^4}\chi(0)\hat{P}_0^a(\mu) = 0.$$

In virtue of (6.1.46), equivalently

$$\frac{1}{2}\chi(0)\left(\Phi_{0y} - iP_{0x} - \frac{1}{\sigma}\Phi_{0x}P_0 - i\sigma\Phi_{0y}P_0\right)$$

$$+ i\chi_x(0)P_0 - i\sigma\chi_y(0) - V_1\chi(0) + \frac{1}{\sigma}V_2\chi(0) = 0. \qquad (6.1.56)$$

One can easily show that this equation is satisfied if

$$V_1 = \Phi_{0y} - i\sigma\chi_y(0)\chi^{-1}(0) - i\frac{1}{2}\chi(0)P_{0x}\chi^{-1}(0),$$

$$V_2 = \Phi_{0x} - i\sigma \chi_x(0)\chi^{-1}(0) + i\sigma^2 \frac{1}{2}\chi(0)P_{0y}\chi^{-1}(0). \qquad (6.1.57)$$

Denoting

$$\Phi = \Phi_0 - i\sigma \ln \det \chi(0), \qquad (6.1.58)$$

one can represent $V_1$ and $V_2$ in the form

$$V_1 = \frac{1}{2}(\Phi_y - iP_x),$$
$$V_2 = \frac{1}{2}(\Phi_x + i\sigma^2 P_y) \qquad (6.1.59)$$

where $P$ is given by (6.1.50).

Further, (6.1.59) obviously obeys the condition (2.3.16). Hence, the second linear problem is

$$D_3\chi + i\sigma D_1 D_2 \chi + \frac{1}{2}(\Phi_y - iP_x)D_1\chi + \frac{1}{2}(\Phi_x + i\sigma^2 P_y)D_2\chi = 0 \qquad (6.1.60)$$

where $P$ and $\Phi$ are given by (6.1.50) and (6.1.58) respectively.

Thus in the case of nontrivial background we have the linear system

$$\sigma D_2\chi + PD_1\chi = 0,$$
$$D_3\chi + i\sigma D_1 D_2 \chi + \frac{1}{2}(\Phi_y - iP_x)D_1\chi + \frac{1}{2}(\Phi_x + i\sigma^2 P_y)D_2\chi = 0. \qquad (6.1.61)$$

This system is of the same form as that of (6.1.10), (6.1.23) in the case of trivial background. The only difference is in the form of the commuting operators $D_i$.

This circumstance makes obvious the fact that the system (6.1.61) gives rise to the Ishimori equation (6.0.1). One can obtain the Ishimori equation directly from the system (6.1.61) evaluating the second equation (6.1.61) at $\lambda = 0$.

The formulae

$$P(x,y,t) = \chi(0)P_0(x,y,t)\chi^{-1}(0) \qquad (6.1.62a)$$

and

$$\Phi(x,y,t) = \Phi_0 - i\sigma \ln \det \chi(0) \qquad (6.1.62b)$$

provide us the $\bar{\partial}$-dressing for the Ishimori equation on the nontrivial background $P_0$ and $\Phi_0$. Note that, due to (6.1.7), (6.1.22) and (6.1.62) the $\bar{\partial}$-dressing produces the multiplication of the values of the function $\chi$ at origin: $\chi_0 \to \chi(0)\chi_0$.

Now, similar to the DS equation, we will consider the special, almost trivial background $\vec{S} = (0,0,1)$ or $P_0 = \sigma_3$. In virtue of the Ishimori equation (6.0.1) the general form of the function $\Phi$ in this case is

$$\Phi_0(x,y,t) = V_1(y - \sigma x, t) + V_2(y + \sigma x, t) \qquad (6.1.63)$$

where $V_1$ and $V_2$ are arbitrary functions.

The corresponding operators $\tilde{\tilde{B}}_i$ are of the form

$$\tilde{\tilde{B}}_1 = \partial_s, \quad \tilde{\tilde{B}}_2 = -\frac{1}{\sigma}\sigma_3 \partial_s,$$

$$\tilde{\tilde{B}}_3 = i\sigma_3 \partial_s^2 - \begin{pmatrix} V_1'(y - \sigma x, t) & 0 \\ 0 & V_2'(y + \sigma x, t) \end{pmatrix} \partial_s \quad (6.1.64)$$

where $V'(\xi) \doteq \frac{dV}{d\xi}$.

The Fourier transform $\tilde{R}$ of the kernel $R$ defined by

$$\tilde{R}(s, \tilde{s}; x, y, t)$$
$$= \frac{1}{(2\pi)^2} \iint_C \frac{d\mu \wedge d\bar{\mu}}{|\mu|^4} \iint_C \frac{d\lambda \wedge d\bar{\lambda}}{|\lambda|^4} R(\mu, \bar{\mu}; \lambda, \bar{\lambda}; x, y, t) e^{i\left(\frac{s}{\mu} + \frac{\bar{s}}{\bar{\mu}}\right) - i\left(\frac{\tilde{s}}{\lambda} + \frac{\tilde{s}}{\bar{\lambda}}\right)} \quad (6.1.65)$$

obeys the equations

$$\tilde{R}_x(s, \tilde{s}; x, y, t) - \tilde{R}_s - \tilde{R}_{\tilde{s}} = 0,$$

$$\sigma \tilde{R}_y(s, \tilde{s}; x, y, t) + \sigma_3 \tilde{R}_s + \tilde{R}_{\tilde{s}}\sigma_3 = 0, \quad (6.1.66)$$

$$\tilde{R}_t(s, \tilde{s}; x, y, t) - i\sigma_3 \tilde{R}_{ss} + i\tilde{R}_{\tilde{s}\tilde{s}}\sigma_3 + C_0(s + x, y, t)\tilde{R}_s + (\tilde{R}C_0(\tilde{s} + x, y, t))_{\tilde{s}} = 0$$

where

$$C_0(z, y, t) = \begin{pmatrix} V_1' & 0 \\ 0 & V_2' \end{pmatrix}. \quad (6.1.67)$$

The system (6.1.66) is solvable by separation of variables

$$\tilde{R}(s, \tilde{s}; x, y, t) = \sum_{i,k=1}^{N} \rho_{ik} X_{0i}(s, x, y, t) Y_{0k}^*(\tilde{s}, x, y, t). \quad (6.1.68)$$

The matrix functions $X_i$ and $Y_k$ obey the system of equations

$$X_{0x} - X_{0s} = 0,$$

$$\sigma X_{0y} + \sigma_3 X_{0s} = 0, \quad (6.1.69)$$

$$iX_{0t} + \sigma_3 X_{0ss} + iC_0(s + x, y, t) X_{0s} = 0.$$

The first two equations (6.1.69) are readily solvable. The third equation is, in fact, the linear problem of the type

$$iX_t \pm X_{0ss} + iV_1'(s + x, y, t) X_{0s} = 0 \quad (6.1.70)$$

for each component of $X_0$. The problem (6.1.70) is nothing but the first linear problem associated with the mKP-I equation. So, the results of section (4.2) provide us the wide class of exact solutions of equation (6.1.70) and, consequently, the explicit kernels $R$. Then, solving the $\bar{\partial}$-problem, one obtains the solutions of the Ishimori equation by the formula

$$P = \chi_0 \sigma_3 \chi_0^{-1},$$

$$\Phi = -i\sigma \ln \det \chi_0 + V_1(y - \sigma x, t) + V_2(y + \sigma x, t). \tag{6.1.71}$$

for given background $\Phi_0 = V_1 + V_2$.

Note that for the component $R_{12}$ equations (6.1.66) are of the form

$$R_{12x} - R_{12s} - R_{12\tilde{s}} = 0,$$

$$\sigma R_{12y} + R_{12s} - R_{12\tilde{s}} = 0, \tag{6.1.72}$$

$$iR_{12t} + R_{12ss} + R_{12\tilde{s}\tilde{s}} + iV_1'(y - \sigma(x+s), t)R_{12s} + (iV_2'(y + \sigma(x+\tilde{s}), t)R_{12\tilde{s}})_{\tilde{s}} = 0.$$

Introducing $S$ defined by $S_{\tilde{s}} \doteq R_{12}$, one rewrites the third equation (6.1.72) as follows

$$iS_t + S_{ss} + S_{\tilde{s}\tilde{s}} + iV_1'(y - \sigma(x+s), t)S_s + iV_2'(y + \sigma(x+s), t)S_{\tilde{s}} = 0. \tag{6.1.73}$$

The separation of variables

$$S(s, \tilde{s}) = \sum_{i,k} \rho_{ik} X_i(s) Y_k(\tilde{s}) \tag{6.1.74}$$

in (6.1.73) leads to the equations

$$iX_t + X_{ss} + iV_1'(y - \sigma x - \sigma s, t)X_s = 0,$$

$$iY_t + Y_{\tilde{s}\tilde{s}} + iV_2'(y + \sigma x + \sigma s, t)Y_{\tilde{s}} = 0 \tag{6.1.75}$$

that are the linear problems for the mKP-I equation.

Since $V_1$ and $V_2$ are the asymptotics values of gradients of $\Phi$ then the $\bar{\partial}$-dressing described above provides us the solutions of the Ishimori equation with nontrivial boundaries of $\Phi$.

In the case $P_0 = \sigma_3$ and $\Phi$ of the form (6.1.63) the linear problems (6.1.61) looks simpler. In terms of the function

$$\tilde{\Psi}(x, y, t, \lambda) = \chi \, e^{\frac{ix}{\lambda} - \frac{i\sigma_3}{\sigma\lambda}y}, \tag{6.1.76}$$

one has the problem

$$\sigma \tilde{\Psi}_y + P\tilde{\Psi}_x = 0,$$

$$\tilde{\Psi}_t + i\sigma\tilde{\Psi}_{xy} + \frac{1}{2}(\Phi_y - iP_x)\tilde{\Psi}_x + \frac{1}{2}(\Phi_x + i\sigma^2 P_y)\tilde{\Psi}_y$$

$$-\frac{i}{2\lambda}\iint_C \frac{d\mu \wedge d\bar{\mu}}{|\mu|^4} \tilde{\Psi}\left(\frac{\lambda\mu}{\lambda+\mu}\right)\tilde{C}_0(\mu) = 0 \qquad (6.1.77)$$

where

$$\tilde{C}_0(\lambda) \doteq \frac{1}{2\pi}\iint_C ds \wedge d\bar{s}\, e^{-i(\frac{s}{\lambda}+\frac{\bar{s}}{\bar{\lambda}})} C_0(s+x,y,t). \qquad (6.1.78)$$

The modified second linear problem (6.1.77) for the Ishimori equation with nontrivial boundaries has been derived first by a different method in [245].

We conclude with the remark that, similar to the DS equation, the cases $\sigma = 1$ and $\sigma^2 = -i$ are essentially different if one considers the bounded functions $V_1'(y - \sigma x, t)$ and $V_2'(y + \sigma x, t)$. For the Ishimori-I equation $V_1'$ and $V_2'$ are arbitrary bounded functions while for the Ishimori-II equation the only admissible bounded $V_1'$ and $v_2'$ are constants.

## 6.2. Exact solutions

According to the results of the previous section the solutions of the Ishimori equation are given by the formulae

$$\vec{S} = \frac{1}{2}\text{tr}(\vec{\sigma}\chi(0)P_0\chi^{-1}(0)),$$

$$\Phi = \Phi_0 - i\sigma \ln \det \chi(0). \qquad (6.2.1)$$

Using the integral equation (2.2.3), one gets for trivial background

$$\chi(0) = 1 + i\Delta = 1 + \frac{1}{2\pi i}\iint_C \frac{d\lambda \wedge d\bar{\lambda}}{\lambda}\iint_C d\mu \wedge d\bar{\mu}\, \chi(\mu,\bar{\mu}) R(\mu,\bar{\mu};\lambda,\bar{\lambda}) \qquad (6.2.2)$$

where

$$R(\mu,\bar{\mu};\lambda,\bar{\lambda};x,y,t) = e^{F(\mu)} R_0(\mu,\bar{\mu};\lambda,\bar{\lambda}) e^{-F(\lambda)} \qquad (6.2.3)$$

and

$$F(\lambda) = \frac{i}{\lambda}x - \frac{i}{\lambda}\sigma_3 y - \frac{i}{\lambda^2}\sigma_3 t. \qquad (6.2.4)$$

In general the solutions $\vec{S}$ and $\Phi$ constructed with the use of these formulae are complex. The conditions of the reality of $\vec{S}$ and $\Phi$ are different for the Ishimori-I and Ishimori-II equations. In the case $\sigma = 1$ the condition $\vec{\bar{S}} = \vec{S}$ (or $P^+ = P$)(symbol + denotes the Hermitian conjugation of matrix) is satisfied if

$$\chi^+(0) \cdot \chi(0) = 1. \qquad (6.2.5)$$

The condition (6.2.5) also implies $|\det \chi(0)| = 1$. So, in this case

$$\Phi = \Phi_0 + \arg \det \chi(0). \qquad (6.2.6)$$

In the case $\sigma = i$ the reality of $\vec{S}$ is reflected in the existence of involution in the corresponding linear problem. Indeed, the condition $\overline{\vec{S}} = \vec{S}$ is equivalent to $\bar{P} = -\sigma_2 P \sigma_2^{-1}$. As a result, one can impose the involution

$$\overline{\chi(\lambda, \bar{\lambda})} = \sigma_2 \chi(-\bar{\lambda}, -\lambda) \sigma_2^{-1} \tag{6.2.7}$$

on the solutions of the linear problem (6.1.61) at $\sigma = i$. The involution (6.2.7) implies

$$\overline{\chi(0)} = \sigma_2 \chi(0) \sigma_2^{-1}, \quad \overline{\det \chi(0)} = \det \chi(0).$$

These conditions obviously lead to real $\vec{S}$ and $\Phi$ given by the formulae (6.2.1).

For the Ishimori-II equation the compatibility of the involution (6.2.7) with the $\bar{\partial}$-problem leads to the following reality constraint for $R$:

$$\overline{R(\mu, \bar{\mu}; \lambda, \bar{\lambda})} = \sigma_2 R(-\bar{\mu}, -\mu; -\bar{\lambda}, -\lambda) \sigma_2^{-1}. \tag{6.2.8}$$

For the Ishimori-I equation one can obtain the necessary condition for reality using the formulae (6.2.1) and (6.2.2). The condition (6.2.5) implies $\Delta^+ = -\Delta$. Then for the small data $R$ one has $\chi \sim 1$ and, consequently,

$$\chi(0) \sim 1 + \frac{1}{2\pi i} \iint_C \frac{d\lambda \wedge d\bar{\lambda}}{\lambda} \iint_C d\mu \wedge d\bar{\mu}\, R(\mu, \bar{\mu}; \lambda, \bar{\lambda}). \tag{6.2.9}$$

So one has the following necessary reality condition

$$\frac{1}{\bar{\lambda}} R^+(\mu, \bar{\mu}; \lambda, \bar{\lambda}) = \frac{1}{\mu} R(\bar{\lambda}, \lambda; \bar{\mu}, \mu) \tag{6.2.10}$$

for the Ishimori-I equation.

We start with the case of the trivial background and consider at first the **solutions with functional parameters**. They typically correspond to the degenerate kernel $R_0$

$$R_0(\mu, \bar{\mu}; \lambda, \bar{\lambda}) = \sum_{k=1}^{N} f_{0k}(\mu, \bar{\mu}) g_{0k}(\lambda, \bar{\lambda}) \tag{6.2.11}$$

where $f_{0k}$ and $g_{0k}$ are arbitrary matrix functions. Using the formulae (2.2.20) - (2.2.21) and (6.2.2), one gets

$$\chi_0 = 1 + \frac{1}{2\pi i} \sum_{l,k=1}^{N} \xi_k (1+A)^{-1}_{kl} \eta_l \tag{6.2.12}$$

where

$$\xi_k(x, y, t) = \iint_C d\lambda \wedge d\bar{\lambda}\, e^{\frac{ix}{\lambda} - \frac{i\sigma_3}{\lambda\sigma}y - \frac{i}{\lambda^2}\sigma_3 t} f_{0k}(\lambda, \bar{\lambda}),$$

$$\eta_l(x, y, t) = \iint_C \frac{d\lambda \wedge d\bar{\lambda}}{\lambda} g_{0l}(\lambda, \bar{\lambda}) e^{-\frac{ix}{\lambda} + \frac{i\sigma_3}{\lambda\sigma}y + \frac{i}{\lambda^2}\sigma_3 t} \tag{6.2.13}$$

and

$$A_{lk}(x,y,t) = \frac{1}{2\pi i} \iint_C d\mu \wedge d\bar{\mu} \iint \frac{d\lambda \wedge d\bar{\lambda}}{\lambda - \mu}$$

$$\times g_{0l}(\mu,\bar{\mu}) \exp\left\{i\left(\frac{1}{\lambda}-\frac{1}{\mu}\right)x - \frac{i}{\sigma}\left(\frac{1}{\lambda}-\frac{1}{\mu}\right)\sigma_3 y - i\left(\frac{1}{\lambda^2}-\frac{1}{\mu^2}\right)\sigma_3 t\right\} f_{0k}(\lambda,\bar{\lambda}). \quad (6.2.14)$$

Using the formal identity

$$\int^x dx' e^{i\left(\frac{(\mu-\lambda)}{\mu\lambda}\right)x'} = \frac{\mu\lambda}{i(\mu-\lambda)} e^{\left(\frac{\mu-\lambda}{\mu\lambda}\right)x}, \quad (6.2.15)$$

one can represent $A_{lk}$ as

$$A_{lk} = \frac{i}{2\pi} \int^x dx' \eta_l(x',y,t) \xi_{kx'}(x',y,t) \quad (6.2.16)$$

where $\xi_k$ and $\eta_l$ are given by (6.2.13).

So the formulae (6.2.1) with $\chi_0$ given by (6.2.12), (6.2.16), (6.2.13) describe the solutions of the Ishimori equation with functional parameters $f_{0k}(\lambda,\bar{\lambda})$ and $g_{0k}(\mu,\bar{\mu})$.

The functions $\xi_k$ and $\eta_l$ obey the system of equations

$$\sigma \xi_y + \sigma_3 \xi_x = 0,$$

$$\xi_t - i\sigma_3 \xi_{xx} = 0 \quad (6.2.17)$$

and

$$\sigma \eta_y + \eta_x \sigma_3 = 0,$$

$$\eta_t + i\eta_{xx}\sigma_3 = 0. \quad (6.2.18)$$

It is easy to check that both $\xi$ and $\eta$ are also solutions of the linearized Ishimori equation.

For the Ishimori-I equation the condition (6.2.10) of the reality of $\vec{S}$ is satisfied for arbitrary $N$ if

$$g_{0k}(\lambda,\bar{\lambda}) = \lambda f_{0k}^+(\bar{\lambda},\lambda) \quad (6.2.19)$$

and for $N = 2M$ where $M$ is an arbitrary integer if

$$f_{0k+M}(\lambda,\bar{\lambda}) = \frac{1}{\lambda} g_{0k}^+(\bar{\lambda},\lambda),$$

$$g_{0k+M}(\lambda,\bar{\lambda}) = \lambda f_{0k}^+(\bar{\lambda},\lambda) \quad (k=1,\ldots,M). \quad (6.2.20)$$

In terms of $\xi_k$ and $\eta_k$ these conditions look like

$$\eta_k(x,y,t) = \xi_k^+(x,y,t) \quad (6.2.21)$$

and
$$\xi_{k+M}(x,y,t) = \eta_k^+(x,y,t),$$
$$\eta_{k+M}(x,y,t) = \xi_k^+(x,y,t) \quad (k=1,\ldots,M) \tag{6.2.22}$$
respectively.

For the Ishimori-II equation the condition (6.2.8) is satisfied if
$$\overline{f_{0k}(\lambda,\bar{\lambda})} = \sigma_2 f_{0k}(-\bar{\lambda},-\lambda)\sigma_2^{-1},$$
$$\overline{g_{0k}(\lambda,\bar{\lambda})} = \sigma_2 g_{0k}(-\bar{\lambda},-\lambda)\sigma_2^{-1} \tag{6.2.23}$$
and, consequently,
$$\overline{\xi_k(x,y,t)} = \sigma_2 \xi_k(x,y,t)\sigma_2^{-1},$$
$$\overline{\eta_k(x,y,t)} = \sigma_2 \eta_k(x,y,t)\sigma_2^{-1}. \tag{6.2.24}$$

The **line solitons**, as usual, correspond to the functions $f_{0k}$ and $g_{0k}$ chosen as the Dirac delta-functions.

For the DS-I equation, taking into account (6.2.19) and (6.2.20), one has
$$R_0(\mu,\bar{\mu};\lambda,\bar{\lambda}) = \sum_{k=1}^{N} \lambda_k \tilde{f}_{0k} \tilde{f}_{0k}^+ \delta(\mu - \lambda_k)\delta(\lambda - \bar{\lambda}_k) \tag{6.2.25}$$
or
$$R_0(\mu,\bar{\mu};\lambda,\bar{\lambda}) = \sum_{k=1}^{N} (\lambda_k \tilde{f}_{0k} \tilde{g}_{0k}^+ \delta(\mu-\mu_k)\delta(\lambda-\lambda_k) + \bar{\lambda}_k \tilde{g}_{0k}^+ \tilde{f}_{0k}^+ \delta(\mu-\bar{\lambda}_k)\delta(\lambda-\bar{\mu}_k)) \tag{6.2.26}$$
where $\tilde{f}_{0k}, \tilde{g}_{0k}$ are arbitrary constant matrices. The function $\xi_k$ in the case (6.2.25) is of the form
$$\xi_k(x,y,t) = -2i \exp\left(\frac{ix}{\lambda_k} - \frac{iy}{\lambda_k}\sigma_3 - \frac{i}{\lambda_k^2}\sigma_3 t\right) \tilde{f}_{0k} \tag{6.2.27}$$
and the matrix $A$ is given by
$$A_{lk} = \frac{2i}{\pi} \frac{1}{\lambda_k - \bar{\lambda}_l} \tilde{f}_{0l}^+ \exp\left(i\left(\frac{1}{\lambda_k} - \frac{1}{\bar{\lambda}_l}\right) - i\left(\frac{1}{\lambda_k} - \frac{1}{\bar{\lambda}_l}\right)\sigma_3 y - i\left(\frac{1}{\lambda_k^2} - \frac{1}{\bar{\lambda}_l^2}\right)\sigma_3 t\right) \tilde{f}_{0k}. \tag{6.2.28}$$

Thus, the line solitons of the Ishimori-I equation are given by the formula
$$\vec{S} = \frac{1}{2}\operatorname{tr}(\vec{\sigma}\chi_0 \sigma_3 \chi_0),$$
$$\Phi = -i\sigma \ln \det \chi_0 \tag{6.2.29}$$

Chapter 6

where

$$\chi_0 = 1 + \frac{1}{2\pi i} \sum_{l,k=1}^{N} \xi_k (1+A)_{kl}^{-1} \xi_l^+ \tag{6.2.30}$$

and $\xi_k$ and $A_{lk}$ are of the form (6.2.27) and (6.2.28).

The line solitons of the Ishimori equations interact elastically. They are the solutions which do not decrease in certain directions. These solutions are the 2+1-dimensional generalization of the solitons of the Heisenberg ferromagnet model equation (1.1.19).

The solutions of the Ishimori-I equation associated with the kernel $R_0$ of the form (6.2.26) are of the breather type. The reader can easily derive the general formula for these solutions.

For the Ishimori-II equation the line solitons, due to (6.2.23), correspond to the choice

$$R_0(\mu,\bar{\mu},\lambda,\bar{\lambda}) = \sum_{k=1}^{N} \tilde{f}_{0k} \tilde{g}_{0k} \delta(\mu - i\mu_k)\delta(\lambda - i\lambda_k) \tag{6.2.31}$$

where $\mu_k$ and $\lambda_k$ are arbitrary real constants and the constant matrices $\tilde{f}_{0k}$ and $\tilde{g}_{0k}$ obey the constraints

$$\overline{\tilde{f}_{0k}} = \sigma_2 \tilde{f}_{0k} \sigma_2^{-1}, \quad \overline{\tilde{g}_{0k}} = \sigma_2 \tilde{g}_{0k} \sigma_2^{-1}. \tag{6.2.32}$$

In this case

$$\xi_k = -2i \exp\left\{\frac{x}{\lambda_k} + i\frac{\sigma_3 y}{\lambda_k} + i\frac{1}{\lambda_k^2}\sigma_3 t\right\} \tilde{f}_{0k},$$

$$\eta_k = -2i\tilde{g}_{0k} \exp\left\{-\frac{x}{\lambda_k} - i\frac{\sigma_3 y}{\lambda_k} - \frac{i}{\lambda_k^2}\sigma_3 t\right\}, \tag{6.2.33}$$

and

$$A_{lk} = \frac{2}{\pi}\frac{1}{\lambda_l - \mu_k}\tilde{g}_{0l}\exp\left\{\left(\frac{1}{\lambda_l}-\frac{1}{\mu_k}\right)x - i\left(\frac{1}{\lambda_l}-\frac{1}{\mu_k}\right)\sigma_3 y + i\left(\frac{1}{\lambda_l^2}-\frac{1}{\mu_k^2}\right)\sigma_3 t\right\}\tilde{f}_{0k}. \tag{6.2.34}$$

The general line solitons of the Ishimori-II equation are of the form (6.2.29) where $\chi_0, \xi_k, \eta_k, A_{lk}$ are given by (6.2.12), (6.2.32) - (6.2.34).

Now we proceed to the **rational solutions (lumps)** of the Ishimori equation. For matrix problems, as we have seen, the rational-exponential exact solutions are typical. To construct pure rational solutions one should choose a rather specific kernel $R_0$ of the $\bar{\partial}$-problem.

Here we present another approach and confine ourselves to the Ishimori-II equation. First we note that the Fourier transform (6.1.65) of the kernel $R$ obeys the system of equations (6.1.66) with $C_0 \equiv 0$. This system can be solved by separation of variables (6.1.68) and the matrix $Y^*$ obeys the system of equations

$$Y_x^* - Y_s^* = 0,$$

$$iY_y^* + Y_s^*\sigma_3 = 0, \tag{6.2.35}$$
$$iY_t^* - Y_{ss}^*\sigma_3 = 0.$$

In virtue of (6.2.23) for real $\vec{S}$, one has

$$\overline{Y^*(s,x,y,t)} = \sigma_2 Y^*(s,x,y,t)\sigma_2^{-1}. \tag{6.2.36}$$

As a result the matrix $Y^*$ is of the form

$$Y^* = \begin{pmatrix} \tilde{f} & -\bar{\tilde{g}} \\ \tilde{g} & \bar{\tilde{f}} \end{pmatrix}. \tag{6.2.37}$$

The system (6.2.35) implies that the functions $\tilde{f}$ and $\tilde{g}$ obey the following equations

$$\begin{aligned} \tilde{f}_x - \tilde{f}_{\tilde{s}} = 0, \quad i\tilde{f}_y + \tilde{f}_{\tilde{s}} = 0, \quad i\tilde{f}_t - \tilde{f}_{\tilde{s}\tilde{s}} = 0, \\ \tilde{g}_x - \tilde{g}_{\tilde{s}} = 0, \quad i\tilde{g}_y + \tilde{g}_{\tilde{s}} = 0, \quad i\tilde{g}_t - \tilde{g}_{\tilde{s}\tilde{s}} = 0. \end{aligned} \tag{6.2.38}$$

Introducing the complex variables $z = \frac{1}{2}(y - ix), \bar{z} = \frac{1}{2}(y + ix)$, one gets

$$\tilde{f} = \tilde{f}(\tilde{s} + 2iz), \quad \tilde{g} = \tilde{g}(\tilde{s} + 2iz)$$

and

$$i\tilde{f}_t + \frac{1}{4}\tilde{f}_{zz} = 0, \quad i\tilde{g}_t + \frac{1}{4}\tilde{g}_{zz} = 0 \tag{6.2.39}$$

and obviously $\tilde{f}_{\bar{z}} = \tilde{g}_{\bar{z}} = 0$. So $\tilde{f}$ and $\tilde{g}$ are the analytic functions of $z$, obeying the linear equations (6.2.39).

Then we should find $\chi_0$. For the degenerate kernel $R = F(\mu, \bar{\mu})G(\lambda, \bar{\lambda})$ one has

$$\chi_0 = 1 + h(x,y,t)\eta(x,y,t) \tag{6.2.40}$$

where

$$h(x,y,t) = \iint_C d\mu \wedge d\bar{\mu}\chi(\mu,\bar{\mu})F(\mu,\bar{\mu}), \tag{6.2.41}$$

$$\eta(x,y,t) = \iint_C \frac{d\lambda \wedge d\bar{\lambda}}{\lambda}G(\lambda,\bar{\lambda}). \tag{6.2.42}$$

It is not difficult to show that $\eta(x,y,t) = Y_x^*(x,y,t)$. Thus, the matrix $\eta$ is representable in the form

$$\eta = \begin{pmatrix} f-1 & -\bar{g} \\ g & \bar{f}-1 \end{pmatrix} \tag{6.2.43}$$

where $f = 1 + f(2iz) = 1 + \tilde{f}_{\tilde{s}}(\tilde{s} + 2iz)|_{\tilde{s}=0}$, $g = g(2iz) = \tilde{g}_{\tilde{s}}(\tilde{s} + 2iz)|_{\tilde{s}=0}$ and $f$ and $g$ obey the linear equations

$$if_t + \frac{1}{4}f_{zz} = 0, \quad ig_t + \frac{1}{4}g_{zz} = 0. \tag{6.2.44}$$

Now let us choose
$$F(\mu,\bar{\mu}) = \frac{1}{2i}\delta\left(\frac{1}{\mu}\right). \tag{6.2.45}$$

Since $\chi(\mu) \to 1$ as $\mu \to \infty$ then one has $h = 1$. The choice (6.2.45) is admissible for the equations of $f(\mu,\bar{\mu},x,y,t)$ since

$$\frac{1}{\mu^n}\delta\left(\frac{1}{\mu}\right) = 0$$

for positive integer $n$.

Thus, for such a choice we have

$$\chi_0(x,y,t) = \begin{pmatrix} f(z,t) & -\overline{g(z,t)} \\ g(z,t) & \overline{f(z,t)} \end{pmatrix}. \tag{6.2.46}$$

So, we obtain the solution of the Ishimori-II equation given by

$$S_+ = S_1 + iS_2 = \frac{2f\bar{g}}{f\bar{f}+g\bar{g}}, \quad S_- = S_1 - iS_2 = \frac{2g\bar{f}}{f\bar{f}+g\bar{g}},$$

$$S_3 = \frac{f\bar{f}-g\bar{g}}{f\bar{f}+g\bar{g}}, \tag{6.2.47}$$

$$\Phi = \ln(f\bar{f}+g\bar{g})$$

where the functions $f$ and $g$ are the solutions of equations (6.2.44).

Pure rational solutions, obviously, correspond to the polynomial in $z$ solutions of equations (6.2.44). The general polynomial solutions of these equations are of the form [92]

$$f = f_N = \sum_{j=0}^{N}\sum_{m+2n=j}\frac{a_j}{m!n!}(2z)^m(2it)^n,$$

$$g = g_M = \sum_{j=0}^{M}\sum_{m+2n=j}\frac{b_j}{m!n!}(2z)^m(2it)^n \tag{6.2.48}$$

where $a_j$ and $b_j$ are arbitrary complex constants and $N$ and $M$ are arbitrary integers. The symbol $\sum_{m+2n=j}$ means the summation over all possible combinations of the non-negative integers $m$ and $n$ such that $m+2n = j$.

For arbitrary $N$ and $M$ the solutions (6.2.47) may increase as $|z| \to \infty$. But if one chooses $|N - M| = 1$, for instance, $M = N - 1$ ($a_N \neq 0$), then for the solutions (6.2.47) one has $\vec{S} \to (0,0,1)$ as $|z| \to \infty$.

The simplest solution of this type ($N = 1$) for which $f = a_0 + 2a_1z$, $g = b_0$ is of the form

$$S_1 + iS_2 = \frac{2\beta(\bar{\alpha}+(x+iy))}{|\alpha-x+iy|^2+|\beta|^2} = \frac{2\beta(x-x_0+i(y-y_0))}{|x-x_0-i(y-y_0)|^2+|\beta|^2},$$

$$S_3 = -\frac{|\alpha - x + iy|^2 - |\beta|^2}{|\alpha - x + iy|^2 + |\beta|^2} = -\frac{|x - x_0 - i(y - y_0)|^2 - |\beta|^2}{|x - x_0 - i(y - y_0)|^2 + |\beta|^2} \tag{6.2.49}$$

where $\alpha = (x_0 - iy_0) = -\frac{b_0}{a_1}$ and $\beta = \frac{a_0}{a_1}$. Equivalently

$$S_1 + iS_2 = \frac{2|\beta|\rho e^{-i(\varphi - \varphi_0)}}{\rho^2 + |\beta|^2}, \quad S_3 = \frac{|\beta|^2 - \rho^2}{|\beta|^2 + \rho^2} \tag{6.2.50}$$

where $\rho e^{i\varphi} \doteq (x - x_0) - i(y - y_0)$ and $|\beta| = |\beta|e^{i\varphi_0}$.

The solution (6.2.49) or (6.2.50) is nonsingular and rational. So it is just the lump of the Ishimori-II equation.

This one-lump solution is the configuration of the vortex-type for the spin field $\vec{S}(x, y, t)$ with the center at the point $(x_0, y_0)$. This solution is a static. The two-lump solution corresponds to $N = 2$ and

$$f_2 = a_0 + a_1 2z + a_2 \left(\frac{1}{2}(2z)^2 + 2it\right), \quad g_1 = b_0 + 2b_1 z.$$

This two-lump solution is the nonstatic and describes the collision of the two vortices of the type (6.2.49). The general solution (6.2.47), (6.2.48) with $M = N - 1$ describes the scattering of $N$ vortices (6.2.49). The analysis of this solution shows that the collision of the vortices is absolutely elastic and the phase shift is absent as in other cases.

The remarkable feature of the vortex solutions is that they have the nontrivial topological properties. The $N$-lump ($N$-vortices) solutions are topologically nontrivial configurations of the field $\vec{S}$. To convince ourselves of this fact let us calculate the topological charge $Q$ (6.0.3) for the $N$-vortex solution. Taking into account (6.0.1) ($\sigma = i$), one has

$$Q = \frac{1}{4\pi} \iint dxdy \left(\Phi_{xx} + \Phi_{yy}\right) = \frac{1}{4\pi} \lim_{|z| \to \infty} \oint_S (\Phi_x dy - \Phi_y dx). \tag{6.2.51}$$

Then it follows from (6.1.40) that for the $N$-vortex solutions $\Phi = \ln(\bar{f}f + \bar{g}g)$. Hence, for the $N$-vortex solutions $\Phi \to 2\ln|a_N(2z)^{2N}| + O(|z|^{-1})$ as $|z| \to \infty$. Substituting this asymptotics into (6.2.51), we obtain

$$Q = N. \tag{6.2.52}$$

Thus, the $N$-vortex solution of the Ishimori-II equation (6.0.1) is characterized by the definite value of the topological charge $Q = N$. By this reason only the topologically nontrivial fluctuations of the field $\vec{S}$ can deform the solutions with the different number of vortices to each other. Note that, by virtue of (6.2.5), the scalar field $\Phi(x, y, t)$ plays a role of the potential created by the distribution of the topological charge density $\frac{1}{4\pi}\vec{S}(\vec{S}_x \times \vec{S}_y)$.

In the completely similar manner one can construct the vortex-type solutions with $Q = -N$. The corresponding solutions differ from those constructed above by the change of variable $z \to \bar{z}$. So these solutions are the anti-vortex solutions.

For the first time these vortex solutions of the Ishimori-II equation have been constructed in the paper [92] by the Hirota method. The result (6.2.52) has been obtained in [92] too. Note that the Ishimori equation is the only basic 2+1-dimensional integrable equation which possesses the topological charge.

Note also that in a manner described above one can construct the polynomial solutions of other 2+1-dimensional integrable equations.

Next important class of exact solutions are **exponentially localized solutions** of the Ishimori-I equation. Similar to the DS-I equation they are associated with the almost trivial background $P_0 = \sigma_3, (\vec{S} = (0,0,1))$ and $\Phi = V_1(y-x,t) + V_2(y+x,t)$ (see (6.1.63)). Analogously to the DS case for the bounded solutions the nonlocal $\bar{\partial}$-problem is reduced to the nonlocal Riemann—Hilbert problem with the jump across the real axis. Consequently the integrals over the complex plane $\lambda$ associated with the dressing on nontrivial background are converted into integrals over the real axis.

We start with the given boundary functions $V_1$ and $V_2$. Our intermediate goal is to calculate the kernel $R$. We will look for the solutions within the class of degenerate kernels $R$ of the form

$$R(\mu, \bar{\mu}; \lambda, \bar{\lambda}; x, y, t)$$
$$= \sum_{i,k=1}^{N} \left\{ \rho_{0ik} \begin{pmatrix} f_i(\mu) \\ 0 \end{pmatrix} \otimes \left(0, \frac{1}{\lambda} g_k(\lambda)\right) + \bar{\rho}_{0ik} \begin{pmatrix} 0 \\ \frac{1}{\bar{\mu}} \bar{g}_i(\mu) \end{pmatrix} \otimes (\bar{f}_k(\lambda), 0) \right\}$$
$$= \begin{pmatrix} 0 & \sum_{i,k}^{N} \rho_{0ik} \frac{1}{\lambda} f_i(\mu) g_k(\lambda) \\ \sum_{i,k=1}^{N} \overline{\rho_{0ik}} \frac{1}{\bar{\mu}} \bar{g}_i(\mu) \bar{f}_k(\lambda) & 0 \end{pmatrix}. \tag{6.2.53}$$

The Fourier transform $\tilde{R}(s, \tilde{s})$ of the kernel $R(\mu, \lambda)$ is defined by

$$\tilde{R}(s, \tilde{s}) = \frac{1}{2\pi} \int_R \int_R d\lambda d\mu \, e^{\frac{2i}{\mu} s - \frac{2i\tilde{s}}{\lambda}} R(\mu, \lambda). \tag{6.2.54}$$

The Fourier transform of the kernel (6.2.53) is

$$\tilde{R}(s, \tilde{s}, x, y, t) = \sum_{i,k=1}^{N} \begin{pmatrix} 0 & \rho_{0ik} X_i(s) \tilde{Y}_k(\tilde{s}) \\ \bar{\rho}_{0ik} \tilde{\bar{Y}}_i(s) \bar{X}_k(\tilde{s}) & 0 \end{pmatrix}. \tag{6.2.55}$$

The quantity $\sum_{i,k=1}^{N} \rho_{0ik} X_i(s) \tilde{Y}_k(\tilde{s})$ obeys the system of equations (6.1.72) while $X_i$ and $\tilde{Y}_k$ solve equations

$$iX_{it} + \frac{1}{4} X_{iss} + iV_1'(y - x - 2s, t) X_{is} = 0 \tag{6.2.56}$$

and
$$i\tilde{Y}_{kt} + \frac{1}{4}\tilde{Y}_{k\tilde{s}\tilde{s}} + (iV_2'(y+(x+2\tilde{s}),t)\tilde{Y}_k)_{\tilde{s}} = 0. \tag{6.2.57}$$

For the matrix $Y_k$ defined by $Y_{k\tilde{s}} \doteq \tilde{Y}_k$ one has
$$iY_{kt} + \frac{1}{4}Y_{kss} + iV_2'(y+(x+2\tilde{s}),t)Y_{k\tilde{s}} = 0. \tag{6.2.58}$$

In the characteristic variables $\xi = \frac{1}{2}(y-x) - \frac{s}{2}$ and $\eta = \frac{1}{2}(y+x) + \frac{s}{2}$ the problems (6.2.56) and (6.2.58) are
$$iX_{it} + X_{i\xi\xi} + iu_1(\xi,t)X_{i\xi} = 0,$$
$$iY_{kt} + Y_{k\eta\eta} + iu_2(\eta,t)Y_{k\eta} = 0 \tag{6.2.59}$$
where $u_1(\xi) \equiv V_1'(2\xi), u_2 = V_2'(2\eta)$.

The problems (6.2.59) are just the linear problems for the mKP-I equation.

So, solving the linear problems (6.2.59) for given $u_1$ and $u_2$, one can construct the degenerate kernel
$$\tilde{R}(s,\tilde{s},x,y,t) = \sum_{i,k=1}^N \begin{pmatrix} 0 & \rho_{0ik} X_i(s) Y_{\tilde{s}k}(\tilde{s}) \\ \bar{\rho}_{0ik} \bar{Y}_{is}(s) \bar{X}_k(\tilde{s}) & 0 \end{pmatrix} \tag{6.2.60}$$
of the $\bar{\partial}$-problem.

Since in our case
$$f_k(\mu) = \int_R ds\, e^{-2i\frac{s}{\mu}} X_k(s+x,y,t),$$
$$\frac{1}{\lambda}g_k(\lambda) = \int_R d\tilde{s}\, e^{-2i\frac{\tilde{s}}{\lambda}} Y_k(\tilde{s}+x,y,t) \tag{6.2.61}$$
one has
$$\int d\mu\, f_k(\mu) = X_k(x,y,t),$$
$$\int d\lambda\, \frac{g_k(\lambda)}{\lambda} = Y_k(x,y,t). \tag{6.2.62}$$

Next step is to calculate $\chi(0)$ for such a kernel $R$. The straightforward but cumbersome calculations with the use of formulae (6.2.2) and (6.2.62) give [245]
$$\chi_0 = \begin{pmatrix} 1 - \langle X, (1-\rho_0 a\rho_0^+ b)^{-1}\rho_0 a\rho_0^+ \bar{X}\rangle & \langle \bar{Y}, \rho_0^+(1-b\rho_0 a\rho_0^+)^{-1}\bar{X}\rangle \\ -\langle X, (1-\rho_0 a\rho_0^+ b)^{-1}\rho_0 Y\rangle & 1 + \langle \bar{Y}, \rho_0^+(1-b\rho_0 a\rho_0^+)^{-1}b\rho_0 Y\rangle \end{pmatrix} \tag{6.2.63}$$
where $\langle X,Y\rangle \doteq \sum_i X_i Y_i$ $(\rho^+)_{ij} = \bar{\rho}_{ij}$ and
$$a_{lm} = \int_{-\infty}^\xi d\xi'\, \bar{X}_m(\xi',t) X_{\xi'l}(\xi',t),$$

Chapter 6                                                                                                         185

$$b_{ik} = -\int_{-\infty}^{\eta} d\eta' Y_k(\eta', t)\bar{Y}_{i\eta'}(\eta', t). \tag{6.2.64}$$

Finally, the formulae

$$\vec{S} = \frac{1}{2}\operatorname{tr}(\vec{\sigma}\chi(0)\sigma_3\chi^{-1}(0)),$$

$$\Phi = u_1(\xi, t) + u_2(\eta, t) - i\sigma\ln\det\chi(0) \tag{6.2.65}$$

give us the solution of the Ishimori-I equation on the background $\vec{S} = (0, 0, 1)$ and $\Phi = u_1(\xi, t) + u_2(\eta, t)$. For the first time the formula (6.2.63) has been derived in [245] by another method.

To construct the solutions (6.2.63) explicitly we should use the solvable cases for the linear problems (6.2.59). The proper solutions of (6.2.59) which describe the moving boundaries are given by line solitons, line lumps and line breathers of the mKP-I equation constructed in section (4.2). Dropping the time dependence in these solutions and changing $y \to t$, we obtain the explicit exact solutions of the problem (6.2.59). Since we can choose the solutions of the two problems (6.2.59) independently (line solitons, lumps, breathers), we have 9 classes of exact solutions of the Ishimori equation. The solutions from these classes behave differently as $\xi^2 + \eta^2 \to \infty$ but all of them are localized.

Here we present only three simplest localized solutions of the Ishimori-I equation.

The first example is the lump-lump boundaries solution. The corresponding boundaries are given by

$$u_1(\xi, t) = \frac{\alpha}{(\xi - \frac{t}{\alpha} + C_1)^2 + \frac{\alpha^2}{4}}, \quad u_2(\eta, t) = \frac{\beta}{(\eta - \frac{t}{\beta} + C_2)^2 + \frac{\beta^2}{4}}. \tag{6.2.66}$$

The formula (6.2.63) provides us the rationally localized solution [246]

$$S_1^{rr}(\xi, \eta, t) = S_\perp^{rr}(\hat{\xi}, \hat{\eta})\cos\left[\frac{1}{\alpha}\hat{\xi} + \frac{1}{\beta}\hat{\eta} + \frac{t}{2}\left(\frac{1}{\alpha^2} + \frac{1}{\beta^2}\right) + \Phi_{rr}(\hat{\xi}, \hat{\eta})\right],$$

$$S_2^{rr}(\xi, \eta, t) = S_\perp^{rr}(\hat{\xi}, \hat{\eta})\sin\left[\frac{1}{\alpha}\hat{\xi} + \frac{1}{\beta}\hat{\eta} + \frac{t}{2}\left(\frac{1}{\alpha^2} + \frac{1}{\beta^2}\right) + \Phi_{rr}(\hat{\xi}, \hat{\eta})\right], \tag{6.2.67}$$

$$S_3^{rr}(\xi, \eta, t) = 1 - \frac{2}{\left[(\hat{\xi} + C_1)(\hat{\eta} + C_2) + \frac{\alpha\beta}{4} + \frac{1}{\alpha\beta}\right]^2 + \left[\frac{\alpha}{2}(\hat{\eta} + C_2) - \frac{\beta}{2}(\hat{\xi} + C_1)\right]^2}$$

where

$$\hat{\xi} = \xi - \frac{t}{\alpha}, \quad \hat{\eta} = \eta - \frac{t}{\beta}$$

and

$$S_\perp^{rr}(\hat{\xi}, \hat{\eta}) = -2\frac{\sqrt{\left[(\hat{\xi} + C_1)(\hat{\eta} + C_2) - \frac{\alpha\beta}{4} + \frac{1}{\alpha\beta}\right]^2 + \left[\frac{\alpha}{2}(\hat{\eta} + C_2) + \frac{\beta}{2}(\hat{\xi} + C_1)\right]^2}}{\left[(\hat{\xi} + C_1)(\hat{\eta} + C_2) + \frac{\alpha\beta}{4} + \frac{1}{\alpha\beta}\right]^2 + \left[\frac{\alpha}{2}(\hat{\eta} + C_2) - \frac{\beta}{2}(\hat{\xi} + C_1)\right]^2},$$

$$\Phi_{rr}(\hat{\xi},\hat{\eta}) = \text{arctg}\frac{\frac{\alpha}{2}(\hat{\eta}+C_2)+\frac{\beta}{2}(\hat{x}i+C_1)}{(\hat{\xi}+C_1)(\hat{\eta}+C_2)-\frac{\alpha\beta}{4}+\frac{1}{\alpha\beta}}. \qquad (6.2.68)$$

The scalar function $\Phi^{rr}(\xi,\eta,t)$ is given by

$$\Phi^{rr}(\xi,\eta,t) = 4\text{arctg}\frac{\frac{\alpha}{2}(\hat{\eta}+C_2)-\frac{\beta}{2}(\hat{\xi}+C_1)}{(\hat{\xi}+C_1)(\hat{\eta}+C_2)+\frac{\alpha\beta}{4}+\frac{1}{\alpha\beta}} + 4\text{arctg}\frac{\beta}{2(\hat{\eta}+C_2)}$$

$$-4\text{arctg}\frac{\alpha}{2(\hat{\xi}+C_1)} + 4\text{arctg}\frac{\alpha}{2}(\hat{\xi}+C_1) + 4\text{arctg}\frac{2}{\beta}(\hat{\eta}+C_2) + 2\pi(\text{sgn}\alpha+\text{sgn}\beta) \quad (6.2.69)$$

and the density of the topological charge is

$$\partial_\xi \partial_\eta \ln \det g = -2i\partial_\xi\partial_\eta \left\{ \text{arctg}\frac{\frac{\alpha}{2}(\hat{\eta}+C_2)-\frac{\beta}{2}(\hat{\xi}+C_1)}{(\hat{\xi}+C_1)(\hat{\eta}+C_2)+\frac{\alpha\beta}{4}+\frac{1}{\alpha\beta}} \right.$$

$$\left. + \text{arctg}\frac{\beta}{2(\hat{\eta}+C_2)} - \text{arctg}\frac{\alpha}{2(\hat{\xi}+C_1)} \right\}. \qquad (6.2.70)$$

The soliton (6.2.67) decays as $1/(\xi\eta)$ as $\xi^2+\eta^2 \to \infty$ and moves with the velocity $V = (V_\xi, V_\eta) = (\alpha^{-1}, \beta^{-1})$. Emphasize that the rationally localized soliton (4.1) is the novel phenomena which is absent in the DS-I case.

Next example corresponds to the choice of the boundaries $u_1(\xi,t)$ and $u_2(\eta,t)$ as the plane solitons

$$u_1(\xi,t) = \frac{4\frac{\mu_I}{|\mu|^2}\text{sgn}R_2}{e^{\frac{2\mu_I\xi}{|\mu|^2}} + \left(e^{-\frac{\mu_I\xi}{|\mu|^2}} + \frac{\mu_R}{\mu_I}(\text{sgn}R_2)e^{\frac{\mu_I\xi}{|\mu|^2}}\right)^2},$$

$$u_2(\eta,t) = \frac{4\frac{\lambda_I}{|\lambda|^2}\text{sgn}R_1}{e^{\frac{2\lambda_I\eta}{|\lambda|^2}} + \left(e^{-\frac{\lambda_I\eta}{|\lambda|^2}} + \frac{\lambda_R}{\lambda_I}(\text{sgn}R_1)e^{\frac{\lambda_I\eta}{|\lambda|^2}}\right)^2}$$

where $\hat{\eta} = \eta - \frac{\lambda_R}{|\lambda|^2}t + \eta_0$, $\hat{\xi} = \xi - \frac{\mu_R}{|\mu|^2}t + \xi_0$. The corresponding soliton of the Ishimori-I equation is of the form [246]:

$$S_1^{ss}(\xi,\eta,t) = S_\perp^{ss}(\hat{\xi},\hat{\eta})\cos\left[\frac{\lambda_R}{|\lambda|^2}\hat{\eta} + \frac{\mu_R}{|\mu|^2}\hat{\xi} + \frac{t}{2}\left(\frac{1}{|\lambda|^2}+\frac{1}{|\mu|^2}\right) + \Phi_{ss}(\hat{\xi},\hat{\eta})\right],$$

$$S_2^{ss}(\xi,\eta,t) = S_\perp^{ss}(\hat{\xi},\hat{\eta})\sin\left[\frac{\lambda_R}{|\lambda|^2}\hat{\eta} + \frac{\mu_R}{|\mu|^2}\hat{\xi} + \frac{t}{2}\left(\frac{1}{|\lambda|^2}+\frac{1}{|\mu|^2}\right) + \Phi_{ss}(\hat{\xi},\hat{\eta})\right], \qquad (6.2.71)$$

$$S_3^{ss}(\xi,\eta,t) = 1 - \frac{2\exp\left[\frac{\mu_I}{|\mu|^2}\hat{\xi} + \frac{\lambda_I}{|\lambda|^2}\hat{\eta}\right]}{A^2+B^2}$$

where

$$S_{\perp}^{ss}(\hat{\xi},\hat{\eta}) = -\frac{2\sqrt{C^2+D^2}}{A^2+B^2}\exp\left\{\frac{\lambda_I}{|\lambda|^2}\hat{\eta} + \frac{\mu_I}{|\mu|^2}\hat{\xi}\right\},$$

$$\Phi_{ss}(\hat{\xi},\hat{\eta}) = \arctg(D/C) \qquad (6.2.72)$$

and

$$A = 1 + \frac{1}{4R_1R_2} + \frac{R_1\lambda_R}{\lambda_I}e^{\frac{\lambda_I\hat{\eta}}{|\lambda|^2}} + \frac{R_2\mu_R}{\mu_I}e^{\frac{\mu_I\hat{\xi}}{|\mu|^2}} + \frac{R_1R_2}{\lambda_I\mu_I}(\lambda_R\mu_R + \lambda_I\mu_I)\exp\left\{\frac{\lambda_I\hat{\eta}}{|\lambda|^2} + \frac{\mu_I\hat{\xi}}{|\mu|^2}\right\},$$

$$B = R_2 e^{\frac{\mu_I\hat{\xi}}{|\mu|^2}} - R_1 e^{\frac{\lambda_I\hat{\eta}}{|\lambda|^2}} + R_1R_2\left(\frac{\lambda_R}{\lambda_I} - \frac{\mu_R}{\mu_I}\right)\exp\left\{\frac{\lambda_I\hat{\eta}}{|\lambda|^2} + \frac{\mu_I\hat{\xi}}{|\mu|^2}\right\},$$

$$C = 1 + \frac{1}{4R_1R_2} + \frac{R_1\lambda_R}{\lambda_I}e^{\frac{\lambda_I\hat{\eta}}{|\lambda|^2}} + \frac{R_2\mu_R}{\mu_I}e^{\frac{\mu_I\hat{\xi}}{|\mu|^2}} + \frac{R_1R_2}{\lambda_I\mu_I}(\lambda_R\mu_R - \lambda_I\mu_I)\exp\left\{\frac{\lambda_I\hat{\eta}}{|\lambda|^2} + \frac{\mu_I\hat{\xi}}{|\mu|^2}\right\},$$

$$D = R_1 e^{\frac{\lambda_I\hat{\eta}}{|\lambda|^2}} + R_2 e^{\frac{\mu_I\hat{\xi}}{|\mu|^2}} + R_1R_2\left(\frac{\lambda_R}{\lambda_I} + \frac{\mu_R}{\mu_I}\right)\exp\left\{\frac{\lambda_I\hat{\eta}}{|\lambda|^2} + \frac{\mu_I\hat{\xi}}{|\mu|^2}\right\}.$$

The soliton (6.2.71) decays exponentially in all directions on the plane $\xi,\eta$ similar to the localized soliton of the DS-I equation and moves with the velocity $V = (V_\xi, V_\eta) = \left(\frac{\mu_R}{|\mu|^2}, \frac{\lambda_R}{|\lambda|^2}\right)$.

Our last example here corresponds to the choice of the rational lump as the boundary $u_1(\xi,t)$ and of the plane soliton as the boundary $u_2(\eta,t)$:

$$u_1(\xi,t) = -\frac{\beta}{(\hat{\xi}+C)^2 + \frac{\beta^2}{4}},$$

$$u_2(\eta,t) = -\frac{4\frac{\mu_I}{|\mu|^2}\mathrm{sgn}\,R}{e^{\frac{2\mu_I\hat{\eta}}{|\mu|^2}} + \left(e^{-\frac{\mu_I\hat{\eta}}{|\mu|^2}} + \frac{\mu_R}{\mu_I}(\mathrm{sgn}\,R)e^{\frac{\mu_I\hat{\eta}}{|\mu|^2}}\right)^2}$$

where $\hat{\xi} = \xi - \frac{t}{\beta}$, $\hat{\eta} = \eta - \frac{\mu_R}{|\mu|^2}t + \xi_0$. In this case one has for the soliton of the Ishimori-I equation [246]:

$$S_1^{rs}(\xi,\eta,t) = S_\perp^{rs}(\hat{\xi},\hat{\eta})\cos\left[\frac{\hat{\xi}}{\beta} + \frac{\mu_R}{|\mu|^2}\hat{\eta} + \frac{t}{2}\left(\frac{1}{\beta^2} + \frac{1}{|\mu|^2}\right) + \Phi_{rs}(\hat{\xi},\hat{\eta})\right],$$

$$S_2^{rs}(\xi,\eta,t) = S_\perp^{rs}(\hat{\xi},\hat{\eta})\sin\left[\frac{\hat{\xi}}{\beta} + \frac{\mu_R}{|\mu|^2}\hat{\eta} + \frac{t}{2}\left(\frac{1}{\beta^2} + \frac{1}{|\mu|^2}\right) + \Phi_{rs}(\hat{\xi},\hat{\eta})\right], \qquad (6.2.73)$$

$$S_3^{rs}(\xi,\eta,t) = 1 - \frac{2\exp\left\{\frac{\mu_I}{|\mu|^2}\hat{\eta}\right\}}{A^2+B^2}$$

where
$$S^{rs}(\hat{\xi},\hat{\eta}) = -2\frac{\sqrt{C^2+D^2}}{A^2+B^2}\exp\left\{\frac{\mu_I}{|\mu|^2}\hat{\eta}\right\},$$

$$\Phi_{rs}(\hat{\xi},\hat{\eta}) = \operatorname{arctg}\frac{D}{C}$$

and
$$A = (\hat{\xi}+C)\left(1+\frac{R\mu_R}{\mu_I}e^{\frac{2\mu_I\hat{\eta}}{|\mu|^2}}\right)+\frac{\beta R}{2}e^{\frac{2\mu_I\hat{\eta}}{|\mu|^2}}+\frac{1}{2R\beta},$$

$$B = (\hat{\xi}+C)Re^{\frac{2\mu_I\hat{\eta}}{|\mu|^2}}-\frac{\beta}{2}\left(1+\frac{R}{\mu_I}\mu_R e^{\frac{2\mu_I\hat{\eta}}{|\mu|^2}}\right),$$

$$C = (\hat{\xi}+C)\left(1+\frac{R\mu_R}{\mu_I}e^{\frac{2\mu_I\hat{\eta}}{|\mu|^2}}\right)-\frac{\beta R}{2}e^{\frac{2\mu_I\hat{\eta}}{|\mu|^2}}+\frac{1}{2R\beta},$$

$$D = (\hat{\xi}+C)R_1 e^{\frac{2\mu_I\hat{\eta}}{|\mu|^2}}+\frac{\beta}{2}\left(1+\frac{R\mu_R}{\mu_I}e^{\frac{2\mu_I\hat{\eta}}{|\mu|^2}}\right),$$

The formula (6.2.73) present the mixed rational-exponential decreasing soliton of the Ishimori equation.

We see that the family of the localized solitons for the Ishimori equation is much richer than the DS-I equation. In addition to the exact solutions mentioned above there are other classes of exact solutions which correspond to the multiple-poles line lumps and line solitons and so on.

The exact explicit solutions of the Ishimori equation constructed in this section provide, of course, the exact solutions of equation (6.0.4). Equation (6.0.4) has an interesting symmetry. Namely, it is invariant under the inversions

$$q \to \tilde{q} = \pm\frac{1}{q}. \qquad (6.2.74)$$

This invariance is connected with the invariance of the Ishimori equation (6.0.1) under the inversion of sign of any pair of components $S_1, S_2, S_3$, for instance $(S_1, S_2, S_3) \to (-S_1, -S_2, S_3)$.

The invariance (6.2.74) is the useful tool for the construction of the localized solutions of equation (6.0.4).

Here we will present several exact solutions of equation (6.0.4). First of them corresponds to the vortex solutions (6.2.46), (6.2.44) of the Ishimori-II equation. The corresponding solutions of equation (6.0.4) are given by

$$q = \frac{\bar{g}}{f} \qquad (6.2.75)$$

*Chapter 6* 189

where the functions $g$ and $f$ obey equations (6.2.44). The counterpart of the vortex (6.2.49) is

$$q = \frac{\bar{b}_0}{\bar{a}_0 e^{2\bar{a}_1 \bar{z}}}.$$

The localized counterparts of the localized solutions (6.2.67), (6.2.71) and (6.2.73) of the Ishimori-I equation are constructed with the use of the inversion (6.2.74). They are of the form [246]:

1. The rationally localized soliton of (6.0.4) is

$$\tilde{q}(\xi, \eta, t) = \frac{\exp\left\{(-i)\left[\frac{\hat{\xi}}{\alpha} + \frac{\hat{\eta}}{\beta} + \frac{t}{2}\left(\frac{1}{\alpha^2} + \frac{1}{\beta^2}\right)\right]\right\}}{(\hat{\eta} + C_2 + \frac{i\beta}{2})(\hat{\xi} + C_1 + \frac{i\alpha}{2}) + \frac{1}{\alpha\beta}}. \tag{6.2.76}$$

2. The exponentially localized soliton of (6.0.4) is

$$\tilde{q}(\xi, \eta, t) = \frac{\exp\left\{\frac{\lambda_I \hat{\eta}}{|\lambda|^2} + \frac{\mu_I \hat{\xi}}{|\mu|^2}\right\}}{\left(1 + \frac{R_1 \lambda}{\lambda_I} e^{\frac{2\lambda_I \hat{\eta}}{|\lambda|^2}}\right)\left(1 + \frac{R_2 \mu}{\mu_I} e^{\frac{2\mu_I \hat{\xi}}{|\mu|^2}}\right) + \frac{1}{4R_1 R_2}}$$

$$\times \exp\left\{(-i)\left[\frac{\mu_R}{|\mu|^2}\hat{\xi} + \frac{\lambda_R}{|\lambda|^2}\hat{\eta} + \frac{t}{2}\left(\frac{1}{|\lambda|^2} + \frac{1}{|\mu|^2}\right)\right]\right\}. \tag{6.2.77}$$

3. The rationally-exponentially localized soliton of (6.0.4) is

$$\tilde{q}(\xi, \eta, t) = \frac{\exp\left(\frac{\mu_I \hat{\eta}}{|\mu|^2}\right)}{\left(\hat{\xi} + C + \frac{i\beta}{2}\right)\left(1 + \frac{R\mu}{\mu_I} e^{\frac{\mu_I \hat{\eta}}{|\mu|^2}}\right) + \frac{1}{2R\beta}}$$

$$\times \exp\left\{(-i)\left[\frac{\hat{\xi}}{\beta} + \frac{\mu_R}{|\mu|^2}\hat{\eta} + \frac{t}{2}\left(\frac{1}{|\mu|^2} + \frac{1}{\beta^2}\right)\right]\right\}. \tag{6.2.78}$$

## 6.3. Initial value problem for the Ishimori-I equation

In order to linearize the initial value problem for the Ishimori equation we will study the forward and inverse spectral problems for the linear equations (6.1.26). We assume that $\vec{S}$ tends sufficiently rapidly to $(0,0,1)$ as $x^2 + y^2 \to \infty$ and $\Phi \to 0$ as $x^2 + y^2 \to \infty$.

To formulate the spectral problem we consider the solutions of the problems (6.1.26) of the type,

$$\Psi = \mu(x, y, t, \lambda) e^{\frac{i}{\lambda}x - \frac{i\sigma_3 y}{\sigma\lambda}} \tag{6.3.1}$$

where $\lambda$ is an arbitrary complex parameter and $\mu(\lambda) \to 1$ as $\lambda \to \infty$. The function $\mu$ obeys the equations

$$\sigma\mu_y + \sigma\mu_x + i\lambda^{-1}[\sigma_3, \mu] + Q(\partial_x + i\lambda^{-1})\mu = 0, \tag{6.3.2}$$

$$\mu_t + i\sigma\mu_{xy} + \frac{1}{2}(\Phi_y - iP_x)\mu_x + \frac{1}{2}(\Phi_x + i\sigma^2 P_y)\mu_y$$

$$+ \frac{1}{\lambda}\left\{\mu_x\sigma_3 - \mu_y + \frac{i}{2}(\Phi_y - iP_x)\mu - \frac{i}{2}(\Phi_x + i\sigma^2 P_y)\mu\sigma_3\right\} + \frac{i}{\lambda^2}[\mu,\sigma_3] = 0, \quad (6.3.3)$$

where $Q = P - \sigma_3$.

In this section we will consider the Ishimori-I equation. For the first time this initial value problem has been discussed in [247] (see also [245]).

We start with the introduction of the characteristic variables

$$\xi = \frac{1}{2}(y-x), \quad \eta = \frac{1}{2}(y+x).$$

In this variables the problem (6.3.2) is

$$\begin{pmatrix} \partial_\eta & 0 \\ 0 & \partial_\xi \end{pmatrix}\mu + \frac{i}{\lambda}[\sigma_3,\mu] + Q\left(\frac{1}{2}(\partial_\eta - \partial_\xi) + \frac{i}{\lambda}\right)\mu = 0. \quad (6.3.4)$$

The unperturbed operator

$$L_0 = \begin{pmatrix} \partial_\eta & 0 \\ 0 & \partial_\xi \end{pmatrix} + \frac{i}{\lambda}[\sigma_3,\cdot] \quad (6.3.5)$$

for equation (6.3.4) is the same (up to the change $\lambda \to \frac{1}{\lambda}$) as for the DS-I equation. So one can use the results of section (5.3). The operator $\hat{G}$ formally inverse to the operator $\hat{L}_0$ (6.3.5) acts as follows

$$(\hat{G}\Phi)(\xi,\eta) \doteq \begin{pmatrix} \partial_\eta^{-1}(\Phi_{11}(\xi,\eta')) & \partial_\eta^{-1}(e^{-\frac{2i(\eta-\eta')}{\lambda}}\Phi_{12}(\xi,\eta')) \\ \partial_\xi^{-1}(e^{\frac{2i(\xi-\xi')}{\lambda}}\Phi_{21}(\xi',\eta)) & \partial_\xi^{-1}(\Phi_{22}(\xi',\eta)) \end{pmatrix} \quad (6.3.6)$$

where $\Phi = \begin{pmatrix} \Phi_{11} & \Phi_{12} \\ \Phi_{21} & \Phi_{22} \end{pmatrix}$ is $2 \times 2$ matrix. The kernel $Q(\xi-\xi',\eta-\eta')$ of the operator $\hat{G}$ is the Green function for the problem (6.3.4). Since the integrals $\partial_\xi^{-1}$ and $\partial_\eta^{-1}$ can be chosen as $\partial_\xi^{-1} = \int_{\pm\infty}^\xi d\xi'$ and $\partial_\eta^{-1} = \int_{\pm\infty}^\eta d\eta'$, the Green function $G$ is defined nonuniquely. This freedom allows us to construct the bounded analytic Green functions. Here we will use the following Green functions:

$$(G^+(\cdot,\lambda)\Phi)(\xi,\eta) \doteq \begin{pmatrix} \int_{-\infty}^\eta d\eta' \Phi_{11}(\xi,\eta') & \int_{-\infty}^\eta d\eta' e^{-\frac{2i(\eta-\eta')}{\lambda}}\Phi_{12}(\xi,\eta') \\ \int_{+\infty}^\xi d\xi' e^{\frac{2i(\xi-\xi')}{\lambda}}\Phi_{21}(\xi',\eta) & \int_{-\infty}^\xi d\xi' \Phi_{22}(\xi',\eta) \end{pmatrix},$$

$$(G^-(\cdot,\lambda)\Phi)(\xi,\eta) \doteq \begin{pmatrix} \int_{-\infty}^\eta d\eta' \Phi_{11}(\xi,\eta') & \int_{+\infty}^\eta d\eta' e^{-\frac{2i(\eta-\eta')}{\lambda}}\Phi_{12}(\xi,\eta') \\ \int_{-\infty}^\xi d\xi' e^{\frac{2i(\xi-\xi')}{\lambda}}\Phi_{21}(\xi',\eta) & \int_{-\infty}^\xi d\xi' \Phi_{22}(\xi',\eta) \end{pmatrix}. \quad (6.3.7)$$

It is easy to see that the Green function $G^+(\lambda)$ is analytic and bounded at the upper half-plane $\text{Im}\lambda > 0$ while the Green function $G^-(\lambda)$ is analytic and bounded at the lower half-plane $\text{Im}\lambda < 0$. Note that the Green functions (6.3.7) differ from the Green functions used in section (5.3) by the rings of the lower limits of integration in the $\Phi_{11}$ and $\Phi_{22}$ respectively.

Using the Green functions $G^+$ and $G^-$ we define the solutions $\mu^+$ and $\mu^-$ of equation (6.3.4) via the integral equations

$$\mu^\pm(\xi,\eta,t) = 1 - \left\{ G^\pm(\cdot,\lambda)\frac{1}{2}Q\left(\partial' - \tilde{\partial}' + \frac{2i}{\lambda}\right)\mu^\pm(\cdot,\lambda)\right\}(\xi,\eta) \qquad (6.3.8)$$

where $\partial' = \partial/\partial\xi'$, $\tilde{\partial}' = \partial/\partial\eta'$. As far as the Green functions $G^+, G^-$, the solutions $\mu^+, \mu^-$ of the integral equation (6.3.8) are analytic and bounded in the upper and lower half-planes $\text{Im}\lambda > 0$ and $\text{Im}\lambda < 0$ respectively. Then since $G^+ - G^- \neq 0$ at $\text{Im}\lambda = 0$ one has also $\mu^+ - \mu^- \neq 0$ at $\text{Im}\lambda = 0$.

Thus, one can define the function

$$\mu = \begin{cases} \mu^+, & \text{Im}\lambda > 0, \\ \mu^-, & \text{Im}\lambda < 0 \end{cases}$$

which is analytic and bounded at whole complex plane and has a jump across the real axis. So we arrive to the standard Riemann—Hilbert problem. We assume that the homogeneous integral equations (6.3.8) have no nontrivial solutions.

To specify the Riemann—Hilbert problem one must express, according to the standard procedure, the jump $\mu^+ - \mu^-$ at $\text{Im}\lambda = 0$ via $\mu^-$. To do this we firstly obtain the integral equation for $\mu^+ - \mu^-$. Using (6.3.8), one gets

$$(\mu^+ - \mu^-)(\xi,\eta,\lambda) = \Gamma(\xi,\eta,\lambda) - \left(\tilde{\tilde{G}}(\cdot,\lambda)\frac{1}{2}Q\left(\partial' - \tilde{\partial}' + \frac{2i}{\lambda}\right)(\mu^+ - \mu^-)\right)(\xi,\eta) \qquad (6.3.9)$$

where

$$\Gamma(\xi,\eta,\lambda) = \begin{pmatrix} 0 & \int_{-\infty}^{+\infty} d\eta' e^{-\frac{2i(\eta-\eta')}{\lambda}} \frac{1}{2}(Q(\xi,\eta')(\partial_\xi - \partial_{\eta'} + \frac{2i}{\lambda})\mu^+)_{12} \\ -\int_{-\infty}^{+\infty} d\xi' e^{\frac{2i(\xi-\xi')}{\lambda}} \frac{1}{2}(Q(\xi',\eta)(\partial_{\xi'} - \partial_\eta + \frac{2i}{\lambda})\mu^-(\xi',\eta))_{21} & 0 \end{pmatrix} \qquad (6.3.10)$$

and

$$(\tilde{G}(\cdot,\lambda)\Phi)(\xi,\eta) = \begin{pmatrix} \int_{-\infty}^\eta d\eta' \Phi_{11}(\xi,\eta') & \int_{+\infty}^\eta d\eta' e^{-\frac{2i(\eta-\eta')}{\lambda}}\Phi_{12}(\xi,\eta') \\ \int_{+\infty}^\xi d\xi' e^{\frac{2i(\xi-\xi')}{\lambda}}\Phi_{21}(\xi',\eta) & \int_{-\infty}^\xi d\xi' \Phi_{22}(\xi',\eta) \end{pmatrix}. \qquad (6.3.11)$$

Note that the diagonal elements of the quantity $\Gamma$ (6.2.10) are equal to zero. This is due to our choice of the Green functions (6.3.7).

We will look for the expression for $\mu^+ - \mu^-$ via $\mu^-$ in the form

$$(\mu^+ - \mu^-)(\xi,\eta,\lambda) = \int_\infty^{+\infty} \frac{dl}{l^2}\mu^-(l)\Sigma_l(\xi,\eta)f(l,\lambda)\Sigma_\lambda^{-1}(\xi,\eta) \qquad (6.3.12)$$

$$\Sigma_\lambda(\xi,\eta) \doteq \begin{pmatrix} e^{-\frac{2i\xi}{\lambda}} & 0 \\ 0 & e^{\frac{2i\eta}{\lambda}} \end{pmatrix} \tag{6.3.13}$$

and $f(l,\lambda)$ is the $2 \times 2$ matrix which must be found.

Substituting (6.3.12) into the r. h. s. of (6.3.9), we obtain

$$\mu^+ - \mu^- = \Gamma - \int_{-\infty}^{+\infty} \frac{dl}{l^2} \begin{pmatrix} \int_{+\infty}^{\eta} d\eta' \frac{1}{2}(Q(\partial' - \tilde{\partial}' + \frac{2i}{\lambda})\mu^-(l) \Sigma_l f(l,\lambda) \Sigma_\lambda^{-1})_{11} \\ \int_{+\infty}^{\xi} d\xi' e^{-\frac{2i(\xi-\xi')}{\lambda}} \frac{1}{2}(Q(\partial' - \tilde{\partial}' + \frac{2i}{\lambda})\mu^-(l) \Sigma_l f(l,\lambda) \Sigma_\lambda^{-1})_{21} \end{pmatrix}$$

$$\begin{pmatrix} \int_{+\infty}^{\eta} d\eta' e^{-\frac{2i(\eta-\eta')}{\lambda}} \frac{1}{2}(Q(\partial' - \tilde{\partial}' + \frac{2i}{\lambda})\mu^-(l) \Sigma_l f(l,\lambda) \Sigma_\lambda^{-1})_{12} \\ \int_{-\infty}^{\xi} d\xi' \frac{1}{2}(Q(\partial' - \tilde{\partial}' + \frac{2i}{\lambda})\mu^-(l) \Sigma_l f(l,\lambda) \Sigma_\lambda^{-1})_{22} \end{pmatrix}. \tag{6.3.14}$$

On the other hand from (6.3.8) one has

$$\int_{-\infty}^{+\infty} \frac{dl}{l^2} \mu^-(l) \Sigma_l f(l,\lambda) \Sigma_\lambda^{-1} = \mu^+ - \mu^- = \int_{-\infty}^{+\infty} \frac{dl}{l^2} \Sigma_l f(l,\lambda) \Sigma_\lambda^{-1}$$

$$- \int_{-\infty}^{+\infty} \frac{dl}{l^2} \begin{pmatrix} \int_{-\infty}^{\eta} d\eta' \frac{1}{2}(Q(\partial' - \tilde{\partial}' + \frac{2i}{l})\mu^-(l))_{11} \\ \int_{-\infty}^{\xi} d\xi' e^{\frac{2i(\xi-\xi')}{l}} \frac{1}{2}(Q(\partial' - \tilde{\partial}' + \frac{2i}{l})\mu^-(l))_{21} \end{pmatrix}$$

$$\begin{pmatrix} \int_{+\infty}^{\eta} d\eta' e^{-\frac{2i(\eta-\eta')}{l}} \frac{1}{2}(Q(\partial' - \tilde{\partial}' + \frac{2i}{l})\mu^-(l))_{12} \\ \int_{-\infty}^{\xi} d\xi' \frac{1}{2}(Q(\partial' - \tilde{\partial}' + \frac{2i}{l})\mu^-(l))_{22} \end{pmatrix} \Sigma_l f(l,\lambda) \Sigma_\lambda^{-1}. \tag{6.3.15}$$

Transforming (6.3.14) with the use the identity

$$\left( \left( \partial' - \tilde{\partial}' + \frac{2i}{l} \right) \mu^-(l) \Sigma_l f(l,\lambda) \Sigma_\lambda^{-1} \right)(\xi,\eta)$$

$$= \left( \left( \partial' - \tilde{\partial}' + \frac{2i}{l} \right) \mu^-(l) \right)(\xi,\eta) \Sigma_l(\xi,\eta) f(l,\lambda) \Sigma_\lambda^{-1}(\xi,\eta), \tag{6.3.16}$$

then performing the matrix multiplication in (6.3.15) and subtracting the obtained equations, one gets

$$\int_{-\infty}^{+\infty} \frac{dl}{l^2} \Sigma_l f(l,\lambda) \Sigma_\lambda^{-1}$$

$$= \begin{pmatrix} 0 & \int_{-\infty}^{+\infty} d\eta' e^{\frac{2i\eta'}{\lambda}} \frac{1}{2}(Q(\partial' - \tilde{\partial}' + \frac{2i}{\lambda})\mu^+)_{12} \\ -\int_{-\infty}^{+\infty} d\xi' e^{-\frac{2i\xi'}{\lambda}} \frac{1}{2}(Q(\partial' - \tilde{\partial}' + \frac{2i}{\lambda})\mu^-)_{21} & 0 \end{pmatrix} \Sigma_\lambda^{-1}(\xi,\eta)$$

$$- \int_{-\infty}^{+\infty} \frac{dl}{l^2} \begin{pmatrix} 0 & 0 \\ \int_{-\infty}^{+\infty} d\xi' e^{-\frac{2i\xi'}{l}} \frac{1}{2}(Q(\partial' - \tilde{\partial}' + \frac{2i}{l})\mu^-)_{21} & 0 \end{pmatrix} \begin{pmatrix} f_{11} & f_{12} \\ f_{21} & f_{22} \end{pmatrix} \Sigma_\lambda^{-1}(\xi,\eta). \tag{6.3.17}$$

Further, multiplying (6.3.17) by $\Sigma_\lambda$ from the right and acting on (6.3.17) by
$\frac{1}{2\pi}\begin{pmatrix} \int_{-\infty}^{+\infty} d\xi\, e^{\frac{i\xi}{p}} & 0 \\ 0 & \int_{-\infty}^{+\infty} d\eta\, e^{-\frac{i\eta}{p}} \end{pmatrix}$ from the left, we obtain

$$\begin{pmatrix} f_{11}(p,\lambda) & f_{12}(p,\lambda) \\ f_{21}(p,\lambda) & f_{22}(p,\lambda) \end{pmatrix} = \begin{pmatrix} 0 & T_{12}^+(p,\lambda) \\ -T_{21}^-(p,\lambda) & 0 \end{pmatrix}$$

$$- \int_{-\infty}^{+\infty} \frac{dl}{l^2} \begin{pmatrix} 0 & 0 \\ T_{21}^-(p,l) & 0 \end{pmatrix} \begin{pmatrix} f_{11}(l,\lambda) & f_{12}(l,\lambda) \\ f_{21}(l,\lambda) & f_{22}(l,\lambda) \end{pmatrix} \quad (6.3.18)$$

where

$$T_{12}^+(p,\lambda) \doteq \frac{1}{2\pi} \iint_{-\infty}^{+\infty} d\xi d\eta\, e^{\frac{2i\xi}{p}+\frac{2i\eta}{\lambda}} \frac{1}{2}\left(Q\left(\partial' - \tilde\partial' + \frac{2i}{\lambda}\right)\mu^+\right)_{12},$$

$$T_{21}^-(p,\lambda) \doteq \frac{1}{2\pi} \iint_{-\infty}^{+\infty} d\xi dp\, e^{-\frac{2i\eta}{p}-\frac{2i\xi}{\lambda}} \frac{1}{2}\left(Q\left(\partial' - \tilde\partial' + \frac{2i}{\lambda}\right)\mu^-\right)_{21}. \quad (6.3.19)$$

In the matrix form equation (6.3.18) looks like

$$f(p,\lambda) + \int_{-\infty}^{+\infty} \frac{dl}{l^2} T^-(p,l) f(l,\lambda) = T^+(p,\lambda) - T^-(p,\lambda) \quad (6.3.20)$$

where

$$T^+ = \begin{pmatrix} 0 & T_{12}^+ \\ 0 & 0 \end{pmatrix}, \quad T^- = \begin{pmatrix} 0 & 0 \\ T_{21}^- & 0 \end{pmatrix}. \quad (6.3.21)$$

Thus, we have proved that the jump $\mu^+ - \mu^-$ is indeed given by (6.3.12) with the matrix $f(l,\lambda)$ which obeys equation (6.3.20).

Note that $f$ is easily expressed via $T_{12}^+$ and $T_{21}^-$

$$f(p,\lambda) = \begin{pmatrix} 0 & T_{12}^+(p,\lambda) \\ -T_{21}^-(p,\lambda) & -\int_{-\infty}^{+\infty} \frac{dl}{l^2} T_{21}^-(p,l) T_{12}^+(l,\lambda) \end{pmatrix}. \quad (6.3.22)$$

So, we have arrived at the standard regular nonlocal Riemann—Hilbert problem. Its solution is given by the standard linear singular equation

$$\mu^-(\lambda) = 1 + \frac{1}{2\pi i} \iint_{-\infty}^{+\infty} \frac{dp}{p^2} \frac{dk}{k^2} \frac{\mu^-(p) \Sigma_p f(p,k) \Sigma_k^{-1}}{k - \lambda + i0}. \quad (6.3.23)$$

Equation (6.3.23) is the inverse problem equation for the linear problem (6.3.4). The functions $T_{12}^+(\lambda,\mu)$ and $T_{21}^-(\lambda,\mu)$ are the inverse problem data. The reconstruction formulae are

$$P = \mu_0 \sigma_3 \mu_0^{-1}, \quad (6.3.24)$$

$$\Phi = \arg\det \mu_0$$

where

$$\mu_0 = \mu(\xi,\eta,t,\lambda=0) = 1 + \frac{1}{2\pi i}\iint_{-\infty}^{+\infty}\frac{dldk}{l^2k^2}\mu^-(\xi,\eta,l)\Sigma_l(\xi,\eta)f(l,k)\Sigma_k^{-1}(\xi,\eta). \quad (6.3.25)$$

For the real valued $\vec{S}$, by virtue of (6.2.10) the data $T_{12}^+$ and $T_{21}^-$ obey the condition

$$\overline{\frac{1}{\lambda}T_{21}^-(\mu,\lambda)} = \frac{1}{\mu}T_{12}^+(\lambda,\mu). \quad (6.3.26)$$

Now to apply these results for the initial value problem of the Ishimori-I equation one should find the time evolution of the inverse problem data. Using equation (6.3.3), in a standard manner one finds

$$T_{12}^+(\lambda,\mu,t) = T_{12}^+(\lambda,\mu,0)e^{i\left(\frac{1}{\lambda^2}+\frac{1}{\mu^2}\right)t},$$
$$T_{21}^-(\lambda,\mu,t) = T_{21}^-(\lambda,\mu,0)e^{-i\left(\frac{1}{\lambda^2}+\frac{1}{\mu^2}\right)t}. \quad (6.3.27)$$

Thus, the formulae (6.3.27) and (6.3.23), (6.3.24) allow us to linearize the initial value problem for the Ishimori-I equation via the standard IST scheme.

Similar to the DS-I equation one can also linearize the special initial-boundary value problem for the Ishimori-I equation [245, 246]. Indeed, equation (6.0.1) defines the function $\Phi$ up to the two arbitrary functions $V_1(\xi)$ and $V_2(\eta)$, i. e.

$$\Phi_\xi = \int_{-\infty}^{\eta} d\eta' \vec{S}(\vec{S}_\xi \times \vec{S}_{\eta'}) + u_1(\xi,t),$$
$$\Phi_\eta = \int_{-\infty}^{\xi} d\xi' \vec{S}(\vec{S}_{\xi'} \times \vec{S}_\eta) + u_2(\eta,t) \quad (6.3.28)$$

where $u_1 = V_1', u_2 = V_2'$. So the functions $u_1$ and $u_2$ are the boundary values

$$u_1(\xi,t) = \lim_{\eta\to-\infty}\Phi_\xi,$$
$$u_2(\eta,t) = \lim_{\xi\to-\infty}\Phi_\eta. \quad (6.3.29)$$

The initial-boundary value problem is to calculate $\vec{S}(x,y,t)$ for given $\vec{S}(x,y,0)$, $u_1(y-x,t)$, $u_2(y+x,t)$. The first linear problem (6.3.2) allows us to introduce the inverse problem data $T(\lambda,\mu,t)$. To find the time evolution of $T(\lambda,\mu,t)$ one should now use the modified second equation (6.1.77). One can prove that the Fourier transform $\hat{S}(\xi,\eta,t)$ of the data $T(\lambda,\mu)$ obeys the linear equation [245]

$$i\hat{S}_t + (\hat{S}_{\xi\xi} + \hat{S}_{\eta\eta}) + u_1(\xi,t)\hat{S}_\xi + u_2(\eta,t)\hat{S}_\eta = 0. \quad (6.3.30)$$

The separation of variables in equation (6.3.30) leads to the linear equation which coincide with the first linear problem for the mKP-I equation. Solving these last one can solve, in principle, the initial-boundary value problem mentioned above for the Ishimori-I equation. This problem has been considered in [246].

## 6.4. Initial value problem for the Ishimori-II equation

The initial value problem for the Ishimori-II equation has been studied in [248, 247, 249]. We will follow mainly to these papers.

For the Ishimori-II equation it is convenient to introduce the complex coordinates $z = \frac{1}{2}(y - ix)$ and $\bar{z} = \frac{1}{2}(y + ix)$. In this variables the linear problem (6.3.2) looks like

$$\begin{pmatrix} \partial_{\bar{z}} & 0 \\ 0 & \partial_z \end{pmatrix} \mu + \frac{1}{\lambda}[\sigma_3, \mu] + \frac{1}{2i}Q\left(\partial_z - \partial_{\bar{z}} + \frac{2i}{\lambda}\right)\mu = 0. \tag{6.4.1}$$

For the real $\vec{S}$ equation (6.4.1) admits the involution

$$\overline{\mu(\lambda, \bar{\lambda})} = \sigma_2 \mu(-\bar{\lambda}, -\lambda)\sigma_2^{-1}. \tag{6.4.2}$$

Again the unperturbed operator

$$\hat{L}_0 = \begin{pmatrix} \partial_{\bar{z}} & 0 \\ 0 & \partial_z \end{pmatrix} + \frac{1}{\lambda}[\sigma_3, \cdot] \tag{6.4.3}$$

is, in essence, the same as for the DS-II equation. So we can use the results of section (5.5).

So, we are looking for the bounded solutions of the problem (6.4.1). Such solutions are given by the bounded solutions of the linear integral equation

$$\mu(z, \bar{z}, \lambda) = 1 - \frac{1}{2}\left(G(\cdot, \cdot, \lambda)Q\left(\partial - \bar{\partial} + \frac{2i}{\lambda}\right)\mu(\cdot, \cdot, \lambda)\right)(z, \bar{z}) \tag{6.4.4}$$

where $G(z, \bar{z}; z', \bar{z}')$ is the kernel of the operator $\hat{L}_0^{-1}$. This operator is $2 \times 2$ operator which acts as follows

$$(\hat{L}_0^{-1}\Phi)(z, \bar{z})$$

$$= \frac{1}{2\pi i} \iint_C dz' \wedge d\bar{z}' \begin{pmatrix} \frac{\Phi_{11}(z',\bar{z}')}{z'-z} & \frac{\Phi_{12}(z',\bar{z}')}{z'-z}e^{-\frac{2}{\lambda}(\bar{z}-\bar{z}')+\frac{2}{\lambda}(z-z')} \\ \frac{\Phi_{21}(z',\bar{z}')}{\bar{z}'-\bar{z}}e^{-\frac{2}{\lambda}(z-z')+\frac{2}{\lambda}(\bar{z}-\bar{z}')} & \frac{\Phi_{22}(z',\bar{z}')}{(\bar{z}'-\bar{z})} \end{pmatrix} \tag{6.4.5}$$

where $\Phi(z, \bar{z})$ is an arbitrary $2 \times 2$ matrix.

It is easy to see that the Green function $G(z, \bar{z}; \lambda, \bar{\lambda})$ is analytic nowhere in $\lambda$. As a result, the solution $\mu(z, \bar{z}; \lambda, \bar{\lambda})$ of equation (6.4.4) is analytic nowhere too.

Following the $\bar{\partial}$-method one must find the corresponding $\bar{\partial}$-equation for $\mu$. Differentiating equation (6.4.4) over $\bar{\lambda}$ and taking into account the explicit form of the Green function (6.4.5), we obtain

$$\frac{\partial \mu(z, \bar{z}; \lambda, \bar{\lambda})}{\partial \bar{\lambda}} = \begin{pmatrix} 0 & F_1(\lambda, \bar{\lambda})e^{-\frac{2}{\lambda}\bar{z}+\frac{2}{\lambda}z} \\ F_2(\lambda, \bar{\lambda})e^{-\frac{2}{\lambda}z+2\frac{\bar{z}}{\lambda}} & 0 \end{pmatrix}$$

$$-\frac{1}{2}\left(GQ\left(\partial - \bar{\partial} + \frac{2i}{\lambda}\right)\frac{\partial \mu(\cdot,\cdot)}{\partial \bar{\lambda}}\right)(z,\bar{z}) \tag{6.4.6}$$

where

$$F_1(\lambda,\bar{\lambda}) = \frac{1}{4\pi\lambda^2}\int dz \wedge d\bar{z}\, e^{\frac{2}{\lambda}\bar{z}-\frac{2}{\lambda}z}\sum_{k=1}^{2} Q_{1k}(z,\bar{z})\left(\bar{\partial}-\partial-\frac{2i}{\lambda}\right)\mu_{k2}(z,\bar{z};\lambda,\bar{\lambda}),$$

$$F_2(\lambda,\bar{\lambda}) = -\frac{1}{4\pi\lambda^2}\int dz \wedge d\bar{z}\, e^{-\frac{2}{\lambda}z+\frac{2}{\lambda}\bar{z}}\sum_{k=1}^{2} Q_{2k}(z,\bar{z})\left(\bar{\partial}-\partial-\frac{2i}{\lambda}\right)\mu_{k1}(z,\bar{z};\lambda,\bar{\lambda}). \tag{6.4.7}$$

The terms proportional to $\delta(\lambda)$ which could appear on the r.h.s. of (6.4.6) are equal to zero due to the special matrix structure of $Q$ and the vanishing at $\lambda = 0$ of the integrals which contain the highly oscillating exponents similar to (6.4.7).

Then we introduce another solution $N(z,\bar{z};\lambda,\bar{\lambda})$ of equation (6.4.1) which also obeys the integral equation

$$N(z,\bar{z};\lambda,\bar{\lambda}) = \Sigma_\lambda(z,\bar{z}) - \frac{1}{2}\left(GQ\left(\partial-\bar{\partial}+\frac{2i}{\lambda}\right)N(\cdot,\cdot;\lambda,\bar{\lambda})\right)(z,\bar{z}) \tag{6.4.8}$$

where

$$\Sigma_\lambda(z,\bar{z}) = \begin{pmatrix} 0 & e^{-\frac{2}{\lambda}\bar{z}+\frac{2}{\lambda}z} \\ e^{\frac{2}{\lambda}z-\frac{2}{\lambda}\bar{z}} & 0 \end{pmatrix}. \tag{6.4.9}$$

Comparing equations (6.4.6) and (6.4.8) and assuming that the homogeneous equation (6.4.8) has no nontrivial solutions, one obtains

$$\frac{\partial \mu(z,\bar{z};\lambda,\bar{\lambda})}{\partial \bar{\lambda}} = N(z,\bar{z};\lambda,\bar{\lambda})\begin{pmatrix} F_2 & 0 \\ 0 & F_1 \end{pmatrix}. \tag{6.4.10}$$

Now it is necessary to establish the relation between functions $N$ and $\mu$. Using the integral equations (6.4.4) and (6.4.8) and the identity

$$G(z,z';\lambda,\bar{\lambda})Q(z',\bar{z}')\left(\partial'-\bar{\partial}'+\frac{2i}{\lambda}\right)\Phi(z',\bar{z}')\Sigma_\lambda(z',\bar{z}')$$

$$= G(z,z';-\bar{\lambda},-\lambda)Q(z',\bar{z}')\left(\partial'-\bar{\partial}'-\frac{2i}{\lambda}\right)\Phi(z',\bar{z}')\Sigma_\lambda(z,\bar{z}), \tag{6.4.11}$$

one finds

$$N(z,\bar{z};\lambda,\bar{\lambda}) = \mu(z,\bar{z};-\bar{\lambda},-\lambda)\Sigma_\lambda(z,\bar{z}). \tag{6.4.12}$$

Substituting the expression (6.4.12) into (6.4.10), we finally arrive at the linear $\bar{\partial}$-equation

$$\frac{\partial \mu(z,\bar{z};\lambda,\bar{\lambda})}{\partial \bar{\lambda}} = \mu(z,\bar{z};-\bar{\lambda},-\lambda)F(\lambda,\bar{\lambda};z,\bar{z}) \tag{6.4.13}$$

Chapter 6

where
$$F(\lambda, \bar{\lambda}; z, \bar{z}) = \begin{pmatrix} 0 & F_1(\lambda, \bar{\lambda})e^{-\frac{2}{\lambda}\bar{z}+\frac{2}{\lambda}z} \\ F_2(\lambda, \bar{\lambda})e^{\frac{2}{\lambda}z-\frac{2}{\lambda}\bar{z}} & 0 \end{pmatrix} \qquad (6.4.14)$$
and $F_1$ and $F_2$ are given by (6.4.7).

In order to complete equation (6.4.13) one should also add the information about the singular points of the function $\mu$. We will assume that the homogeneous equation (6.4.4) has a finite number of nontrivial solutions in a finite number of distinct points $\lambda_1, \ldots, \lambda_n$. This implies that the solutions of equation (6.4.4) have a form

$$\mu(z, \bar{z}; \lambda, \bar{\lambda}) = \sum_i \Phi_i^{(s)} + \tilde{\mu}(z, \bar{z}; \lambda, \bar{\lambda}) \qquad (6.4.15)$$

where $\Phi_i$ are the solutions of the homogeneous equation (6.4.4) and $\tilde{\mu}$ is a function bounded in $\lambda$. A structure of the singular part of the function $\mu$ can be determined by the use of the following three properties of equation (6.4.4). Firstly, each column of the $2 \times 2$ matrix $\Phi_i^{(s)}$ obeys the homogeneous equation (6.4.4) separately. Therefore, the columns $\begin{pmatrix} \Phi_{i11} \\ \Phi_{i21} \end{pmatrix}$ and $\begin{pmatrix} \Phi_{i12} \\ \Phi_{i22} \end{pmatrix}$ can be the solutions of the homogeneous equation (6.4.4) in different points $\lambda_i$ and $\mu_k$. Secondly, it follows from the identity (6.4.11) that if the matrix $\begin{pmatrix} \Phi_{11} & 0 \\ \Phi_{21} & 0 \end{pmatrix}$ is the solution of the homogeneous equation (6.4.4) at the point $\lambda_i$ then the matrix $a \begin{pmatrix} 0 & \Phi_{11} \\ 0 & \Phi_{21} \end{pmatrix} \exp\left(i\frac{\bar{z}}{\lambda} + i\frac{z}{\lambda}\right)$ where $a$ is arbitrary constant is the solution of the homogeneous equation (6.4.4) at the point $-\bar{\lambda}_i$.

The third point is that for real $\vec{S}$ the function $\mu$ possesses the involution (6.4.2). This involution implies that

$$\Phi_i^{(s)} = \begin{pmatrix} \Phi_{i11} & -\bar{\Phi}_{i21} \\ \Phi_{i21} & \bar{\Phi}_{i11} \end{pmatrix}. \qquad (6.4.16)$$

In what follows we confine ourselves to the case of real $\vec{S}$.

The nontrivial $\Phi^{(s)}$ which obeys the involution (6.4.16) corresponds to the choice $a = 0$. So we have

$$\Phi^{(s)} = \sum_{i=1}^N \frac{C_i}{\lambda - \lambda_i} \begin{pmatrix} \Phi_{11i} & 0 \\ \Phi_{21i} & 0 \end{pmatrix} + \sum_{i=1}^N \frac{\bar{C}_i}{\lambda + \bar{\lambda}_i} \begin{pmatrix} 0 & -\overline{\Phi_{21i}} \\ 0 & \overline{\Phi_{11i}} \end{pmatrix}. \qquad (6.4.17)$$

So the full $\bar{\partial}$-problem is of the form

$$\frac{\partial \mu}{\partial \bar{\lambda}} = \mu \begin{pmatrix} 0 & \overline{\lambda F(\lambda, \bar{\lambda})} e^{-2\lambda^{-1}z + 2\bar{\lambda}^{-1}\bar{z}} \\ \lambda F(\lambda, \bar{\lambda}) e^{2\lambda^{-1}z + 2\bar{\lambda}^{-1}\bar{z}} & 0 \end{pmatrix}$$

$$+ \sum_{i=1}^N C_i \pi \delta(\lambda - \lambda_i) \begin{pmatrix} \Phi_{11i} & 0 \\ \Phi_{21i} & 0 \end{pmatrix} + \sum_{i=1}^N \bar{C}_i \pi \delta(\lambda + \bar{\lambda}_i) \begin{pmatrix} 0 & -\overline{\Phi_{21i}} \\ 0 & \overline{\Phi_{11i}} \end{pmatrix}. \qquad (6.4.18)$$

In virtue of (6.4.16) one can consider only the equation for the first column $\mu^{(1)} = \begin{pmatrix} \mu_{11} \\ \mu_{21} \end{pmatrix}$. It is

$$\frac{\partial \mu^{(1)}(\lambda, \bar\lambda)}{\partial \bar\lambda} = \pi \sum_{i=1}^{N} \delta(\lambda - \lambda_i)\Phi_i^{(1)} - i\lambda F(\lambda, \bar\lambda) e^{-2\lambda^{-1}z + 2\bar\lambda^{-1}\bar z}\sigma_2 \overline{\mu^{(1)}(\lambda, \bar\lambda)}, \qquad (6.4.19)$$

where we normalize $\Phi_i^{(1)}$ as $\Phi_i^{(1)} \xrightarrow[|z|\to\infty]{} \frac{i}{z\lambda_i}\begin{pmatrix}1\\0\end{pmatrix}$, i. e. $C_i = \frac{i}{\lambda_i^2}$. The generalized Cauchy formula gives

$$\mu^{(1)}(\lambda, \bar\lambda) = \begin{pmatrix} 1 \\ 0 \end{pmatrix} + \sum_{i=1}^{N} \frac{i\Phi_i^{(1)}(z, \bar z)}{\lambda - \lambda_i}$$

$$-\frac{1}{2\pi}\iint_C \frac{d\lambda' \wedge d\bar\lambda'}{\lambda' - \lambda} \lambda' F(\lambda', \bar\lambda') e^{2\lambda'^{-1}z - 2\bar\lambda'^{-1}\bar z} \sigma_2 \overline{\mu^{(1)}(\lambda', \bar\lambda')}. \qquad (6.4.20)$$

Equation (6.4.20) is the part of the inverse problem equations. To derive the rest of such equations one should proceed in equation (6.4.20) to the limit $\lambda \to \lambda_k$ to obtain the system of equations for $\Phi_i^{(1)}$. To do this we must calculate the limit

$$\lim_{\lambda \to \lambda_i}\left(\mu^{(1)} - \frac{i\Phi_i^{(1)}}{\lambda - \lambda_i}\right). \qquad (6.4.21)$$

Completely similar to the DS-II equation (section 5.4) one can show that the quantity

$$\lim_{\lambda \to \lambda_i}\left(\mu^{(1)} - \frac{i\Phi_i^{(1)}}{\lambda - \lambda_i}\right) - \frac{z\Phi_i^{(1)}}{\lambda_i^2} \qquad (6.4.22)$$

is also the solution of the homogeneous equation (6.4.4). Therefore the expression (6.4.22) is the linear superposition of the two independent solutions of the homogeneous equation (6.4.4), i. e.

$$\lim_{\lambda \to \lambda_i}\left(\mu^{(1)} - \frac{i\Phi_i^{(1)}}{\lambda - \lambda_i}\right) - \frac{z\Phi_i^{(1)}}{\lambda_i^2} = \gamma_i \Phi_i^{(1)} - i\tilde\gamma_i \sigma_2 \overline{\Phi_i^{(1)}}, \qquad (6.4.23)$$

where $\gamma_i$ and $\tilde\gamma_i$ are some constants.

Further, proceeding to the limit $\lambda \to \lambda_k$ in equation (6.4.20) and taking into account the identity (6.4.23), we obtain

$$\left(\frac{z}{\lambda_k^2} + \gamma_k\right)\Phi_k^{(1)} - \tilde\gamma_k \sigma_2 \bar\Phi_k^{(1)} e^{-2\lambda_k^{-1}z + 2\bar\lambda_k^{-1}\bar z} = \begin{pmatrix} 1 \\ 0 \end{pmatrix} + \sum_{j\neq k}^{N} \frac{i\mu_j^{(1)}}{\lambda_j - \lambda_k}$$

$$-\frac{1}{2\pi}\iint_C \frac{d\mu \wedge d\bar\mu}{\mu - \lambda_k}\mu F(\mu, \bar\mu) e^{-2\mu^{-1}z + 2\bar\mu^{-1}\bar z}\sigma_2 \overline{\mu^{(1)}(\mu, \bar\mu)} \quad (k=1,\ldots,N). \qquad (6.4.24)$$

The formulae (6.4.20), (6.4.24) form the complete set of the inverse problem equations for the linear problem (6.4.1). The quantities $\{F(\lambda,\bar{\lambda}), \lambda_i, \gamma_i, \tilde{\gamma}_i, (i=1,\ldots,N)\}$ are the inverse problem data.

Finally the reconstruction formula is $P = \chi_0 \sigma_3 \chi_0^{-1}$ where $\chi_0 = \chi^{(1)}(0) - i\sigma_2 \overline{\chi^{(1)}(0)}$ and

$$\chi^{(1)}(0) = \begin{pmatrix} 1 \\ 0 \end{pmatrix} - \sum_{i=1}^{N} \frac{i\Phi_i^{(1)}}{\lambda_i} - i \iint_C d\lambda \wedge d\bar{\lambda} F(\lambda,\bar{\lambda}) e^{\frac{2z}{\lambda} - \frac{2\bar{z}}{\bar{\lambda}}} \sigma_2 \overline{\mu^{(1)}(\lambda,\bar{\lambda})}. \qquad (6.4.25)$$

The evolution of the inverse problem data can be found by the standard procedure from equation (6.3.3). One has

$$F(\lambda,\bar{\lambda},t) = F(\lambda,\bar{\lambda},0) e^{2(\frac{1}{\lambda^2} - \frac{1}{\bar{\lambda}^2})t},$$

$$\lambda_k(t) = \lambda_k(0),$$

$$\gamma_k(t) = -\frac{2i}{\lambda_k^3} t + \gamma_k(0), \qquad (6.4.26)$$

$$\tilde{\gamma}_k(t) = \tilde{\gamma}_k(0) e^{2\left(\frac{1}{\lambda_k^2} - \frac{1}{\bar{\lambda}_k^2}\right)t}.$$

The formulae (6.4.20), (6.4.24) – (6.4.26) provide us the solution (linearization) of the initial value problem for the Ishimori-II equation.

These formulae allow us also to construct exact decaying solutions of the Ishimori-II equation. In the case of the pure discrete inverse problem data ($F \equiv 0$) the inverse problem equations are reduced to the following

$$\chi(0) = 1 - i\sum_{i=1}^{N} \left( \frac{\Phi_i^{(1)}}{\lambda_i} + i\sigma_2 \frac{\overline{\Phi_i^{(1)}}}{\overline{\lambda_i}} \right),$$

$$\left( \frac{z}{\lambda_k^2} - \frac{2i}{\lambda_k^3} t + \gamma_k(0) \right) \Phi_k^{(1)} - i\tilde{\gamma}_k(0) \sigma_2 \overline{\Phi_k^{(1)}} e^{2\frac{z}{\lambda_k} - \frac{2\bar{z}}{\bar{\lambda}_k} + 2\left(\frac{1}{\lambda_k^2} - \frac{1}{\bar{\lambda}_k^2}\right)t} = \begin{pmatrix} 1 \\ 0 \end{pmatrix}. \qquad (6.4.27)$$

The corresponding solutions of the Ishimori-II equation are the rational-exponential one. The simplest solution of this type ($N = 1$) looks like

$$S_t = S_1 + iS_2$$

$$= \frac{-2i\left( \left|\frac{z}{\lambda_1^2} - \frac{2i}{\lambda_1^3} t + \gamma_{10}\right|^2 + |\tilde{\gamma}_{10}|^2 - i\left(\frac{\bar{z}}{\lambda_1^2} + \frac{2i}{\lambda_1^3} t + \bar{\gamma}_{10}\right)\right) \tilde{\gamma}_{10} e^{\frac{2z}{\lambda_1} - \frac{2\bar{z}}{\bar{\lambda}_1} + 2\left(\frac{1}{\lambda_1^2} - \frac{1}{\bar{\lambda}_1^2}\right)t}}{\left|\left|\frac{z}{\lambda_1^2} - \frac{2i}{\lambda_1^3} t + \gamma_{10}\right|^2 + |\tilde{\gamma}_{10}|^2 - i\left(\frac{\bar{z}}{\lambda_1^2} + \frac{2i}{\lambda_1^3} t + \bar{\gamma}_{10}\right)\right|^2 + |\tilde{\gamma}_{k0}|^2},$$

$$S_3 = \frac{\left|\left|\frac{z}{\lambda_1^2} - \frac{2i}{\lambda_1^3}t + \gamma_{10}\right|^2 + |\tilde{\gamma}_{10}|^2 - i\left(\frac{\bar{z}}{\lambda_1^2} + \frac{2i}{\lambda_1^3}t + \bar{\gamma}_{10}\right)\right|^2 - |\tilde{\gamma}_{10}|^2}{\left|\left|\frac{z}{\lambda_1^2} - \frac{2i}{\lambda_1^3}t + \gamma_{10}\right|^2 + |\tilde{\gamma}_{10}|^2 - i\left(\frac{\bar{z}}{\lambda_1^2} + \frac{2it}{\lambda_1^3} + \bar{\gamma}_{10}\right)\right|^2 + |\tilde{\gamma}_{10}|^2}. \qquad (6.4.28)$$

This solution is bounded and moves with the velocity $V = \left(2\text{Re}\frac{1}{\lambda_1}, 2\text{Im}\frac{1}{\lambda_1}\right)$. General solution (6.4.25) describes the completely trivial scattering of the simple objects (6.4.28).

Within a different approach the solutions of the Ishimori-II equation of this type has been obtained in [250].

## 6.5. Gauge equivalence of the DS and Ishimori equations

The Ishimori and DS equations are, in fact, very closely connected. They are gauge equivalent to each other similar to the KP and mKP equations. This gauge equivalence is the 2+1-dimensional extension of the well-known gauge equivalence of the NLS and Heisenberg ferromagnet equations [14].

Let the problems

$$L_i^{\text{DS}}\Psi^{\text{DS}} = 0 \quad (i = 1, 2) \qquad (6.5.1)$$

and

$$L_i^{\text{Is}}\Psi^{\text{Is}} = 0 \quad (i = 1, 2) \qquad (6.5.2)$$

be the linear auxiliary systems for the DS and Ishimori equations respectively. It is not difficult to verify that they are connected by the transformation [248]

$$\Psi^{\text{Is}} = g(x, y, t)\Psi^{\text{DS}},$$

$$L_i^{\text{Is}} = g L_i^{\text{DS}} g^{-1} \qquad (6.5.3)$$

where $g(x, y, t)$ is the $2 \times 2$ matrix-valued function such that

$$P^{\text{Is}} = g\sigma_3 g^{-1}, \qquad (6.5.4)$$

$$P^{\text{DS}} = \sigma g^{-1} g_y + \sigma_3 g^{-1} g_x. \qquad (6.5.5)$$

The formulae (6.5.4) and (6.5.5) establish the interrelation between the solutions of the DS equation $P^{\text{DS}} = \begin{pmatrix} 0 & q \\ r & 0 \end{pmatrix}$ and the Ishimori equation $P^{\text{Is}} = \sum_{i=1}^{3} \sigma_i S_i$.

For the complex valued general $q, r$ and $\vec{S}$ this transformation is effectively the two-dimensional transformation. The situation changes crucially if one considers the gauge transformation between the DS equation under the reduction $r = -\bar{q}$ and the Ishimori equation with real $\vec{S}$.

Chapter 6

In the case $\sigma = 1$ the formula (6.5.4) for real $\vec{S}$ means (see (6.2.5)) that

$$gg^+ = 1 \tag{6.5.6}$$

while the reduction $P^{\text{DS}} = \begin{pmatrix} 0 & q \\ -\bar{q} & 0 \end{pmatrix}$ implies

$$g_y^+(g^{-1})^+ + g_x^+(g^{-1})^+\sigma_3 = -g^{-1}g_y - \sigma_3 g^{-1}g_x. \tag{6.5.7}$$

Combining (6.5.6) and (6.5.7), one gets $[\sigma_3, g^{-1}g_x] = 0$. So, in the case $\sigma = 1$ the gauge equivalence takes place only between the very restricted classes of solutions (effectively, one-dimensional).

It is easy to see that this restriction is absent in the case $\sigma = i$. So the transformation (6.5.3), (6.5.4) establishes the gauge equivalence between the real-valued solutions of the Ishimori-II equation and solutions $P^{\text{DS}} = \begin{pmatrix} 0 & q \\ -\bar{q} & 0 \end{pmatrix}$ of the DS-II equation [248]. Note that such property of the gauge transformation between the Ishimori and DS equations is completely similar to that for the mKP and KP equations (section 4.5). For both pair of equations the gauge transformation interrelates the "real" solutions of these equations only in the case $\sigma^2 = -1$.

Using a suitable parametrization of $P^{\text{Is}}$, one can convert the formulae (6.5.4), (6.5.5) in a more explicit form and finds the relations between the integrals of motion, including the topological charge $Q$, for the DS-II equation and Ishimori-II equation [251]. The comparison of the formulae (6.5.4) and (6.2.12) leads to the identification $g = \chi(0)$. So, using the results of section (6.2), one can establish the direct interrelation between the solutions of the Ishimori-II and DS-II equations.

We see that the function $\chi(0) = \chi^{\text{Is}}(\lambda = 0)$ plays an important role in the gauge equivalence between Ishimori and DS equations. This is not an accidental fact. As we mentioned in section (6.1) the function $\chi(0)$ is, in fact, the primary quantity in the $\bar{\partial}$-dressing for the Ishimori equation. This function obeys the nonlinear equation (6.1.31). This equation is of principal importance not only for the Ishimori equation but also for the DS equation. Indeed, given the solution $\chi_0$, $C$ of the system (6.1.31) the combinations

$$P^{\text{Is}} = \chi_0 \sigma_3 \chi_0^{-1},$$
$$\Phi^{\text{Is}} = -i\sigma \ln \det \chi_0 \tag{6.5.8}$$

solves the Ishimori equation while the combinations

$$P^{\text{DS}} = \sigma \chi_0^{-1} \chi_{0y} + \sigma_3 \chi_0^{-1} \chi_{0x},$$

$$C^{DS} = (\sigma \partial_y + \sigma_3 \partial_x) C \qquad (6.5.9)$$

solve the DS equation [152].

An interesting fact is that equation (6.1.31) is just nothing but the equation for the DS wave function $\Psi^{DS}$ which can be obtained by the elimination of $P^{DS}$ from the system (5.1.32). Equation (6.1.31) has been studied by the IST method in [152]. This equation is the representative of the class of nonlinear integrable equations for soliton wave functions [104].

The gauge equivalence between the Ishimori and DS equations can be formulated on the general level of the $\bar{\partial}$-dressing. In addition to the scheme of the $\bar{\partial}$-dressing for the DS equation described in section (5.1), one can derive the DS equation also if one starts with the $2 \times 2$ matrix $\bar{\partial}$-problem

$$\frac{\partial \chi^{DS}(\lambda, \bar{\lambda})}{\partial \bar{\lambda}} = \pi \delta(\lambda) + \iint_C d\mu \wedge d\bar{\mu}\, \chi^{DS}(\mu, \bar{\mu}) R^{DS}(\mu, \bar{\mu}; \lambda, \bar{\lambda}) \qquad (6.5.10)$$

with the normalization

$$\chi^{DS} \to \frac{1}{\lambda} \quad \text{as} \quad \lambda \to 0 \qquad (6.5.11)$$

and the choice

$$\hat{B}_1 = \frac{i}{\lambda}, \quad \hat{B}_2 = -\frac{i}{\lambda}\sigma_3, \quad \hat{B}_3 = -\frac{i}{\lambda^2}\sigma_3. \qquad (6.5.12)$$

So the kernel $R^{DS}$ in (6.5 10) is of the form

$$R^{DS}(\mu, \bar{\mu}; \lambda, \bar{\lambda}) = e^{\frac{i}{\mu}x - \frac{i}{\mu}\sigma_3 y - \frac{i}{\mu^2}\sigma_3 t} R_0(\mu, \bar{\mu}; \lambda, \bar{\lambda}) e^{-\frac{i}{\lambda}x + \frac{i}{\lambda}\sigma_3 y + \frac{i}{\lambda^2}\sigma_3 t}. \qquad (6.5.13)$$

Remind that for the Ishimori equation the function $\chi^{Is}$ solves the $\bar{\partial}$-problem

$$\frac{\partial \chi^{Is}(\lambda, \bar{\lambda})}{\partial \bar{\lambda}} = 1 + \iint_C d\mu \wedge d\bar{\mu}\, \chi^{Is}(\mu, \bar{\mu}) R^{Is}(\mu, \bar{\mu}; \lambda, \bar{\lambda}) \qquad (6.5.14)$$

where the kernel $R^{Is}$ have the same structure as $R^{DS}$ (6.5.13).

It is not difficult to verify that the function $\tilde{\chi} = \frac{1}{\lambda}\chi^{Is}$ is the solution of the problem

$$\frac{\partial \tilde{\chi}(\lambda, \bar{\lambda})}{\partial \bar{\lambda}} = \pi \delta(\lambda)\chi_0^{Is} + \iint_C d\mu \wedge d\bar{\mu}\, \tilde{\chi}(\mu, \bar{\mu})\frac{\mu}{\lambda} R^{Is}(\mu, \bar{\mu}; \lambda, \bar{\lambda}) \qquad (6.5.15)$$

where $\chi_0^{Is} \doteq \chi^{Is}(\lambda = 0)$. Then comparing (6.5.15) and (6.5.10), one finds that the function $\frac{1}{\lambda}\chi_0^{Is^{-1}}\chi^{Is}(\lambda, \bar{\lambda})$ solves the same $\bar{\partial}$-problem as the function $\chi^{DS}(\lambda, \bar{\lambda})$ with the kernel $\frac{\mu}{\lambda}R^{Is}(\mu, \bar{\mu}; \lambda, \bar{\lambda})$. So one can identify

$$\chi^{DS}(\lambda, \bar{\lambda}) = \frac{1}{\lambda}\chi_0^{Is^{-1}}\chi^{Is}(\lambda, \bar{\lambda}) \qquad (6.5.16)$$

and
$$R^{DS}(\mu,\bar{\mu};\lambda,\bar{\lambda}) = \frac{\mu}{\lambda}R^{Is}(\mu,\bar{\mu};\lambda,\bar{\lambda}). \qquad (6.5.17)$$
These formulae describe the gauge equivalence between the Ishimori and DS equations within the framework of the $\bar{\partial}$-dressing method.

Note in conclusion the papers [252, 253] in which the noncompact versions of the Ishimori equation and its gauge equivalence to the noncompact DS equation have been considered.

# Chapter 7

# The 2+1-dimensional integrable sine-Gordon equation

## 7.1. The 2DISG equation and its properties

The sine-Gordon equation (1.1.12) is a very important 1+1-dimensional integrable equation with a number of interesting properties. Thereby a search of its reasonable 2+1-dimensional integrable counterpart is of great interest.

Two different multidimensional integrable extensions of the sine-Gordon equation have been proposed and studied in the papers [254–256]. The disadvantage of these extensions is that one of them [254] has a rather complicated form and contains the spatial variable very unsymmetrically. The other [254] is the system for several dependent variables.

A simple and symmetric 2+1-dimensional integrable generalization of the sine-Gordon equation has been found recently in [105]. This equation has been derived within the analysis of the class of 2+1-dimensional nonlinear equations introduced by C. Loewner in the year 1952. He studied the problem of construction of solutions of the plane potential gas dynamic [106]. For this purpose he considered the infinitesimal generalized Bäcklund transformations which are defined by the system of equations [106]

$$\sigma \Psi_y + P \Psi_x = 0,$$

$$\Psi_{tx} + U \Psi_x + V \Psi_t + W \Psi = 0,$$

$$\Psi_{ty} + \tilde{U} \Psi_y + \tilde{V} \Psi_t + \tilde{W} \Psi = 0 \qquad (7.1.1)$$

where $P, U, V, W, \tilde{U}, \tilde{V}, \tilde{W}$ are $2 \times 2$ matrices.

The compatibility condition of the system (7.1.1) gives rise to the system of partial differential equations for $P, U, V, W, \tilde{U}, \tilde{V}, \tilde{W}$. C. Loewner has managed to represent this system in a compact form. Considering the special 1+1-dimensional case of the system (7.1.1) he demonstrated that the sine-Gordon equation (1.1.12) is among the nonlinear equations associated with (7.1.1).

It should be emphasized that C. Loewner got these results 25 years before the discovery of the IST method.

The $N \times N$ matrix extension of the Loewner system (7.1.1) has been studied in [105, 112]. The compact form of the general equations and several interesting particular cases have been obtained in [105, 112]. The 2+1-dimensional integrable generalizations of several well-known 1+1-dimensional integrable equations are among them.

One of the simplest system of this type is of the form [105, 112]

$$\theta_t = \phi_1 - \phi_2,$$
$$(e^{i\theta}\phi_{1x})_x - \sigma^2(e^{i\theta}\phi_{1y})_y = 0, \qquad (7.1.2)$$
$$(e^{-i\theta}\phi_{2x})_x - \sigma^2(e^{-i\theta}\phi_{2y})_y = 0$$

where $\theta(x,y,t)$, $\phi_1(x,y,t)$ and $\phi_2(x,y,t)$ are scalar functions. Introducing the variables $\varphi$ and $\tilde{\varphi}$ by $\varphi_t = \phi_1$, $\tilde{\varphi}_t = -\phi_2$, one represents the system (7.1.2) as follows

$$(e^{i(\varphi+\tilde{\varphi})}\varphi_{tx})_x - \sigma^2(e^{i(\varphi+\tilde{\varphi})}\varphi_{ty})_y = 0,$$
$$(e^{-i(\varphi+\tilde{\varphi})}\tilde{\varphi}_{tx})_x - \sigma^2(e^{-i(\varphi+\tilde{\varphi})}\tilde{\varphi}_{ty})_y = 0. \qquad (7.1.3)$$

The system (7.1.3) is equivalent to the compatibility condition for the linear system [112]

$$\sigma\phi_y - \sigma \begin{pmatrix} 0 & e^{-i(\varphi+\tilde{\varphi})} \\ e^{i(\varphi+\tilde{\varphi})} & 0 \end{pmatrix} \phi_x = 0,$$

$$\phi_{tx} + \frac{i}{2}\begin{pmatrix} \varphi_t + \tilde{\varphi}_t & 0 \\ 0 & -\varphi_t - \tilde{\varphi}_t \end{pmatrix}\phi_x + \frac{i}{2}\begin{pmatrix} \varphi_{tx} & e^{-i(\varphi+\tilde{\varphi})}\tilde{\varphi}_{ty} \\ e^{-i(\varphi+\tilde{\varphi})}\varphi_{ty} & -\tilde{\varphi}_{tx} \end{pmatrix}\phi = 0, \qquad (7.1.4)$$

$$\phi_{ty} + \frac{i}{2}\begin{pmatrix} \varphi_t + \tilde{\varphi}_t & 0 \\ 0 & -\varphi_t - \tilde{\varphi}_t \end{pmatrix}\phi_y + \frac{i}{2}\begin{pmatrix} \varphi_{ty} & \frac{1}{\sigma}e^{-i(\varphi+\tilde{\varphi})}\tilde{\varphi}_{ty} \\ \frac{1}{\sigma}e^{-i(\varphi+\tilde{\varphi})}\varphi_{ty} & -\tilde{\varphi}_{ty} \end{pmatrix}\phi = 0.$$

Further, in the variables

$$\theta = \varphi + \tilde{\varphi}, \quad \tilde{\theta} = i(\varphi - \tilde{\varphi})$$

the system (7.1.3) looks like

$$(e^{i\theta}(\theta + \tilde{\theta})_{tx})_x - \sigma^2(e^{-i\theta}(\theta + \tilde{\theta})_{ty})_y = 0,$$
$$(e^{-i\theta}(\theta - \tilde{\theta})_{tx})_x - \sigma^2(e^{-i\theta}(\theta - \tilde{\theta})_{ty})_y = 0. \qquad (7.1.5)$$

The system (7.1.4) or (7.1.5) represents the 2+1-dimensional integrable generalization of the sine-Gordon equation. We will refer to this system as the 2DISG equation.

One of the important features of the 2DISG equation (7.1.4) is that it is invariant under the rotations in $(x,y)$ plane, namely, under the hyperbolic rotations in the case $\sigma^2 = 1$ and under the usual rotations at $\sigma^2 = -1$.

In terms of the characteristic variables
$$\xi = \frac{1}{2}(y - \sigma x), \quad \eta = \frac{1}{2}(y + \sigma x)$$

the 2DISG equation looks like
$$\theta_{t\xi\eta} + \frac{1}{2}\theta_\eta \cdot \tilde{\theta}_{\xi t} + \frac{1}{2}\theta_\xi \cdot \tilde{\theta}_{\eta t} = 0,$$

$$\tilde{\theta}_{\xi\eta} - \frac{1}{2}\theta_\xi \cdot \theta_\eta = \alpha(\xi,\eta) \tag{7.1.6}$$

where $\alpha(\xi,\eta)$ is an arbitrary function. Introducing the variable $\rho \equiv \tilde{\theta}_t$, one gets the system
$$\theta_{t\xi\eta} + \frac{1}{2}\theta_\eta \cdot \rho_\xi + \frac{1}{2}\theta_\xi \cdot \rho_\eta = 0,$$

$$\rho_{\xi\eta} - \frac{1}{2}(\theta_\xi \cdot \theta_\eta)_t = 0. \tag{7.1.7}$$

The system (7.1.7) has a broad group of usual Lie symmetries. In addition to the obvious translational invariance it is invariant under independent scaling of each variable $\xi$, $\eta$, $t$
$$\xi \to \xi' = \lambda_1 \xi, \quad \theta \to \theta' = \theta,$$

$$\eta \to \eta' = \lambda_2 \eta, \quad \rho \to \rho' = \lambda_3^{-1} \rho, \tag{7.1.8}$$

$$t \to t' = \lambda_3 t$$

where $\lambda_1$, $\lambda_2$ are arbitrary parameters. The group of transformations (7.1.8) includes the rotations in $(x,y)$, $(x,t)$ and $(y,t)$ planes. Note that in the case $\sigma^2 = -1$ the variables $\xi$, $\eta$ are complex conjugate to each other and $\lambda_2 = \bar{\lambda}_1$.

The system (7.1.7) is integrable with the use of the linear system (7.1.4) rewritten in the characteristic variables $\xi$ and $\eta$. One can also find another auxiliary linear system for (7.1.7). Combining equations (7.1.4) and performing the gauge transformation
$$\phi = \begin{pmatrix} e^{-\frac{i}{2}\theta} & e^{-\frac{i}{2}\theta} \\ -e^{\frac{i}{2}\theta} & e^{\frac{i}{2}\theta} \end{pmatrix} \psi, \tag{7.1.9}$$

one obtains the simpler and more convenient linear system for equation (7.1.7), namely, the system [257]
$$L_1 \psi \equiv \begin{pmatrix} \partial_\eta & -\frac{i}{2}\theta_\eta \\ -\frac{i}{2}\theta_\xi & \partial_\xi \end{pmatrix} \psi = 0, \tag{7.1.10a}$$

Chapter 7    207

$$L_2\psi \equiv \begin{pmatrix} \partial_{t\xi}^2 + \frac{1}{2}\rho_\xi, & -\frac{i}{2}\theta_\xi \cdot \partial_t \\ -\frac{i}{2}\theta_\eta \cdot \partial_t, & \partial_{t\eta}^2 + \frac{1}{2}\rho_\eta \end{pmatrix}\psi = 0. \tag{7.1.10b}$$

The operator form of the compatibility condition for the system (7.1.10), i. e. the operator form of the system (7.1.7) is the following [257]

$$[L_1, L_2] = A_1 L_1 + A_2 L_2, \tag{7.1.11}$$

where

$$A_1 = \begin{pmatrix} 0 & \frac{i}{2}(\theta_\eta + \theta_\xi) \cdot \partial_t + \frac{i}{2}\theta_{\eta t} \\ \frac{i}{2}(\theta_\eta + \theta_\xi) \cdot \partial_t + \frac{i}{2}\theta_{\xi t} & 0 \end{pmatrix}, \tag{7.1.12}$$

$$A_2 = -\frac{i}{2}(\theta_\xi + \theta_\eta)\begin{pmatrix} 0 & 1 \\ 1 & 0 \end{pmatrix}. \tag{7.1.13}$$

The linear system (7.1.8) will be our basic tool for the study of the 2DISG equation (7.1.7).

Similarly for the DS and Ishimori equations one can eliminate the variable $\rho$ from the system (7.1.7). As a result, one obtains the following single equation for $\theta$:

$$\theta_{t\xi\eta} + m_1(t) \cdot \theta_\eta + m_2(\eta, t) \cdot \theta_\xi$$
$$+ \frac{1}{4}\theta_\eta \cdot \int_{-\infty}^{\eta} d\eta'(\theta_\xi \cdot \theta_{\eta'})_t + \frac{1}{4}\theta_\xi \int_{-\infty}^{\xi} d\xi'(\theta_{\xi'} \cdot \theta_\eta)_t = 0. \tag{7.1.14}$$

where

$$m_1(\xi, t) = \frac{1}{2}\lim_{\eta \to -\infty} \rho_\xi,$$

$$m_2(\eta, t) = \frac{1}{2}\lim_{\xi \to -\infty} \rho_\eta. \tag{7.1.15}$$

Note that in the case $\sigma^2 = -1$ for real $\theta$ and $\rho$ one has $m_2 = \bar{m}_1$.

So the solution of equation (7.1.14) with the fixed functions $m_1(\xi, t)$ and $m_2(\eta, t)$ gives the solution of the 2DISG equation (7.1.7) with the boundary values of $\rho$ given by (7.1.15).

The properties of the 2DISG equation essentially depend on the boundaries $m_1$ and $m_2$. In the case $m_1 = m_2 = 0$ equation (7.1.14) is the dispersiveless one with the linear part $\theta_{t\xi\eta} = 0$. In this case equation (7.1.14) possesses the symmetry group (7.1.8).

In the case $m_2 = 0$ equation (7.1.4) has the solution $\theta = \theta_1(\xi, t)$ where $\theta_1$ is an arbitrary function while in the case $m_1 = 0$ it has the solution $\theta = \theta_2(\eta, t)$ where $\theta_2$ is an arbitrary function. This is an obvious consequence of the general structure of the 2DISG equation.

An important case corresponds to the constant boundary values $m_1$ and $m_2$. The dispersion law for the corresponding 2DISG equation

$$\theta_{t\xi\eta} + m_1 \theta_\eta + m_2 \theta_\xi$$
$$+ \frac{1}{4}\theta_\eta \int_{-\infty}^{\eta} d\eta'(\theta_\xi \cdot \theta_{\eta'})_t + \frac{1}{4}\theta_\xi \int_{-\infty}^{\xi} d\xi'(\theta_{\xi'} \cdot \theta_\eta)_t = 0, \tag{7.1.16}$$

in the characteristic variables is

$$\omega(p_1, p_2) = \frac{m_1}{p_1} + \frac{m_2}{p_2}. \tag{7.1.17}$$

Equation (7.1.16) does not possess the full symmetry group (7.1.8). At $m_1 \neq 0, m_2 \neq 0$ it possesses the scale invariance

$$\xi \to \xi' = \lambda \xi,$$

$$\eta \to \eta' = \lambda \eta, \tag{7.1.18}$$

$$t \to t' = \lambda^{-1} t,$$

where $\lambda$ is an arbitrary parameter. In the case $m_1 = 0, m_2 \neq 0$ it is invariant under the transformation

$$\xi \to \xi' = \lambda_1 \xi,$$

$$\eta \to \eta' = \lambda_2 \eta, \tag{7.1.19}$$

$$t \to t' = \lambda_2^{-1} t,$$

where $\lambda_1$ and $\lambda_2$ are arbitrary parameters. At $m_1 \neq 0$, $m_2 = 0$ it has the symmetry (7.1.19) with the exchange $\xi \leftrightarrow \eta$. The scale symmetries (7.1.8), (7.1.18), (7.1.19) imply the existence of various types of the similarity solutions for equations (7.1.7) and (7.1.16).

Note that in the case $m_1 \neq 0$, $m_2 \neq 0$ equation (7.1.16) is not invariant under the rotations in $(x, y)$ plane while in the case $m_1 = 0$ or $m_2 = 0$ it possesses such a symmetry group which is the subgroup of transformations (7.1.19) with $\lambda_1 \lambda_2 = 1$.

The constant boundaries $m_1$ and $m_2$ are the 2+1-dimensional analog of the mass $m$ (more precisely, squared mass) in the 1+1-dimensional sine-Gordon equation. Indeed, in the 1+1-dimensional limit $\theta_\xi = \theta_\eta$ equation (7.1.16) looks like

$$\theta_{t\xi\xi} + (m_1 + m_2) \cdot \theta_\xi + \theta_\xi \int_{-\infty}^{\xi} d\xi' \theta_{\xi'} \cdot \theta_{\xi' t} = 0. \tag{7.1.20}$$

Assuming that

$$\theta_{t\xi} = F'(\theta), \tag{7.1.21}$$

where $F(\theta)$ is some function and $F'(\theta) = dF/d\theta$, one gets from (7.1.18)

$$F''(\theta) + F(\theta) = F(\theta(\xi = -\infty)) - (m_1 + m_2). \tag{7.1.22}$$

The particular solution of (7.1.22) is

$$F(\theta) = (m_1 + m_2) \cdot \cos \theta$$

for which one has the sine-Gordon equation

$$\theta_{t\xi} = -(m_1 + m_2) \cdot \sin\theta \qquad (7.1.23)$$

with the squared mass term $m = -(m_1 + m_2)$.

Emphasize also that the invariance of the 2DISG equation under rotations on the plane $(x, y)$ makes the introduction of the polar coordinates $r, \varphi$ on the plane $(x, y)$ very natural. In such independent variables the 2DISG equations look like a deformation of the sine-Gordon equation in $(r, t)$. Indeed, considering the limit $\theta_\varphi = \hat\theta_\varphi = 0'$, one gets from the 2DISG equation the sine-Gordon equation $\theta_{rt} = \gamma\sin\theta$ where $r$ is the radius on the plane.

So the 2DISG equation for the class of solutions with the asymptotic behaviour $\theta \to 2\pi n$ ($n$ is arbitrary integer) and $\rho \to 2(m_1\eta + m_2\xi)$ as $\xi^2 + \eta^2 \to \infty$ is an adequate 2+1-dimensional analog of the sine-Gordon equation (1.1.12).

An interesting property of the 2DISG equation is that it is closely connected both with the DS equation and Ishimori equation. Namely, it is the first negative member for both the DS and Ishimori hierarchies under certain reductions.

The DS hierarchy is an infinite family of equations which are associated with the $\bar\partial$-problem (5.1.1) for the choice

$$B_1 = i\lambda, \quad B_2 = -\frac{i\lambda\sigma_3}{\sigma}, \quad B_3 \sim -i\lambda^n\sigma_3 \qquad (7.1.24)$$

where $n$ is an arbitrary integer.

The corresponding linear problems and nonlinear integrable equations can be derived similar to the DS equation case $n = 2$. For arbitrary $n$ the auxiliary linear system is of the form

$$\sigma\psi_y + \sigma_3\psi_x + P\psi = 0,$$

$$\psi_t - i\sigma_3\partial_x^n\psi + \ldots = 0. \qquad (7.1.25)$$

Comparing (7.1.8) and (7.1.25), we see that the 2DISG equation is the $n = -1$ member of the DS hierarchy under the reduction

$$P = -\frac{i}{2}\begin{pmatrix} 0 & \theta_\eta \\ \theta_\xi & 0 \end{pmatrix}. \qquad (7.1.26)$$

The Ishimori hierarchy is an infinite family of equations which are generated by the 2×2 matrix $\bar\partial$-problem with

$$B_1 = \frac{i}{\lambda}, \quad B_2 = -\frac{i}{\sigma\lambda}\sigma_3, \quad B_3 \sim \frac{i\sigma_3}{\lambda^n} \qquad (7.1.27)$$

where $n$ is arbitrary integer. The Ishimori equation corresponds to $n = 2$. Taking into account (7.1.4), it is not difficult to see that the 2DISG equation is the member of this hierarchy with $n = -1$ and the reduction

$$P^{IST} = \begin{pmatrix} 0 & e^{-i\theta} \\ e^{i\theta} & 0 \end{pmatrix}, \tag{7.1.28}$$

i. e. $S_1 + iS_2 = e^{-i\theta}$, $S_3 = 0$.

So the 2DISG equation is the point of coincidence of DS and Ishimori hierarchies.

This fact also indicates the gauge invariance of the 2DISG equation. In section (6.5) it was shown that the DS and Ishimori equations are gauge equivalent to each other. In general, the whole DS and Ishimori hierarchies are transformed to each other by the gauge transformations (6.5.3). However the formula (7.1.9) clearly shows us that such gauge transformation converts the 2DISG equation into the same 2DISG equation while it converts the linear problem (7.1.4) of the Ishimori type into the linear problem (7.1.10) of the DS type.

The gauge invariance is an important feature of the 2DISG equation. One can show that all the members of the DS and Ishimori hierarchies for $n = 2m + 1$ where $m$ is an arbitrary integer under the reductions (7.1.26) and (7.1.28) possess the similar gauge invariance.

Note in conclusion that under the reduction (7.1.28) the topological charge (6.0.3) is apparently identical zero.

## 7.2. $\bar{\partial}$-dressing

Similar to the DS and Ishimori equations we start with the 2×2 $\bar{\partial}$-problem

$$\frac{\partial \chi(\lambda, \bar{\lambda})}{\partial \bar{\lambda}} = \iint_C d\mu \wedge d\bar{\mu}\, \chi(\mu, \bar{\mu})\, R(\mu, \bar{\mu}; \lambda, \bar{\lambda}) \tag{7.2.1}$$

which is assumed to be uniquely solvable with the canonical normalization

$$\chi \to 1 + \frac{1}{\lambda}\chi_{-1} + \frac{1}{\lambda^2}\chi_{-2} + \ldots \tag{7.2.2}$$

as $\lambda \to \infty$. For the trivial background we choose

$$\hat{B}_1 = i\lambda, \quad \hat{B}_2 = -\frac{i\lambda}{\sigma}\sigma_3, \quad \hat{B}_3 = -\frac{i\sigma}{2\lambda}m \tag{7.2.3}$$

where

$$m \equiv \begin{pmatrix} m_1 & 0 \\ 0 & -m_2 \end{pmatrix}.$$

Chapter 7

So the operator $D_i$ are [257]

$$D_1 f = \partial_x f + i\lambda f,$$
$$D_2 f = \partial_y f - \frac{i\lambda}{\sigma} f \sigma_3, \qquad (7.2.4)$$
$$D_3 f = \partial_t f - \frac{i}{2\lambda} f m$$

and the kernel $R$ is of the form

$$R(\mu, \bar{\mu}; \lambda, \bar{\lambda}; x, y, t) = e^{F(\mu)} R_0(\mu, \bar{\mu}; \lambda, \bar{\lambda}) e^{-F(\lambda)} \qquad (7.2.5)$$

where

$$F(\lambda, x, y, t) = i\lambda x - \frac{i\lambda}{\sigma}\sigma_3 y - \frac{i}{2\lambda} m t. \qquad (7.2.6)$$

Note that in the case under consideration the operators $D_i$ have the singularities both at the point and infinity in contrast to the previous cases with the single point singularities.

The operators $D_1$ and $D_2$ (7.2.4) are exactly the same as for the DS equation. Hence, the first linear problem is given by (5.1.11), i. e.

$$\sigma D_2 \chi + \sigma_3 D_1 \chi + P\chi = 0 \qquad (7.2.7)$$

where

$$P(x, y, t) = \begin{pmatrix} 0 & q \\ r & -0 \end{pmatrix} = -i[\sigma_3, \chi_{-1}]. \qquad (7.2.8)$$

Note that equation (7.2.7) also implies

$$P = (\sigma \chi_{0y} + \sigma_3 \chi_{0x}) \chi_0^{-1} \qquad (7.2.9)$$

where $\chi_0 = \chi(\lambda = 0)$.

The construction of the second linear problem is completely different from the DS case. First we note that there is the combination

$$\sigma D_2 \chi - \sigma_3 D_1 \chi = \sigma \chi_y - \sigma_3 \chi_x - i\lambda \chi \sigma_3 - i\lambda \sigma_3 \chi \qquad (7.2.10)$$

which has the first order singularity as $\lambda \to \infty$ in contrast to $\sigma D_2 \chi + \sigma_3 D_1 \chi$. So it is rather obvious that the quantity $D_3(\sigma D_2 \chi - \sigma_3 D_1 \chi)$ will not have singularity as $\lambda \to \infty$. Indeed

$$D_3(\sigma D_2 \chi - \sigma_3 D_1 \chi) = \sigma \chi_{yx} - \sigma_3 \chi_{xx} - i\lambda \{\sigma_3, \chi_t\}$$
$$- \frac{i}{2\lambda}(\sigma \chi_y - \sigma_3 \chi_x)m + \frac{i}{2}\{\sigma_3, \chi\}m \qquad (7.2.11)$$

where $\{A, B\} \doteq AB + BA$.

The first order singularity at the origin in (7.2.11) can be compensated by the addition of the term $QD_3\chi$. The requirement of vanishing the residue in the first order pole in the quantity

$$D_3(\sigma D_2\chi - \sigma_3 D_1\chi) + QD_3\chi \qquad (7.2.12)$$

gives

$$Q = -(\sigma\chi_{0y} - \sigma_3\chi_{0x})\chi_0^{-1}. \qquad (7.2.13)$$

Further it is not difficult to check that

$$D_3(\sigma D_2\chi - \sigma_3 D_1\chi) + QD_3\chi + \tilde{Q}\chi \to 0 \qquad (7.2.14)$$

as $\lambda \to \infty$ if

$$\tilde{Q} = i\{\sigma_3, \chi_{-1t}\} - m\sigma_3. \qquad (7.2.15)$$

So, the second linear problem is

$$D_3(\sigma D_2\chi - \sigma_3 D_1\chi) + QD_3\chi + \tilde{Q}\chi = 0 \qquad (7.2.16)$$

where $Q$ and $\tilde{Q}$ are given by (7.2.13) and (7.2.15).

The system (7.2.7), (7.2.16) is the desired linear system associated with the choice (7.2.3). This system is automatically compatible and provides the corresponding nonlinear integrable equation. Substituting the asymptotic expansion (7.2.2) into equation (7.2.16), considering the terms of the order $\bar{\lambda}'$, one gets

$$\sigma\chi_{-1yt} - \sigma_3\chi_{-1xt} - i\{\sigma_3, \chi_{-2t}\} - \frac{1}{2}m[\sigma_3, \chi_{-1}]$$
$$- (\sigma\chi_{0y} - \sigma_3\chi_{0x})\chi_0^{-1}(\chi_{-1t} - \frac{1}{2}m) + i\{\sigma_3, \chi_{-1t}\}\chi_{-1} = 0. \qquad (7.2.17)$$

Using equation (5.1.13), the equality $\{\sigma_3, A\} = 2\sigma_3 A_{diag}$, choosing $Q = P^T$ and denoting $C \doteq 2i(\chi_{-1t})_{diag}$, one obtains from (7.2.17) the system

$$\sigma P_{yt} - \sigma_3 P_{xt} + (m+C)\sigma_3 P + \sigma_3 P^T(m+C) = 0, \qquad (7.2.18a)$$

$$\sigma C_y + \sigma_3 C_x = -\sigma_3(P^2)_t. \qquad (7.2.18b)$$

Equation (7.2.18b) implies that

$$C_1 = \frac{1}{2}(\sigma\partial_y - \partial_x)\tilde{\rho},$$

$$C_2 = -\frac{1}{2}(\sigma\partial_y + \partial_x)\tilde{\rho} \qquad (7.2.19)$$

*Chapter 7* 213

where $\tilde{\rho}$ is the scalar function which obeys the equation

$$(\sigma^2 \partial_y^2 - \partial_x^2)\tilde{\rho} = -2(qr)_t. \tag{7.2.20}$$

Further, denoting $\rho = \tilde{\rho} + 2m_1\eta + 2m_2\xi$ and transiting to the characteristic variables $\xi = \frac{1}{2}(y - \sigma x)$, $\eta = \frac{1}{2}(y + \sigma x)$, one rewrites the system (7.1.8) as

$$q_{\xi t} + \frac{1}{2}\rho_\xi q + \frac{1}{2}\rho_\eta r = 0,$$

$$r_{\eta t} + \frac{1}{2}\rho_\xi q + \frac{1}{2}\rho_\eta r = 0, \tag{7.2.21}$$

$$\sigma^2 \rho_{\xi\eta} = -2(qr)_t.$$

Finally, under the reduction

$$q = -\frac{i\sigma}{2}\theta_\eta, \quad r = -\frac{i\sigma}{2}\theta_\xi \tag{7.2.22}$$

the system (7.2.21) is converted to the following

$$\theta_{t\xi\eta} + \frac{1}{2}\rho_\xi \theta_\eta + \frac{1}{2}\rho_\eta \theta_\xi = 0,$$

$$\rho_{\xi\eta} - \frac{1}{2}(\theta_\xi \theta_\eta)_t = 0. \tag{7.2.23}$$

So, one gets the 2DISG equation in the form (7.1.7).

The corresponding linear system looks like

$$\sigma D_2 \chi + \sigma_3 D_1 \chi - \frac{i\sigma}{2}\begin{pmatrix} 0 & \theta_\eta \\ \theta_\xi & 0 \end{pmatrix}\chi = 0, \tag{7.2.24}$$

$$D_3(\sigma D_2 \chi - \sigma_3 D_1 \chi) - \frac{i\sigma}{2}\begin{pmatrix} 0 & \theta_\xi \\ \theta_\eta & 0 \end{pmatrix} D_3 \chi + \frac{1}{2}\begin{pmatrix} \rho_\xi & 0 \\ 0 & \rho_\eta \end{pmatrix}\chi = 0. \tag{7.2.25}$$

Transiting to the matrix function $\psi$ defined by

$$\psi = \chi \exp\left(i\lambda x - \frac{i\lambda}{\sigma}\sigma_3 y + \frac{i}{2\lambda}mt\right), \tag{7.2.26}$$

one gets the linear system

$$\sigma \psi_y + \sigma_3 \psi_x - \frac{i\sigma}{2}\begin{pmatrix} 0 & \theta_\eta \\ \theta_\xi & 0 \end{pmatrix}\psi = 0,$$

$$\sigma \psi_{yt} - \sigma_3 \psi_{xt} - \frac{i\sigma}{2}\begin{pmatrix} 0 & \theta_\xi \\ \theta_\eta & 0 \end{pmatrix}\psi_t + \frac{1}{2}\begin{pmatrix} \rho_\xi & 0 \\ 0 & \rho_\eta \end{pmatrix}\psi = 0. \tag{7.2.27}$$

In the variables $\xi$ and $\eta$ it is exactly the problem (7.1.10).

Emphasize that the trivial background for the 2DISG equation is $\theta = 2\pi n$ and $\rho = 2m_1\xi + 2m_2\eta$ where $n$ is an arbitrary integer and $m_1$, $m_2$ are arbitrary constants.

The $\bar{\partial}$-dressing on nontrivial background for the 2DISG equation on one hand is simple since the first linear problem is, in fact, the same as for the DS equation. On the other hand it is more complicated than the previous cases since the second linear problem is not of the form (2.5.10). However the general scheme presented in section (2.5) is also effective in this case.

So we choose the operators $\hat{B}_i$ ($i = 1, 2, 3$) as the integral operators with the same leading singularities as in the trivial background case, namely

$$\hat{B}_1 \sim i\lambda, \quad \hat{B}_2 \sim -\frac{i\lambda}{\sigma}\sigma_3 \quad \text{as} \quad \lambda \to \infty \qquad (7.2.28)$$

and

$$\hat{B}_3 \sim \frac{i}{2\lambda}m \quad \text{as} \quad \lambda \to 0 \qquad (7.2.29)$$

The dual space is introduced via the Fourier transform

$$X(s, \bar{s}, x, y, t) = \frac{1}{2\pi}\iint_C d\lambda \wedge d\bar{\lambda}\, e^{i(\lambda s + \bar{\lambda}\bar{s})} f(\lambda, \bar{\lambda}; x, y, t). \qquad (7.2.30)$$

According to the general scheme we would like to have in the dual space the linear system of the 2DISG equation for the background $(\theta_0, \rho_0)$:

$$\sigma\psi_{0y} + \sigma_3\psi_{0s} + P_0\psi_0 = 0,$$

$$\sigma\psi_{0yt} - \sigma_3\psi_{0st} + P_0^T\psi_{0x} + \Gamma_0\psi_0 = 0 \qquad (7.2.31)$$

where

$$P_0 = -\frac{i\sigma}{2}\begin{pmatrix} 0 & \theta_{0\eta} \\ \theta_{0\xi} & 0 \end{pmatrix}, \quad \Gamma_0 = \frac{1}{2}\begin{pmatrix} \rho_{0\eta} & 0 \\ 0 & \rho_{0\xi} \end{pmatrix}. \qquad (7.2.32)$$

Here for simplicity we will confine ourselves by the almost trivial background $\theta_0 = 0$. In this case $\rho_{\xi\eta} = 0$, i. e. $\rho = 2v_1(\xi, t) + 2v_2(\eta, t)$ where $v_1$ and $v_2$ are arbitrary functions. So our background system is

$$\sigma\psi_{0y} + \sigma_3\psi_{0s} = 0,$$

$$\sigma\psi_{0yt} - \sigma_3\psi_{0st} + \Gamma_0\psi_0 = 0 \qquad (7.2.33)$$

where

$$\Gamma_0 = \begin{pmatrix} -2u_1(\xi, t) & 0 \\ 0 & -2u_2(\eta, t) \end{pmatrix} \qquad (7.2.34)$$

and

$$\sigma u_1(\xi, t) \doteq -\frac{\partial}{2\partial\xi}v_1(\xi, t), \quad \sigma u_2(\eta, t) \doteq -\frac{\partial}{2\partial\eta}v_2(\eta, t).$$

Chapter 7

The system (7.2.33) is equivalent to the following

$$\psi_{0y} = -\frac{1}{\sigma}\sigma_3 \partial s \psi_0,$$

$$\psi_{0t} = -\frac{m}{2}\partial_s^{-1}\psi_0 + \frac{1}{2}\partial_s^{-1}((\sigma_3\Gamma_0 + m)\psi_0). \tag{7.2.35}$$

The system (7.2.35) determines the choice of the operators $\tilde{\hat{B}}_i$ as

$$\tilde{\hat{B}}_1 = \partial_s,$$

$$\tilde{\hat{B}}_2 = -\frac{1}{\sigma}\sigma_3 \partial_s, \tag{7.2.36}$$

$$\tilde{\hat{B}}_3 = -\frac{m}{2}\partial_s^{-1} + \frac{1}{2}\partial_s^{-1}(\sigma_3\Gamma_0 + m).$$

In the spectral space these operators act as follows

$$\hat{B}_1 = i\lambda,$$

$$\hat{B}_2 = -\frac{i\lambda}{\sigma}\sigma_3, \tag{7.2.37}$$

$$(\hat{B}_3 f)(\lambda) = \frac{im}{2\lambda}f(\lambda) - \frac{i}{2\lambda}\iint_C d\mu \wedge d\bar{\mu}\tilde{\Gamma}_0(\lambda - \mu)f(\mu)$$

where

$$\tilde{\Gamma}_0(\lambda, \bar{\lambda}, s, x, y, t) \doteq \frac{1}{2\pi}\iint_C ds \wedge d\bar{s}\, e^{-i(\lambda s + \bar{\lambda}\bar{s})}(\sigma_3\Gamma_0 + m)(s + x, y, t). \tag{7.2.38}$$

So the corresponding operators $D_i$ are

$$(D_1\chi)(\lambda) = (\partial_x + i\lambda)\chi(\lambda),$$

$$(D_2\chi)(\lambda) = \frac{1}{\sigma}\partial_y\chi - \frac{i\lambda}{\sigma}\chi(\lambda)\sigma_3, \tag{7.2.39}$$

$$(D_3\chi)(\lambda) = \partial_x\chi(\lambda) + \frac{i}{2\lambda}\chi(\lambda)m - \frac{i}{2}\iint_C d\mu \wedge d\bar{\mu}\frac{1}{\lambda + \mu}\chi(\lambda + \mu)\tilde{\Gamma}_0(\mu).$$

Using these operators $D_i$ ($i = 1, 2, 3$), one should construct the linear problems. The first problem is, obviously, the same as in the case of trivial background (7.2.24). To find the second problem we again start with the quantity

$$D_3(\sigma D_2\chi - \sigma_3 D_1\chi) = \sigma\chi_{yt} - \sigma_3\chi_{xt} - i\lambda\{\sigma_3, \chi_t\}$$

$$+ \frac{i}{2\lambda}(\sigma\chi_y - \sigma_3\chi_x - i\lambda\{\sigma_3, \chi\})m - \frac{i}{2}\iint_C d\mu \wedge d\bar{\mu}\,\tilde{\Gamma}_0(\mu)\frac{1}{\lambda + \mu}(\sigma\chi_y(\lambda + \mu)$$

$$- \sigma_3\chi_x(\lambda + \mu) - i(\lambda + \mu)\{\sigma_3, \chi(\lambda + \mu)\}) = 0. \tag{7.2.40}$$

It is clear that the compensation of the singularity at $\lambda = 0$ is achieved by adding the term $QD_3\chi$ where $Q$ is given by (7.2.13). Finally, the requirement that

$$D_3(\sigma D_2\chi - \sigma_3 D_1\chi) + QD_3\chi + \tilde{Q}\chi \to 0 \tag{7.2.41}$$

as $\lambda \to \infty$ gives

$$\tilde{Q} = i\{\sigma_3, \chi_{-1t}\} + \iint_C d\mu \wedge d\bar{\mu}\tilde{\Gamma}_0(\mu)\sigma_3. \tag{7.2.42}$$

Taking into account (7.2.38), one gets

$$\tilde{Q} = i\{\sigma_3, \chi_{-1t}\} + \Gamma_0\sigma_3 + \sigma_3 m. \tag{7.2.43}$$

So we have the second problem

$$D_3(\sigma D_2\chi - \sigma_3 D_1\chi) + P^T D_3\chi + \tilde{Q}\chi = 0 \tag{7.2.44}$$

where $\tilde{Q}$ is given by (7.2.43). Denoting

$$\frac{\sigma}{2}\begin{pmatrix} \rho_\eta & 0 \\ 0 & \rho_\xi \end{pmatrix} = -2i\sigma_3(\chi_{-1t})_{diag} + \Gamma_0 + m\sigma_3, \tag{7.2.45}$$

we finally get the linear system

$$\sigma D_2\chi + \sigma_3 D_1\chi - \frac{i\sigma}{2}\begin{pmatrix} 0 & \theta_\eta \\ \theta_\xi & 0 \end{pmatrix}\chi = 0,$$

$$D_3(\sigma D_2 - \sigma_3 D_1)\chi - \frac{i\sigma}{2}\begin{pmatrix} 0 & \theta_\xi \\ \theta_\eta & 0 \end{pmatrix}D_3\chi + \frac{\sigma}{2}\begin{pmatrix} \rho_\eta & 0 \\ 0 & \rho_\xi \end{pmatrix}\chi = 0 \tag{7.2.46}$$

in the case of the nontrivial background (7.2.34). This system is of the same form as (7.2.25) for the trivial background and obviously provides the 2DISG equation for $\theta$ and $\rho$ given by

$$\theta_\eta = 4(\chi_{-1})_{12},$$

$$\rho_\eta = -\frac{4i}{\sigma}(\chi_{-1t})_{11} - 2u_1(\eta,t) + m_1. \tag{7.2.47}$$

The method of construction of exact solutions of the 2DISG equation is the same as that used in the previous chapters.

The Fourier transform $\tilde{R}(s,\tilde{s},x,y,t)$ of the kernel $R$ defined by (2.5.15) obeys the system of equations

$$\tilde{R}_x(s,\tilde{s},x,y,t) - \tilde{R}_s - \tilde{R}_{\tilde{s}} = 0,$$

$$\sigma\tilde{R}_y(s,\tilde{s},x,y,t) + \sigma_3\tilde{R}_s + \tilde{R}_{\tilde{s}}\sigma_3 = 0, \tag{7.2.48}$$

$$\tilde{R}_t(s,\tilde{s},x,y,t) + \frac{\sigma_3}{2}\partial_s^{-1}(\Gamma_0(s')R(s',\tilde{s},x,y,t)) + \frac{1}{2}\partial_{\tilde{s}}^{-1}(R(s,\tilde{s}',x,y,t))\Gamma_0(\tilde{s}')\sigma_3) = 0.$$

This system is solvable by separation of variables. In what follows we will consider the particular case of the kernels $R$ of the form

$$R = \begin{pmatrix} 0 & R_1(\mu, \bar{\mu}; \lambda, \bar{\lambda}) \\ R_2(\mu, \bar{\mu}; \lambda, \bar{\lambda}) & 0 \end{pmatrix} \quad (7.2.49)$$

or, equivalently,

$$\tilde{R}(\tilde{s}, s, \ldots) = \begin{pmatrix} 0 & \tilde{R}_1(s, \tilde{s}, \ldots) \\ \tilde{R}_2(s, \tilde{s}, \ldots) & 0 \end{pmatrix}. \quad (7.2.50)$$

For such a kernel the system (7.2.48) is reduced to

$$\tilde{R}_{1x} - \tilde{R}_{1s} - \tilde{R}_{1\tilde{s}} = 0,$$

$$\sigma \tilde{R}_{1y} + \tilde{R}_{1s} - \tilde{R}_{1\tilde{s}} = 0, \quad (7.2.51)$$

$$\tilde{R}_{1t} - \sigma \partial_s^{-1}(u_1(y - \sigma(x + s'), t) R(s', \tilde{s})) + \sigma \partial_{\tilde{s}}^{-1}(u_2(y + \sigma(x + \tilde{s}'), t) R(s, \tilde{s}')) = 0.$$

The separation of variables in (7.2.51) via

$$\tilde{R}_1(s, \tilde{s}) = \sum_{i,k=1}^{N} \rho_{0ik} X_i(s, x, y, t) Y_k(\tilde{s}, x, y, t) \quad (7.2.52)$$

leads to the equations

$$X_{ix} - X_{is} = 0,$$

$$\sigma X_{iy} + X_{is} = 0, \quad (7.2.53)$$

$$X_{it} - \sigma \partial_s^{-1}(u_1(y - \sigma(x + s'), t)) X(s') = 0$$

and

$$Y_{ix} - Y_{i\tilde{s}} = 0,$$

$$\sigma Y_{iy} - Y_{i\tilde{s}} = 0, \quad (7.2.54)$$

$$Y_{it} + \sigma \partial_{\tilde{s}}^{-1}(u_2(y + \sigma(x + \tilde{s}'), t)) Y(\tilde{s}') = 0.$$

The system (7.2.53) is equivalent to the following equation

$$X_{izt} + u_1(z, t) X_i = 0$$

where $z = \xi - \sigma s$ and

$$X_i = X_i(z, t). \quad (7.2.55)$$

Respectively the system (7.2.54) is equivalent to

$$Y_{i\tilde{z}t} + u_2(\tilde{z},t)Y_i = 0 \tag{7.2.56}$$

where $\tilde{z} = \eta + \sigma\tilde{s}$ and

$$Y_i = Y_i(\tilde{z},t). \tag{7.2.57}$$

Note that

$$u_i(z,t) \underset{|z|\to\infty}{\longrightarrow} m_i \tag{7.2.58}$$

where $m_1$ and $m_2$ are constants.

Solving equations (7.2.54) and (7.2.56), one can construct the kernel $R$.

Introducing the variable

$$\tilde{\psi} = \chi(\lambda,\bar{\lambda};x,y,t)\exp(i\lambda x - \frac{i\lambda}{\sigma}\sigma_3 y), \tag{7.2.59}$$

one represents the linear system (7.2.46) in the form

$$\sigma\tilde{\psi}_y + \sigma_3\tilde{\psi}_x - \frac{i\sigma}{2}\begin{pmatrix} 0 & \theta_\eta \\ \theta_\xi & 0 \end{pmatrix}\tilde{\psi} = 0,$$

$$\sigma\tilde{\psi}_{yt} - \sigma_3\tilde{\psi}_{xt} - \frac{i\sigma}{2}\begin{pmatrix} 0 & \theta_\xi \\ \theta_\eta & 0 \end{pmatrix}\tilde{\psi}_t + \frac{\sigma}{2}\begin{pmatrix} \rho_\eta & 0 \\ 0 & \rho_\xi \end{pmatrix}\psi$$

$$-\frac{i\sigma}{2}\iint_C d\mu\wedge d\bar{\mu}\,\tilde{\Gamma}_0(\mu)\frac{1}{\lambda+\mu}(\sigma\tilde{\psi}_y(\lambda+\mu)$$

$$-\sigma_3\tilde{\psi}_x(\lambda+\mu) + (\lambda+\mu)\{\sigma_3,\tilde{\psi}(\lambda+\mu)\})\exp\left(-i\mu x + \frac{i\mu}{\sigma}\sigma_3 y\right)$$

$$+\frac{1}{4}\begin{pmatrix} 0 & \theta_\xi \\ \theta_\eta & 0 \end{pmatrix}\iint_C d\mu\wedge d\bar{\mu}\,\tilde{\Gamma}_0(\mu)\frac{1}{\lambda+\mu}\tilde{\psi}(\lambda+\mu)\exp\left(-i\mu x + \frac{i\mu}{\sigma}\sigma_3 y\right) = 0.$$

For the first time this system has been derived in [257] by different method.

## 7.3. Exact solutions

At first we will consider the case of the trivial background $\theta_0 = 0$ $\rho_0 = 2m_1\xi + 2m_2\eta$. We will follow mainly the paper [257].

We will confine ourselves to the solutions of the 2DISG equation which correspond to a class of off-diagonal kernels $R$ of the $\bar{\partial}$-problem (7.2.1)

$$R(\mu,\bar{\mu};\lambda,\bar{\lambda}) = \begin{pmatrix} 0 & R_1(\mu,\bar{\mu};\lambda,\bar{\lambda}) \\ R_2(\mu,\bar{\mu};\lambda,\bar{\lambda}) & 0 \end{pmatrix}. \tag{7.3.1}$$

The $\xi$, $\eta$ and $t$ dependence of the functions $R_1$, $R_2$ is given by

$$R_1(\mu,\bar{\mu},\lambda,\bar{\lambda};\xi,\eta,t) = R_{10}(\mu,\bar{\mu};\lambda,\bar{\lambda})\exp\left(-\frac{2i\mu}{\sigma}\xi - \frac{2i\lambda}{\sigma}\eta - \frac{i}{2}\left(\frac{m_1}{\mu} + \frac{m_2}{\lambda}\right)t\right),$$

$$R_2(\mu,\bar{\mu},\lambda,\bar{\lambda};\xi,\eta,t) = R_{20}(\mu,\bar{\mu};\lambda,\bar{\lambda})\exp\left(\frac{2i\mu}{\sigma}\eta + \frac{2i\lambda}{\sigma}\xi + \frac{i}{2}\left(\frac{m_1}{\lambda} + \frac{m_2}{\mu}\right)t\right), \quad (7.3.2)$$

where $R_{10}$ and $R_{20}$ are arbitrary functions.

For the kernels $R$ of the form (7.3.1) the reconstruction formulae (7.2.8) and (7.2.19) give

$$q = \frac{1}{\pi}\iint_C d\lambda \wedge d\bar{\lambda} \iint_C d\mu \wedge d\bar{\mu} R_1(\mu,\bar{\mu};\lambda,\bar{\lambda})\chi_{11}(\mu,\bar{\mu}),$$

$$r = -\frac{1}{\pi}\iint_C d\lambda \wedge d\bar{\lambda} \iint_C d\mu \wedge d\bar{\mu} R_2(\mu,\bar{\mu};\lambda,\bar{\lambda})\chi_{22}(\mu,\bar{\mu}),$$

$$\rho_\xi = 2m_1 - \frac{2}{\pi}\iint_C d\mu \wedge d\bar{\mu} \iint_C d\lambda \wedge d\bar{\lambda} R_2(\mu,\bar{\mu};\lambda,\bar{\lambda})\chi_{12}(\mu,\bar{\mu}). \quad (7.3.3)$$

For small $q$ and $r$ one has

$$q = \frac{i}{\pi}\iint_C d\lambda \wedge d\bar{\lambda} \iint_C d\mu \wedge d\bar{\mu} R_{10}(\mu,\bar{\mu};\lambda,\bar{\lambda})\exp\left(\frac{-2i\mu}{\sigma}\xi - \frac{2i\lambda}{\sigma}\eta - \frac{i}{2}\left(\frac{m_1}{\mu} + \frac{m_2}{\lambda}\right)t\right),$$
$$\quad (7.3.4)$$

$$r = -\frac{i}{\pi}\iint_C d\lambda \wedge d\bar{\lambda} \iint_C d\mu \wedge d\bar{\mu} R_{20}(\mu,\bar{\mu};\lambda,\bar{\lambda})\exp\left(\frac{2i\mu}{\sigma}\eta + \frac{2i\lambda}{\sigma}\xi + \frac{i}{2}\left(\frac{m_1}{\lambda} + \frac{m_2}{\mu}\right)t\right).$$
$$\quad (7.3.5)$$

A necessary condition for the reduction

$$q = -i\sigma/2\theta_\eta, \quad r = -i\sigma/2\theta_\xi \quad (7.3.6)$$

can also be obtained from the formulae (7.3.4). In the case $\sigma = 1$ this condition is of the form

$$R_{10}(\mu,\bar{\mu};\lambda,\bar{\lambda}) = i\bar{\lambda}\cdot\tilde{R}(\mu,\bar{\mu};\lambda,\bar{\lambda}),$$

$$R_{20}(\mu,\bar{\mu};\lambda,\bar{\lambda}) = i\lambda\cdot\tilde{R}(-\lambda,-\bar{\lambda};-\mu,-\bar{\mu}) \quad (7.3.7)$$

where $\tilde{R}(\mu,\bar{\mu};\lambda,\bar{\lambda})$ is some function and

$$\theta = \frac{2\sigma^2}{\pi i}\iint_C d\lambda \wedge d\bar{\lambda} \iint_C d\mu \wedge d\bar{\mu} \tilde{R}(\mu,\bar{\mu};\lambda,\bar{\lambda})\exp\left(-\frac{2i}{\sigma}(\mu\xi + \lambda\eta) + \frac{\sigma}{4}\left(\frac{m_1}{\mu} + \frac{m_2}{\lambda}\right)t\right).$$
$$\quad (7.3.8)$$

The reality condition $\theta = \bar{\theta}$, in particular, implies Im $m_1$ = Im $m_2$ = 0 and

$$\overline{\tilde{R}(\mu,\bar{\mu};\lambda,\bar{\lambda})} = \tilde{R}(-\bar{\mu},-\mu;-\bar{\lambda},-\lambda). \quad (7.3.9)$$

In the case $\sigma^2 = -1$, $\bar{\eta} = \xi = \bar{z}$ and the necessary condition for the reduction (7.3.6) is

$$R_{10}(\mu, \bar{\mu}; \lambda, \bar{\lambda}) = i\bar{\lambda} \cdot \tilde{R}(\mu, \bar{\mu}; \lambda, \bar{\lambda}),$$

$$R_{20}(\mu, \bar{\mu}; \lambda, \bar{\lambda}) = i\lambda \cdot \tilde{R}(-\lambda, -\bar{\lambda}; -\mu, -\bar{\mu}), \qquad (7.3.10)$$

while for the real valued $\theta$ one has $m_2 = \bar{m}_1 = m$ and

$$\overline{\tilde{R}(\mu, \bar{\mu}; \lambda, \bar{\lambda})} = \tilde{R}(-\bar{\lambda}, -\lambda; -\bar{\mu}, -\mu). \qquad (7.3.11)$$

In all these formulae nothing is assumed about the asymptotic behaviour of $\theta$ at the infinity $\xi^2 + \eta^2 \to \infty$. The necessary condition for the boundedness of $\theta$ can also be extracted from (7.3.8). It is equivalent to the condition of the pure oscillating character of the exponents in (7.3.8). So at $\sigma = 1$ one has

$$R_{10}(\mu, \bar{\mu}; \lambda, \bar{\lambda}) = \delta(\mathrm{Im}\lambda) \cdot \delta(\mathrm{Im}\mu) \cdot \tilde{R}_1(\mu, \lambda),$$

$$R_{20}(\mu, \bar{\mu}; \lambda, \bar{\lambda}) = \delta(\mathrm{Im}\lambda) \cdot \delta(\mathrm{Im}\mu) \cdot \tilde{R}_2(\mu, \lambda). \qquad (7.3.12)$$

For such $R_{10}$ and $R_{20}$ the $\bar{\partial}$-problem (7.2.1) is reduced to the nonlocal Riemann—Hilbert problem with the jump across the real axis.

In another case $\sigma^2 = -1$ the condition of boundedness is

$$R_{10}(\mu, \bar{\mu}; \lambda, \bar{\lambda}) = \delta(\lambda + \bar{\mu}) \cdot \tilde{R}_1(\lambda, \bar{\lambda}),$$

$$R_{20}(\mu, \bar{\mu}; \lambda, \bar{\lambda}) = \delta(\lambda + \bar{\mu}) \cdot \tilde{R}_2(\lambda, \bar{\lambda}). \qquad (7.3.13)$$

In this case the problem (7.2.1) is reduced to the quasilocal $\bar{\partial}$-problem.

**Solutions with functional parameters.** A first class of the exact solutions corresponds, typically for the dressing method, to the general degenerated kernels $R_1$ and $R_2$. So, having in mind the conditions (7.3.7) and (7.3.10), we put

$$R_{10}(\mu, \bar{\mu}; \lambda, \bar{\lambda}) = i\bar{\lambda} \sum_{n=1}^{N} f_n(\mu, \bar{\mu}) \cdot g_n(\lambda, \bar{\lambda}),$$

$$R_{20}(\mu, \bar{\mu}; \lambda, \bar{\lambda}) = i\lambda \cdot \sum_{n=1}^{N} f_n(-\lambda, -\bar{\lambda}) \cdot g_n(-\mu, -\bar{\mu}). \qquad (7.3.14)$$

The condition of the reality of $\theta$ gives at $\sigma = 1$

$$\sum_{n=1}^{N} \overline{f_n(\mu, \bar{\mu})} \cdot \overline{g_n(\lambda, \bar{\lambda})} = \sum_{n=1}^{N} f_n(-\bar{\mu}, -\mu) \cdot g_n(-\bar{\lambda}, -\lambda). \qquad (7.3.15)$$

Chapter 7

The condition (7.3.15) means that

$$\overline{f_n(\mu,\bar{\mu})} = f_n(-\bar{\mu},-\mu),$$

$$\overline{g_n(\lambda,\bar{\lambda})} = g_n(-\bar{\lambda},-\lambda), \quad (n=1,\ldots,N) \tag{7.3.16}$$

for arbitrary $N$ or $N = 2M$ and

$$\overline{f_n(\mu,\bar{\mu})} = f_{n+M}(-\bar{\mu},-\mu),$$

$$\overline{g_n(\lambda,\bar{\lambda})} = g_{n+M}(-\bar{\lambda},-\lambda), \quad (n=1,\ldots,M), \quad (\text{mod } M). \tag{7.3.17}$$

For the 2DISG-II equation ($\sigma^2 = -1$) the condition (7.3.10) of the reality $\theta$ gives

$$\sum_{n=1}^{N} \overline{f_n(\mu,\bar{\mu})} \cdot \overline{g_n(\lambda,\bar{\lambda})} = \sum_{n=1}^{N} f_n(-\bar{\lambda},-\lambda) \cdot g_n(-\bar{\mu},-\mu). \tag{7.3.18}$$

The condition (7.3.18) means that

$$g_n(\lambda,\bar{\lambda}) = \overline{f_n(-\bar{\lambda},-\lambda)}, \quad (n=1,\ldots,N) \tag{7.3.19}$$

for arbitrary $N$ or $N = 2M$ and

$$g_n(\lambda,\bar{\lambda}) = \overline{f_{n+M}(-\bar{\lambda},-\lambda)}, \quad n=1,\ldots,M \quad (\text{mod } M). \tag{7.3.20}$$

So the kernels $R_{10}$ and $R_{20}$ are of the form

$$R_{10}(\mu,\bar{\mu};\lambda,\bar{\lambda}) = i\bar{\lambda} \sum_{n=1}^{N} f_n(\mu,\bar{\mu}) \cdot \overline{f_n(-\lambda,-\bar{\lambda})},$$

$$R_{20}(\mu,\bar{\mu};\lambda,\bar{\lambda}) = i\lambda \sum_{n=1}^{N} f_n(-\lambda,-\bar{\lambda}) \cdot \overline{f_n(\mu,\bar{\mu})}, \tag{7.3.21}$$

for an arbitrary $N$ or

$$R_{10}(\mu,\bar{\mu};\lambda,\bar{\lambda}) = i\bar{\lambda} \sum_{n=1}^{2M} f_n(\mu,\bar{\mu}) \cdot \overline{f_{n+M}(-\lambda,-\bar{\lambda})},$$

$$R_{20}(\mu,\bar{\mu};\lambda,\bar{\lambda}) = i\lambda \sum_{n=1}^{2M} f_n(-\lambda,-\bar{\lambda}) \cdot \overline{f_{n+M}(\mu,\bar{\mu})},$$

where $f_{n+2M} = f_n$.

To calculate $\theta$, using the formula

$$\theta_\xi = \frac{1}{\pi} \iint_C d\lambda \wedge d\bar{\lambda} \iint_C d\mu \wedge d\bar{\mu} \, \exp\left\{-\frac{2i}{\sigma}\left(\mu\xi + \lambda\eta + \frac{\sigma}{4}\left(\frac{m_1}{\mu} + \frac{m_2}{\lambda}\right)t\right)\right\}$$

$$\times i\lambda \sum_{n=1}^{N} f_n(\mu,\bar{\mu}) \cdot g_n(\lambda,\bar{\lambda}) \cdot \chi_{11}(\mu,\bar{\mu}), \tag{7.3.22}$$

one should find $\chi_{11}(\mu, \bar{\mu})$. From the system (2.2.3) one has

$$\chi_{11}(\lambda, \bar{\lambda}) = 1 - \frac{1}{(2\pi)^2} \iint_C \frac{d\lambda' \wedge d\bar{\lambda}'}{\lambda' - \lambda} \iint_C d\lambda'' \wedge d\bar{\lambda}'' R_2(\lambda'', \bar{\lambda}''; \lambda', \bar{\lambda}')$$

$$\times \iint_C \frac{d\mu' \wedge d\bar{\mu}'}{\bar{\mu}' - \bar{\lambda}''} \iint_C d\mu'' \wedge d\bar{\mu}'' R_1(\mu'', \bar{\mu}''; \mu', \bar{\mu}') \chi_{11}(\mu'', \bar{\mu}''). \quad (7.3.23)$$

Multiplying equation (7.3.17) by

$$\exp\left(-\frac{2i}{\sigma}\left(\lambda\xi + \frac{\sigma m_1}{4\lambda}t\right)\right) f_m(\lambda, \bar{\lambda}),$$

integrating over $\lambda$ and taking into account (7.3.14), we obtain the system

$$F_m = Y_m + \sum_{k=1}^{N} (A \cdot B)_{mk} F_k, \quad m = 1, \ldots, N \quad (7.3.24)$$

where

$$F_m(\xi, \eta, t) \doteq \iint d\lambda \wedge d\bar{\lambda} \exp\left(\frac{2i}{\sigma}(\lambda\xi + \frac{\sigma m_1}{4\lambda}t) \cdot f_k(\lambda, \bar{\lambda}) \cdot \chi_{11}(\lambda, \bar{\lambda}; \xi, \eta, t)\right),$$

$$X_m(\xi, t) \doteq \iint d\lambda \wedge d\bar{\lambda} \exp\left(\frac{-2i}{\sigma}(\lambda\xi + \frac{\sigma m_1}{4\lambda}t) \cdot f_m(\lambda, \bar{\lambda})\right), \quad (7.3.25)$$
$$m = 1, \ldots, N$$

and the matrix elements of the matrices $A$ and $B$ are

$$A_{mn}(\xi, t) \doteq \frac{1}{2\pi} \iint_C d\lambda \wedge d\bar{\lambda} \iint_C d\mu \wedge d\bar{\mu} \frac{\mu}{\mu - \lambda}$$

$$\times \exp\left(\frac{-2i}{\sigma}\left(\lambda\xi - \mu\xi + \frac{\sigma}{4}\left(\frac{m_1}{\lambda}t - \frac{m_1}{\mu}t\right)\right)\right) \cdot f_m(\lambda, \bar{\lambda}) \cdot f_n(-\mu, -\bar{\mu}),$$

$$B_{mn}(\eta, t) \doteq \frac{1}{2\pi} \iint_C d\lambda \wedge d\bar{\lambda} \iint_C d\mu \wedge d\bar{\mu} \frac{\bar{\mu}}{\bar{\mu} - \bar{\lambda}}$$

$$\times \exp\left(\frac{-2i}{\sigma}\left(-\bar{\lambda}\eta + \bar{\mu}\eta - \frac{\sigma m_2}{4\bar{\lambda}}t + \frac{\sigma m_2}{4\bar{\mu}}t\right)\right) \cdot g_m(-\lambda, -\bar{\lambda}) \cdot g_n(\mu, \bar{\mu}) \quad (7.3.26)$$

$$m, n = 1, \ldots, N.$$

Solving the algebraic system (7.3.24) and substituting the results into (7.3.22), one gets

$$\theta_\xi = 2 \cdot \sum_{n,m=1}^{N} Y_{n,\eta}(\eta, t) \cdot (1 - AB)^{-1}_{nm} X_m(\xi, t) \quad (7.3.27)$$

Chapter 7

where
$$Y_n(\eta, t) \doteq \iint d\lambda \wedge d\bar{\lambda} \, \exp\left(\frac{-2i}{\sigma}\left(\lambda\eta + \frac{\sigma m_2}{4\lambda}t\right)\right) \cdot g_n(\lambda, \bar{\lambda}). \tag{7.3.28}$$

From (7.3.25), (7.3.26) and (7.3.28) it is not difficult to see that the matrices $A_{mn}$ and $B_{mn}$ can be expressed in terms of the quantities $X_n(\xi, t)$ and $Y_n(\eta, t)$ by the compact formulae
$$A_{nm} = i \int^{\xi} d\xi' \, X_n(\xi', t) \cdot X_{m\xi'}(\xi', t),$$
$$B_{nm} = i \int^{\eta} d\eta' \, Y_n(\eta', t) \cdot Y_{m\eta'}(\eta', t). \tag{7.3.29}$$

Integration in these expressions is performed over suitable contour.

Note that the functions $X_n$ and $Y_n$ are arbitrary solutions of the linear Klein–Gordon equations
$$X_{\xi t} + m_1 X = 0,$$
and
$$Y_{\eta t} + m_2 Y = 0. \tag{7.3.30}$$

Similar formulae can also be obtained for $\theta_\eta$.

So for arbitrary functions $f_n$ and $g_n$ we have the solutions (7.3.27) which depend on several functional parameters.

The reality conditions (7.3.16), (7.3.17) and (7.3.19), (7.3.20) imply that at $\sigma = 1$:
$$X_n(\xi, t) = \overline{X_n(\xi, t)}, \quad Y_n(\eta, t) = \overline{Y_n(\eta, t)}$$

and at $\sigma^2 = -1$
$$Y_n(z, t) = \overline{X_n(\bar{z}, t)},$$

where $z = \frac{1}{2}(y - ix)$, $\bar{z} = \frac{1}{2}(y + ix)$.

The simplest solution ($N = 1$) of this type for the 2DISG-I equation is of the form:
$$\theta = 4 \arctan \frac{X(\xi, t) \cdot Y(\eta, t)}{2} \tag{7.3.31}$$

where $X(\xi, t) = \overline{X(\xi, t)}$ and $Y(\eta, t) = \overline{Y(\eta, t)}$. In particular, choosing
$$X = \cos(a\xi + \frac{m_1}{a}t - \xi_0), \quad Y = \sin(b\eta + \frac{m_2}{b}t - \eta_0)$$

where $a$ and $b$ are arbitrary constants, one gets the periodic solution
$$\theta = 4 \cdot \arctan\left(\frac{1}{2} \cos\left(a\xi + \frac{m_1}{a}t - \xi_0\right) \cdot \sin\left(b\eta + \frac{m_2}{b}t - \eta_0\right)\right).$$

The simplest solution ($N = 1$) with functional parameters for the 2DISG-II equation ($\sigma^2 = -1$) is of the form:
$$\theta = 4\arctan\frac{|X(\bar{z},t)|^2}{2} \tag{7.3.32}$$
where $z = \frac{1}{2}(Y + ix)$.

**Line solitons (line kinks).** Line solitons of the 2DISG equation are, as usual, the very special case of the solutions with functional parameters considered above. They correspond to the choice
$$f_n(\lambda,\bar{\lambda}) = C_n \cdot \delta(\lambda - \lambda_n),$$
$$g_n(\mu,\bar{\mu}) = \tilde{C}_n \cdot \delta(\mu - \mu_n) \tag{7.3.33}$$
with the constraints (7.3.16) ($\sigma^2 = 1$) and (7.3.19) ($\sigma^2 = -1$).

Let us consider first the case $\sigma^2 = 1$. The constraint (7.3.19) means
$$\bar{\lambda}_n = -\lambda_n, \quad \bar{C}_n = C_n,$$
$$\bar{\mu}_n = -\mu_n, \quad \overline{\tilde{C}_n} = \tilde{C}_n, \quad (n = 1,\ldots,N) \tag{7.3.34}$$
i.e. $\lambda_n = ip_n$, $\mu_n = iq_n$, $\mathrm{Im}\, p_n = \mathrm{Im}\, q_n = 0$ and
$$X_n = -\tilde{C}_n \cdot \exp(2p_n\xi - \frac{m_1}{2p_n}t), \quad Y_n = -C_n \cdot \exp(2q_n\eta - \frac{m_2}{2q_n}t). \tag{7.3.35}$$

The general $N$-line soliton solutions of the 2DISG-I equation is of the form
$$\theta_\xi = 2 \cdot \sum_{n,m} Y_{n\eta}(\eta,t) \cdot (1 - AB)^{-1}_{nm} \cdot X_m(\xi,t) \tag{7.3.36}$$
where
$$A_{nm} = \frac{iq_m}{q_n + q_m} \cdot X_n(\xi,t) \cdot X_m(\xi,t),$$
$$B_{nm} = \frac{ip_m}{p_n + p_m} \cdot Y_n(\eta,t) \cdot Y_m(\eta,t). \tag{7.3.37}$$
The simplest plane soliton looks like
$$\theta(\xi,\eta,t) = 4 \cdot \arctan\frac{X \cdot Y}{2}$$
$$= 4 \cdot \arctan\left(\frac{C \cdot \tilde{C}}{2} \cdot \exp(2p\xi + 2q\eta - \frac{m_1}{2q}t - \frac{m_2}{2p}t)\right). \tag{7.3.38}$$
Note that $\theta \to 2\pi$ as $p\xi + q\eta \to \infty$ and $\theta \to 0$ as $p\xi + q\eta \to -\infty$.

At $N = 2$ one has the two-soliton solution
$$\theta(\xi,\eta,t) = 4 \cdot \arctan\frac{X_1 Y_1 + X_2 Y_2}{2 - \frac{(q_1-q_2)(p_1-p_2)}{2(q_1+q_2)(p_1+p_2)} \cdot X_1 Y_1 X_2 Y_2}. \tag{7.3.39}$$

The general solution (7.3.29) describes an elastic scattering of $N$ line solitons (7.3.38).

Chapter 7

In the case $\sigma^2 = -1$ the reality condition (7.3.19) gives

$$\mu_n = -\bar{\lambda}_n,$$

$$\tilde{C}_n = \bar{C}_n, \quad (n = 1, \ldots, N) \qquad (7.3.40)$$

and

$$X_n(z,t) = \overline{Y_n(\bar{z},t)} = -C_n \cdot \exp\left(-2i\left(\lambda_n z + \frac{i}{4}\frac{m}{\lambda_n}t\right)\right). \qquad (7.3.41)$$

The simplest plane soliton ($n = 1$) of the 2DISG-II equation is

$$\theta(z,\bar{z},t) = 4 \cdot \arctan \frac{|X(z,t)|^2}{2}$$

$$= 4 \arctan \frac{|C|^2}{2} \cdot \exp\left(-2\left(\lambda z + \bar{\lambda}\bar{z} + \frac{im}{4\lambda}t - \frac{i\bar{m}}{4\bar{\lambda}}t\right)\right). \qquad (7.3.42)$$

The general $N$-line soliton solution has the form

$$\theta_z = 2 \cdot \sum_{n,m} \overline{X_{nz}(\bar{z},t)}(1 + D \cdot \bar{D})^{-1}_{nm} \cdot X_m(z,t) \qquad (7.3.43)$$

where

$$D_{nm} = \frac{\lambda_n \bar{X}_n \bar{X}_m}{\lambda_n + \lambda_m}.$$

At $N = 2$ we have the solution

$$\theta = 4 \cdot \arctan \frac{(|X_1|^2 + |X_2|^2)/2}{1 - \frac{|\lambda_1 - \lambda_2|^2}{|\lambda_1 + \lambda_2|^2} \cdot \frac{|X_1 X_2|^2}{4}} \qquad (7.3.44)$$

where $X_n$ are given by the formula (4.3.41).

**Breathers.** The solutions of the 2DISG equation which are the 2+1-dimensional analogs of the well-known breathers of the 1+1-dimensional sine-Gordon equation correspond to the delta-functional $f_n$ and $g_n$ which obey the conditions (7.3.18) ($\sigma^2 = 1$) and (7.3.20) ($\sigma^2 = -1$).

Namely, for $\sigma = 1$

$$R_{10}(\mu,\bar{\mu};\lambda,\bar{\lambda}) = i\bar{\lambda} \cdot \sum_{n=1}^{M}(C_n \delta(\mu - \mu_n)\delta(\lambda - \lambda_n)$$

$$+ \bar{C}_n \delta(\mu + \bar{\mu}_n)\delta(\lambda + \bar{\lambda}_n)), \qquad (7.3.45)$$

$$R_{20}(\mu,\bar{\mu};\lambda,\bar{\lambda}) = i\lambda \cdot \sum_{n=1}^{M}(C_n \delta(\lambda + \mu_n)\delta(\mu + \lambda_n)$$

$$+ \bar{C}_n \delta(\lambda - \bar{\mu}_n)\delta(\mu - \bar{\lambda}_n)) \qquad (7.3.46)$$

and for $\sigma^2 = -1$:

$$R_{10}(\mu, \bar\mu; \lambda, \bar\lambda) = i\bar\lambda \cdot \sum_{n=1}^{M}(C_n \cdot \delta(\lambda + \lambda_n) \cdot \delta(\mu - \bar\mu_n)$$
$$+ \bar C_n \cdot \delta(\lambda + \mu_n) \cdot (\mu - \bar\lambda_n)),$$

$$R_{20}(\mu, \bar\mu; \lambda, \lambda) = i\lambda \cdot \sum_{n=1}^{M}(C_n \cdot \delta(\lambda + \bar\mu_n) \cdot \delta(\mu - \lambda_n)$$
$$+ \bar C_n \cdot (\lambda + \lambda_n) \cdot (\mu - \bar\mu_n)). \qquad (7.3.47)$$

The simplest breather of the 2DISG-I equation looks like ($N = 1$)

$$\theta(\xi, \eta, t) = 4 \cdot \arctan \frac{|C| \cdot e^f \cdot \cos \varphi}{1 - \frac{|C|^2 \lambda_R \mu_R e^{2f}}{4\lambda_I \mu_I}}, \qquad (7.3.48)$$

where

$$C = |C|e^{i\delta},$$

$$f(\xi, \eta, t) = 2\lambda_I \eta - 2\mu_I \xi - \frac{m_1 \lambda_I}{2|\lambda|^2} t + \frac{m_2 \mu_I}{2|\mu|^2} t,$$

$$\varphi(\xi, \eta, t) = 2\lambda_R \eta + 2\mu_R \xi + \frac{m_1 \lambda_R}{2|\lambda|^2} t + \frac{m_2 \mu_R}{2|\mu|^2} t + \delta,$$

$$\lambda = \lambda_R + i\lambda_I, \quad \mu = \mu_R + i\mu_I.$$

The simplest breather of the 2DISG-II equation is ($N = 1$):

$$\theta(z, \bar z, t) = 4\arctan \frac{|C| \cdot e^{-f} \cdot \cos \varphi}{1 + \frac{|\lambda - \mu|^2 |C|^2 e^{-2f}}{4|\lambda + \mu|^2}} \qquad (7.3.49)$$

where

$$C = |C|e^{i\delta},$$

$$f = \mathrm{Re}\left(\lambda z + \bar\mu \bar z + \frac{im\bar\lambda}{|\lambda|^2}t + \frac{i\bar m\mu}{|\mu|^2}t\right),$$

$$\varphi = \mathrm{Im}\left(\lambda z + \bar\mu \bar z + \frac{im\bar\lambda}{|\lambda|^2}t + \frac{i\bar m\mu}{|\mu|^2}t\right) - \delta.$$

The explicit exact solutions of the 2DISG equation enumerated above have been constructed in [257].

## 7.4. Localized solitons

Now let us proceed to the localized solutions associated with nontrivial boundaries. In section (7.2) it has been demonstrated that the construction of such solutions of the 2DISG equation is similar to that of the DS equation. The differences consist in the reduction $q = (-i\sigma)/2\theta_\eta$, $r = (-i\sigma)/2\theta_\xi$ and in the different equations (7.2.54) and (7.2.56) for the functions $X$ and $Y$. Again the consideration of the bounded solutions fixes the 2DISG-I equation ($\sigma = 1$) as the nontrivial case.

Repeating the calculations of section (5.2) (see formulae (5.2.40)-(5.2.51)), one obtains the following formula for exact solutions of the 2DISG-I equation [258]

$$\theta_\xi = 2 \sum_{k,l,m=1}^{N} \rho_{kl} X_{l\xi}(1 + \alpha\rho_0 \beta \rho_0^T)^{-1}_{lm} Y_l \qquad (7.4.1)$$

where $\rho_0$ is an arbitrary matrix and the matrices $\alpha$ and $\beta$ are given by

$$\alpha_{nk} = \int^\eta d\eta' Y_n(\eta', t) Y_{k\eta'}(\eta', t),$$

$$\beta_{nk} = \int^\xi d\xi' X_n(\xi', t) X_{k\xi'}(\xi', t). \qquad (7.4.2)$$

Thus, to construct the solutions (7.4.1) explicitly one should find the exact solutions $X$ and $Y$ of the linear equations (7.2.54) and (7.2.56). They can be constructed by the $\bar{\partial}$-dressing method. In fact, the derivation of the problem of the type (7.2.54) or (7.2.56) has been given in section (2.4) as the second example. With the scalar nonlocal $\bar{\partial}$-problem as the base and the choice (2.4.10) we constructed the linear problem (2.4.26).

$$\psi_{\xi t} + u(\xi, t)\psi = \varepsilon\psi \qquad (7.4.3)$$

where $\varepsilon$ is some constant. The reconstruction formula is of the form (2.4.18)

$$u(\xi, t) = -i\partial \chi_1(\xi, t) \qquad (7.4.4)$$

where the function $\chi$ is the solution of the integral equation

$$\chi(\lambda, \bar\lambda) = 1 + \frac{1}{2\pi i} \iint_C \frac{d\lambda' \wedge d\bar\lambda}{\lambda' - \lambda} \iint_C d\mu \wedge d\bar\mu \, \chi(\mu\bar\mu)$$
$$\times \exp\left(i(\mu - \lambda)\xi - i\varepsilon\left(\frac{1}{\mu} - \frac{1}{\lambda}\right)t\right) R_0(\mu, \bar\mu; \lambda, \bar\lambda) \qquad (7.4.5)$$

normalized as $\chi \to 1$ as $\lambda \to \infty$ with the additional constraint $\chi(\lambda = 0) = 1$. Here for convenience we made a change of variable $\lambda \to i\lambda$.

The condition $\chi(\lambda = 0) = 1$ is equivalent to the following

$$\iint_C \frac{d\lambda \wedge d\bar{\lambda}}{\lambda} \iint_C d\mu \wedge d\bar{\mu}\, \chi(\mu, \bar{\mu})$$
$$\times \exp\left(i(\mu - \lambda)\xi - i\varepsilon(\frac{1}{\mu} - \frac{1}{\lambda})t\right) R_0(\mu, \bar{\mu}; \lambda, \bar{\lambda}) = 0. \qquad (7.4.6)$$

It is also not difficult to show similar to all previous cases, that the necessary condition for reality of $u$ is of the form

$$\overline{R_0(\mu, \bar{\mu}; \lambda, \bar{\lambda})} = -R_0(-\mu, -\bar{\mu}; -\lambda, -\bar{\lambda}). \qquad (7.4.7)$$

Using these formulae one can find the exact solutions of equation (7.4.3). Solutions of (7.4.3) which are relevant for the construction of the localized solutions of the 2DISG-I equation are soliton like solutions. They correspond, similar to the previous cases, to the delta-function kernels $R$. The complexification in this case is due to the condition (7.4.6).

One can show that the kernels $R_0$ which are compatible both with reality and potentiality conditions (7.4.7), (7.4.6) are of the form

$$R_0(\mu, \bar{\mu}; \lambda, \bar{\lambda}) = i\pi^2 \sum_{k=1}^{2N} (S_k \delta(\mu - i\alpha_k)\delta(\lambda - i\beta_k)) \qquad (7.4.8)$$

where $S_k$, $r_k$, $\alpha_k$, $\beta_k$, $\gamma_k$, $\delta_k$ are real constants or

$$R_0(\mu, \bar{\mu}; \lambda, \bar{\lambda}) = i\pi^2 \sum_{k=1}^{N} (S_k \delta(\mu - \mu_k)\delta(\lambda - \lambda_k)$$

$$+ S_k \delta(\mu + \bar{\mu}_k)\delta(\lambda + \bar{\lambda}_k)) \qquad (7.4.9)$$

where $S_k$, $\mu_k$, $\lambda_k$ are complex constants and in both cases $N$ is an arbitrary integer. The corresponding solutions $u(\xi, t)$ are calculated by the same procedure (2.2.23)–(2.2.25) as in the previous cases. In addition the condition (7.4.6) imposes certain constraint on all constants which parametrize the kernels (7.4.8) and (7.4.9). The constructed solutions $u$ are the line solitons [258].

Here we restrict ourselves by the simplest case $N = 1$. For the choice (7.4.8) the function $\chi(\lambda)$ is given by

$$\chi(\lambda, \bar{\lambda}) = 1 + i \sum_{k=1}^{2} \frac{S_k \chi(i\alpha_k) \exp(F(i\alpha_k) - F(i\beta_k))}{i\beta_k - \lambda} \qquad (7.4.10)$$

where

$$F(\lambda) = i\lambda\xi - \frac{\varepsilon}{\lambda}t.$$

In virtue of (7.3.45), one has

$$\sum_{k=1}^{2} \frac{S_k}{\beta_k}\chi(i\alpha_k)\exp(F(i\alpha_k) - F(i\beta_k)) = 0. \qquad (7.4.11)$$

Formula (7.4.10) also implies the system of two equations for $\chi(i\alpha_1)$ and $\chi(i\alpha_2)$. An analysis of this system and equation (7.4.11) shows that they are satisfied if

$$\alpha_1 = -\beta_2, \quad \alpha_2 = -\beta_1, \quad \beta_2 = -\frac{S_2}{S_1}\beta_1, \quad \alpha_1 = -\frac{S_2}{S_1}\alpha_2 \qquad (7.4.12)$$

and

$$\chi(i\alpha_1) = \chi(i\alpha_2) = \frac{1}{1 + \frac{S_2}{2\alpha_1}\frac{(\alpha_1-\alpha_2)}{(\alpha_1+\alpha_2)}\exp(F(i\alpha_1) - F(i\alpha_2))}. \qquad (7.4.13)$$

The corresponding potential $u$ is calculated by the formula (7.4.4). It is of the form [258]

$$u(\xi, t) = \frac{\varepsilon(\alpha_1 + \alpha_2)^2}{2\alpha_2\alpha_1}\text{ch}^{-2}\left[\left(\frac{\alpha_1 + \alpha_2}{2}\right)\left(\xi + \frac{\varepsilon t}{\alpha_1\alpha_2} - \xi_0\right)\right] \qquad (7.4.14)$$

where

$$\xi_0 = \frac{1}{(\alpha_1 + \alpha_2)}\ln\frac{S_2(\alpha_1 - \alpha_2)}{2\alpha_1(\alpha_1 + \alpha_2)}.$$

This solution is bounded if

$$\frac{S_2(\alpha_1 - \alpha_2)}{2\alpha_1(\alpha_1 + \alpha_2)} > 1. \qquad (7.4.15)$$

The solutions $\psi$ of equation (7.4.3) for the potential $u$ (7.4.14) are

$$\psi_1(\xi, t) = \left(-\exp\left(i\lambda\xi - \frac{i\varepsilon t}{\lambda}\right)\chi(\lambda, \bar{\lambda})\right)\bigg|_{\lambda=i\alpha_1}$$

$$= \frac{\exp\left(-\alpha_1\xi - \frac{\varepsilon}{\alpha_1}t\right)}{1 + \frac{S_2(\alpha_1-\alpha_2)}{2\alpha_1(\alpha_1+\alpha_2)}\exp(F(i\alpha_1) - F(i\alpha_2))},$$

$$\psi_2(\xi, t) = \left(\exp\left(i\lambda\xi - \frac{i\varepsilon t}{\lambda}\right)\chi(\lambda, \bar{\lambda})\right)\bigg|_{\lambda=i\alpha_2}$$

$$= \frac{\exp\left(-\alpha_2\xi - \frac{\varepsilon}{\alpha_2}t\right)}{1 + \frac{S_2(\alpha_1-\alpha_2)}{2\alpha_1(\alpha_1+\alpha_2)}\exp(F(i\alpha_1) - F(i\alpha_2))}. \qquad (7.4.16)$$

In a similar manner one can show that for the kernel (7.4.9) at $N = 1$ the condition (7.4.6) is satisfied if

$$\mu_1 = \bar{\lambda}_1, \quad \frac{S_1}{\lambda_1} = \frac{\bar{S}_1}{\bar{\lambda}_1}. \qquad (7.4.17)$$

The corresponding solution of equation (7.4.3) is [258]

$$u(\xi,t) = \frac{2\varepsilon(\lambda_{1Im})^2}{|\lambda_1|^2}\text{ch}^{-2}\left[\lambda_{1Im}\left(\xi + \frac{\varepsilon t}{|\lambda_0|^2} - \xi_0\right)\right], \qquad (7.4.18)$$

$$\psi_1(\xi,t) = \left(\exp\left(i\lambda\xi - \frac{i\varepsilon t}{\lambda}\right)\chi(\lambda,\bar\lambda)\right)\Big|_{\lambda=\lambda_1} = \frac{\exp\left(i\lambda_1\xi - \frac{i\varepsilon t}{\lambda_1}\right)}{1 + \exp\left(2\lambda_{1Im}\left(\xi + \frac{\varepsilon t}{|\lambda_1|^2} - \xi_0\right)\right)}, \qquad (7.4.19)$$

$$\psi_2(\xi,t) = \left(\exp\left(i\lambda\xi - \frac{i\varepsilon t}{\lambda}\right)\chi(\lambda,\bar\lambda)\right)\Big|_{\lambda=-\bar\lambda_1} = \frac{\exp\left(-i\bar\lambda_1 t + \frac{i\varepsilon t}{\bar\lambda_1}\right)}{1 + \exp\left(2\lambda_{1Im}\left(\xi + \frac{\varepsilon t}{|\lambda_1|^2} - \xi_0\right)\right)}, \qquad (7.4.20)$$

where

$$\xi_0 = -\frac{1}{2\lambda_{1Im}}\ln\left(-\frac{S_1 + \bar S_1}{4\lambda_{1Im}}\right) \qquad (7.4.21)$$

and for boundedness it is assumed that

$$\frac{S_1 + \bar S_1}{4\lambda_{1Im}} < -1.$$

So we have now the exact solutions of equations (7.2.54) and (7.2.56). Comparing (7.2.58) with (7.4.3), we see that

$$u_1(z,t) = u(z,t) - \varepsilon. \qquad (7.4.22)$$

and for the function $X$ one should choose $\varepsilon = -m_1$ while for $Y$ the choice is $\varepsilon = -m_2$. Thus, for the boundaries $u_1$ and $u_2$ of the form

$$u_1(\xi,t) = m_1 - \frac{m_1(\alpha_1 + \alpha_2)^2}{2\alpha_2\alpha_1}\text{ch}^{-2}\left\{\frac{(\alpha_1 + \alpha_2)}{2}\left(\xi - \frac{m_1 t}{\alpha_1\alpha_2} - \varepsilon_0\right)\right\},$$

$$u_2(\eta,t) = m_2 - \frac{m_2(\tilde\alpha_1 + \tilde\alpha_2)^2}{2\tilde\alpha_2\tilde\alpha_1}\text{ch}^{-2}\left\{\frac{(\tilde\alpha_1 + \tilde\alpha_2)}{2}\left(\eta - \frac{m_2 t}{\tilde\alpha_1\tilde\alpha_2} - \eta_0\right)\right\}, \qquad (7.4.23)$$

the formulae (7.4.2), (7.4.16) give

$$\alpha = \frac{1}{2}X^2(\xi,t), \quad \beta = \frac{1}{2}Y^2(\eta,t).$$

Choosing $\rho_0 = 2\rho\begin{pmatrix}1 & 0\\ 0 & 0\end{pmatrix}$, one gets

$$\theta_\xi = 2\rho\frac{X_\xi Y}{1 + \frac{1}{4}\rho^2 X^2 Y^2} = 4\partial_\xi \arctan\frac{\rho}{2}XY.$$

Chapter 7

Hence [258]

$$\theta(\xi,\eta,t) = 4\arctan\left(\frac{\rho\exp(-\alpha_1\hat{\xi}-\tilde{\alpha}_1\hat{\eta})\cdot\exp\left[m_1\left(\frac{1}{\alpha_1}-\frac{1}{\alpha_2}\right)+m_2\left(\frac{1}{\tilde{\alpha}_1}-\frac{1}{\tilde{\alpha}_2}\right)\right]t}{[1+\exp(-(\alpha_1+\alpha_2)(\hat{\xi}-\xi_0))][1+\exp(-(\tilde{\alpha}_1+\tilde{\alpha}_2)(\hat{\eta}-\eta_0))]}\right) \tag{7.4.24}$$

where

$$\hat{\xi}=\xi-\frac{m_1 t}{\alpha_1\alpha_2},\quad \hat{\eta}=\eta-\frac{m_2 t}{\tilde{\alpha}_1\tilde{\alpha}_2}. \tag{7.4.25}$$

If one imposes the additional constraint

$$m_1\left(\frac{1}{\alpha_1}-\frac{1}{\alpha_2}\right)+m_2\left(\frac{1}{\tilde{\alpha}_1}-\frac{1}{\tilde{\alpha}_2}\right)=0 \tag{7.4.26}$$

the solution (7.4.24) looks like

$$\theta(\xi,\eta,t)=4\arctan\left(\frac{\rho\exp(-\alpha_1\hat{\xi}-\tilde{\alpha}_1\hat{\eta})}{[1+\exp(-(\alpha_1+\alpha_2)(\hat{\xi}-\xi_0))][1+\exp(-(\tilde{\alpha}_1+\tilde{\alpha}_2)(\hat{\eta}-\eta_0))]}\right). \tag{7.4.27}$$

For $\alpha_1$, $\alpha_2$ and $\tilde{\alpha}_1$, $\tilde{\alpha}_2$ of the same sign the solution (7.4.27) is bounded, exponentially localized solution of the 2DISG-I equation. It moves with the velocity

$$V=(V_\xi, V_\eta)=\left(\frac{m_1}{\alpha_1\alpha_2},\frac{m_2}{\tilde{\alpha}_1\tilde{\alpha}_2}\right).$$

Further let us consider the boundaries of the type (7.4.18). In this case

$$X_1(\xi,t)=\psi_1(\xi,t)\equiv X(\xi,t),\quad X_2(\xi,t)=\overline{X(\xi,t)}$$

$$Y_1(\eta,t)=\psi_1(\eta,t)\doteq Y(\eta,t),\quad Y_2(\eta,t)=\overline{Y(\eta,t)}. \tag{7.4.28}$$

So if one chooses the boundaries $u_1$ and $u_2$ as ($\lambda\equiv\lambda_1$, $\mu\equiv\mu_1$)

$$u_1(\xi,t)=m_1-\frac{2m_1(\lambda_{Im})^2}{|\lambda|^2}\text{ch}^{-2}\left\{\lambda_{Im}\left(\xi-\frac{m_1 t}{|\lambda|^2}-\xi_0\right)\right\},$$

$$u_2(\eta,t)=m_2-\frac{2m_2(\mu_{Im})^2}{|\mu|^2}\text{ch}^{-2}\left\{\mu_{Im}\left(\eta-\frac{m_2 t}{|\mu|^2}-\eta_0\right)\right\}, \tag{7.4.29}$$

the matrix $\rho_0$ as $\rho_0=\rho\begin{pmatrix}1 & 0\\ 0 & 1\end{pmatrix}$, one gets the following exponentially localized solution of the 2DISG-I equation [258]

$$\theta(\xi,\eta,t)=4\arctan\frac{\rho\cos(\varphi_1+\varphi_2)}{\text{ch}f_1\text{ch}f_2+\frac{\lambda_{Re}\mu_{Re}}{\lambda_{Im}\mu_{Im}}\text{sh}f_1\text{sh}f_2} \tag{7.4.30}$$

where
$$\varphi_1 = \lambda_{Re}\left(\xi + \frac{m_1 t}{|\lambda|^2}\right), \quad \varphi_2 = \mu_{Re}\left(\eta + \frac{m_2 t}{|\mu|^2}\right),$$

$$f_1 = \lambda_{Im}\left(\xi - \frac{m_1 t}{|\lambda|^2} - \xi_o\right), \quad f_2 = \mu_{Im}\left(\eta - \frac{m_2 t}{|\mu|^2} - \eta_0\right). \tag{7.4.31}$$

Under the constraint
$$\frac{m_1 \lambda_{Re}}{|\lambda|^2} + \frac{m_2 \mu_{Re}}{|\mu|^2} = 0 \tag{7.4.32}$$

the formula (7.4.30) describes the exponentially localized soliton which moves with the velocity $V = (V_\xi, V_\eta) = \left(\frac{m_1}{|\lambda|^2}, \frac{m_2}{|\mu|^2}\right)$. For the choice $\rho_0 = \rho \begin{pmatrix} 1 & 1 \\ 1 & 1 \end{pmatrix}$ and boundaries (7.4.29) the localized solution of the 2DISG-I equation looks like

$$\theta(\xi, \eta, t) = 4\arctan \frac{\rho \cos\varphi_1 \cos\varphi_2}{\operatorname{ch} f_1 \operatorname{ch} f_2} \tag{7.4.33}$$

where $\varphi_1, \varphi_2, f_1$ and $f_2$ are given by (7.4.31). Choosing different matrices $\rho_0$, one obtains different localized solutions.

For the general boundaries $u_1$ and $u_2$ associated with the kernels $R_0$ (7.4.8) or (7.4.9) for $N > 1$ the corresponding localized solutions of the 2DISG-I equation describe the elastic scattering of the simple localized objects of the type (7.4.27), (7.4.30), (7.4.33). The solutions (7.4.27), (7.4.30), (7.4.33) have been calculated in [258].

In addition to the solutions with bounded boundaries $u_1$ and $u_2$ the $\bar{\partial}$-dressing method described above allows us to construct, similar to the DS equation, the exact solutions for other solvable cases for equations (7.2.54) and (7.2.56). We will present here the three simplest examples.

These examples correspond to the choice of the boundaries in the form $u_1 = u_1(\xi - v_1 t)$ and $u_2 = u_2(\eta - v_2 t)$. In this case equation (7.4.3) is reduced to the stationary Schrödinger equation

$$-\psi_{\hat{\xi}\hat{\xi}} + \frac{1}{v}u(\hat{\xi})\psi = \frac{\epsilon}{v}\psi. \tag{7.4.34}$$

where $\hat{\xi} = \xi - vt$.

The two most well known cases of solvable Schrödinger equation are the harmonic oscillator and delta-function potential [230].

Choosing
$$u(\hat{\xi}) = \frac{v}{2}\hat{\xi}^2 \tag{7.4.35}$$

one has an infinite set of solutions $\psi_n(\hat{\xi})$ of equation (7.4.34) that are the wave functions of the stationary states. So for the boundaries $u_1$ and $u_2$ of the form

$$u_1 = -v_1\left(n + \frac{1}{2}\right) + \frac{v}{2}(\xi - v_1 t)^2,$$

$$u_2 = -v_2\left(m + \frac{1}{2}\right) + \frac{v_2}{2}(\eta - v_2 t)^2, \tag{7.4.36}$$

one gets the following exponentially localized solutions of the 2DISG-I equation with $\rho_0 = 2\rho$ [258]

$$\theta(\xi, \eta, t) = 4\arctan\left\{\rho \exp\left(\frac{-[v_1(\xi - v_1 t)^2 + v_2(\eta - v_2 t)^2]}{2}\right)\right.$$

$$\left. \times H_n(\sqrt{v_1}(\xi - v_1 t)) H_m(\sqrt{v_2}(\eta - v_2 t))\right\} \tag{7.4.37}$$

where $H_n$ are the Hermite polynomials [230]. At $m = n = 0$ one has

$$\theta(\xi, \eta, t) = 4\arctan\left\{\rho \exp\left(\frac{-[v_1(\xi - v_1 t)^2 + v_2(\eta - v_2 t)^2]}{2}\right)\right\}. \tag{7.4.38}$$

In the case of the delta-function potential $u(\hat{\xi}) = \delta(\hat{\xi})$ one has the solution [230]

$$\psi(\hat{\xi}) = \exp\left(\frac{1}{2v}|\hat{\xi}|\right) \tag{7.4.39}$$

for negative $v$ and $\epsilon = -\frac{1}{4|v|}$.

So choosing the boundaries $u_1$ and $u_2$ as

$$u_1 = \frac{1}{4|v_1|} - \frac{1}{|v_1|}\delta(\xi - v_1 t),$$

$$u_2 = \frac{1}{4|v_2|} - \frac{1}{|v_2|}\delta(\eta - v_2 t), \tag{7.4.40}$$

we obtain the exponentially localized solution of the 2DISG-I equation of the form ($\rho_0 = 2\rho$) [258]

$$\theta(\xi, \eta, t) = 4\arctan\left\{\rho \exp\left(-\left(\frac{1}{2|v_1|}|\xi - v_1 t| + \frac{1}{2|v_2|}|\eta - v_2 t|\right)\right)\right\}. \tag{7.4.41}$$

All the explicit exact solutions of the 2DISG equation presented above have one common feature. They are characterized by the boundaries $u_1$ and $u_2$ with the constant asymptotics $m_1$ and $m_2$ at infinities. To construct the solutions of the 2DISG equation for the boundaries which vanish at infinities one should find the exact solutions of equations (7.4.3) with $\epsilon = 0$. This can also be done by the $\bar{\partial}$-dressing method. We will present here only one example.

One can check that equation (7.4.34) with $\epsilon = 0$ has the solvable case [258]

$$u(\hat{\xi}) = \frac{6\xi^2 - 2\alpha^2}{(\xi^2 + \alpha^2)^2}, \tag{7.4.42}$$

and
$$\psi(\hat{\xi}) = \frac{1}{\hat{\xi}^2 + \alpha^2}$$
where $\alpha$ is an arbitrary real constant.

So one has the following localized solution [258]
$$\theta(\xi, \eta, t) = 4 \arctan\left(\frac{\rho}{((\xi - v_1 t)^2 + \alpha_1^2)((\eta - v_2 t)^2 + \alpha_2^2)}\right) \quad (7.4.43)$$
of the 2DISG-I equation with $\rho_0 = 2\rho$ and the boundaries
$$u_1(\xi, t) = \frac{6(\xi - v_1 t)^2 - 2\alpha_1^2}{((\xi - v_1 t)^2 + \alpha_1^2)^2}$$
$$u_2(\eta, t) = \frac{6(\eta - v_2 t)^2 - 2\alpha_2^2}{((\eta - v_2 t)^2 + \alpha_2^2)^2} \quad (7.4.44)$$
where $\alpha_1$ and $\alpha_2$ are arbitrary real constants.

In contrast to the previous examples this solution is localized rationally. Other exact solutions of the 2DISG-I equation have been found in [258].

In conclusion we emphasize that exact solutions of equation (7.4.3) associated with the kernels (7.4.8), (7.4.9) are of great interest by two more reasons. First, without any connection with the 2DISG equation they are the exact solutions of the so-called perturbed telegraph equation (7.4.3) ($\epsilon \neq 0$) which has a number of applications. Second if one change $t \to \eta$ in (7.4.3) and add the dependence on the time $t$ in the quantities $S_k$ from (7.4.8) or (7.4.9) via (see (2.4.23))
$$S_k = S_{k0} \exp\left(i\left[k_1(\lambda_k^3 - \mu_k^3) + \epsilon^3 k_2\left(\frac{1}{\lambda_k^3} - \frac{1}{\mu_k^3}\right)\right] t\right) \quad (7.4.45)$$
then the formula (7.4.4) gives the multi-line solitons of the NVN equation (1.2.19) with $u \to u - \epsilon$.

## 7.5. Initial value problem for the 2DISG equation

We will consider the initial value problem for the 2DISG equation with the constant boundaries $m_1$ and $m_2$, i. e. for equation (7.1.16). We will follow the paper [257]. The corresponding linear problem is of the form (7.2.60) with
$$\Gamma_1^0(\lambda) = \frac{i}{2} m_1 \delta(\lambda), \quad \Gamma_2^0(\lambda) = -\frac{i}{2} m_2 \delta(\lambda). \quad (7.5.1)$$
In the characteristic coordinate one has
$$\begin{pmatrix} \partial_\eta & 0 \\ 0 & \partial_\xi \end{pmatrix} \tilde{\psi} - \frac{i}{2} \begin{pmatrix} 0 & \theta_\eta \\ \theta_\xi & 0 \end{pmatrix} \tilde{\psi} = 0, \quad (7.5.2)$$

Chapter 7

$$\begin{pmatrix} \partial_\xi & 0 \\ 0 & \partial_\eta \end{pmatrix} \partial_t \tilde{\psi} - \frac{i}{2} \begin{pmatrix} 0 & \theta_\xi \\ \theta_\eta & 0 \end{pmatrix} \partial_t \tilde{\psi} + \begin{pmatrix} m_1 + \frac{1}{4}(\partial_\xi^{-1}(\theta_\xi \theta_\eta)_t & 0 \\ 0 & m_2 + \frac{1}{4}\partial_\eta^{-1}(\theta_\xi \theta_\eta)_t \end{pmatrix} \tilde{\psi}$$
$$+ \frac{i}{\lambda} \left( \partial_\xi + \partial_\eta - \frac{i}{2}(\theta_\xi + \theta_\eta) \right) \begin{pmatrix} 0 & 1 \\ 1 & 0 \end{pmatrix} \tilde{\psi} \begin{pmatrix} m_1 & 0 \\ 0 & -m_2 \end{pmatrix} = 0. \qquad (7.5.3)$$

The spectral parameter is introduced in a standard manner by specifying the class of solutions of (7.5.2)–(7.5.3):

$$\tilde{\psi} = \mu(\xi, \eta, t, \lambda) \begin{pmatrix} e^{-\frac{2i\lambda}{\sigma}\xi} & 0 \\ 0 & e^{\frac{2i\lambda\eta}{\sigma}} \end{pmatrix}. \qquad (7.5.4)$$

The function $\mu$ obeys the equation

$$\begin{pmatrix} \partial_\eta & 0 \\ 0 & \partial_\xi \end{pmatrix} \mu + \frac{2i\lambda}{\sigma}[\sigma_3, \mu] - \frac{i}{2} \begin{pmatrix} 0 & \theta_\eta \\ \theta_\xi & 0 \end{pmatrix} \mu = 0 \qquad (7.5.5)$$

and the corresponding equation (7.5.3).

The linear problem (7.5.5) is the same as for the DS equation with an additional reduction

$$q = -\frac{i\sigma}{2}\theta_\eta, \quad r = -\frac{i\sigma}{2}\theta_\xi. \qquad (7.5.6)$$

Hence, one can use the results of sections (5.3) and (5.4). So we will omit most part of the intermediate calculations. In contrast to the DS case we will formulate all equations in terms of the first and second columns $\mu^{(1)}$ and $\mu^{(2)}$ of the $2 \times 2$ matrix $\mu$.

We start with the 2DSIG-I equation ($\sigma = 1$). In terms of the first column $\mu^{(1)}$ and second column $\mu^{(2)}$ the linear problem (7.5.5) looks like

$$\begin{pmatrix} \partial_\eta & 0 \\ 0 & \partial_\xi \end{pmatrix} \mu^{(1)} - 2i\lambda\sigma_-\mu^{(1)} - \frac{i}{2} \begin{pmatrix} 0 & \theta_\eta \\ \theta_\xi & 0 \end{pmatrix} \mu^{(1)} = 0 \qquad (7.5.7)$$

and

$$\begin{pmatrix} \partial_\eta & 0 \\ 0 & \partial_\xi \end{pmatrix} \mu^{(2)} + 2i\lambda\sigma_+\mu^{(2)} - \frac{i}{2} \begin{pmatrix} 0 & \theta_\eta \\ \theta_\xi & 0 \end{pmatrix} \mu^{(2)} = 0. \qquad (7.5.8)$$

where $\sigma_+ = \begin{pmatrix} 1 & 0 \\ 0 & 0 \end{pmatrix}, \sigma_- = \begin{pmatrix} 0 & 0 \\ 0 & 1 \end{pmatrix}$.

The Green functions $G^{(1)}$, $G^{(2)}$ for equations (7.5.7), (7.5.8) which are bounded and analytic except the real axis are the same as for the DS-I case.

These Green functions allow us to construct, in a manner completely similar to section (5.3), the solutions of the problems (7.5.7) and (7.5.8) which are analytic in upper and lower half planes and have the jumps across the real axis given by

$$\mu^{(1)+}(\lambda) - \mu^{(1)-}(\lambda) = \int_R dl T(\lambda, l) e^{2i\lambda\xi + 2il\eta} \mu^{(2)+}(l),$$

$$\mu^{(2)+}(\lambda) - \mu^{(2)-}(\lambda) = \int_R dl S(\lambda, l) e^{-2i\lambda\eta - 2il\xi} \mu^{(1)-}(l) \tag{7.5.9}$$

where $\mu^{\pm}(\lambda) \doteq \mu(\lambda \pm i0)$ and

$$S(\lambda, l) \doteq -\frac{1}{4\pi} \iint_{R^2} d\xi d\eta (-\frac{i}{2}\theta_\xi) \mu_{22}^-(\lambda) e^{2il\xi + 2i\lambda\eta}, \tag{7.5.10}$$

$$T(\lambda, l) \doteq \frac{1}{4\pi} \iint_{R^2} d\xi d\eta (-\frac{i}{2}\theta_\eta) \mu_{11}^+(\lambda) e^{-2il\eta - 2i\lambda\xi}.$$

So we have the nonlocal Riemann—Hilbert problems. Their solutions are given, for instance, by the formulae

$$\mu^{(1)\pm}(\lambda) = \begin{pmatrix} 1 \\ 0 \end{pmatrix} + \left( \int_R dl T(\lambda, l) e^{2i\lambda\xi + 2il\eta} \mu^{(2)+}(l) \right)^{\pm},$$

$$\mu^{(2)\pm}(\lambda) = \begin{pmatrix} 0 \\ 1 \end{pmatrix} + \left( \int_R dl S(\lambda, l) e^{-2il\xi - 2i\lambda\eta} \mu^{(1)-}(l) \right)^{\pm} \tag{7.5.11}$$

where

$$(f(\lambda))^{\pm} = \frac{1}{2\pi i} \int_R \frac{d\lambda' f(\lambda')}{\lambda' - (\lambda \pm i0)}. \tag{7.5.12}$$

The corresponding reconstruction formulae are of the form

$$\theta_\xi = -\frac{2i}{\pi} \iint_{R^2} d\lambda dl S(\lambda, l) e^{-2i\lambda\eta - 2il\xi} \mu_{11}^-(l),$$

$$\theta_\eta = \frac{2i}{\pi} \iint_{R^2} d\lambda dl T(\lambda, l) e^{2i\lambda\xi + 2il\eta} \mu_{22}^+(l). \tag{7.5.13}$$

The reduction $q = -\frac{i}{2}\theta_\xi$, $r = -\frac{i}{2}\theta_\eta$ implies certain constraint on the inverse problem data $S$ and $T$. Considering the small $\theta$, one gets from (7.5.13) that

$$S(\lambda, l) = \lambda \tilde{S}(\lambda, l),$$

$$T(\lambda, l) = \lambda \tilde{S}(-l, -\lambda). \tag{7.5.14}$$

The condition of the reality of $\theta$ implies

$$\overline{\tilde{S}(\lambda, l)} = \tilde{S}(-\lambda, -l).$$

To solve the initial value problem one should find, as usual, the time evolution of the inverse problem data $\tilde{S}(\lambda, l)$. Using equation (7.5.3), one finds

$$\tilde{S}(\lambda, l, t) = \tilde{S}(\lambda, l, 0) \exp\left( \frac{i}{2} \left( \frac{m_1}{l} + \frac{m_2}{\lambda} \right) t \right). \tag{7.5.15}$$

The time evolution law (5.5.15) of the inverse problem data, the formula (7.5.10) and the inverse problem equations (7.5.11) allow us to solve the initial value problem for the 2DISG-I equation by the standard IST scheme

$$\theta(\xi,\eta,0) \to \tilde{S}(\lambda,l,0) \to \tilde{S}(\lambda,l,t) \to \theta(\xi,\eta,t). \tag{7.5.16}$$

These formulae also allow us to construct the exact solutions of the 2DISG-I equation with functional parameters which correspond to the degenerated data $\tilde{S}(\lambda,l,t)$. These solutions are the particular cases of those with functional parameters constructed in section (7.4).

For the 2DISG-II equation for the real $\theta$ one has $m_1 = \bar{m}_2 = m$ and $z = \frac{1}{2}(y - ix)$, $\bar{z} = \frac{1}{2}(y + ix)$. Using the results of section (5.4) one gets the following $\bar{\partial}$-problem

$$\frac{\partial \mu(\lambda,\bar{\lambda})}{\partial \bar{\lambda}} = \mu \begin{pmatrix} 0 & \overline{\lambda F(\lambda,\bar{\lambda})}e^{-2i\lambda z - 2i\bar{\lambda}\bar{z}} \\ \lambda F(\lambda,\bar{\lambda})e^{2i\lambda z + 2i\bar{\lambda}\bar{z}} & 0 \end{pmatrix} \tag{7.5.17}$$

where

$$F(\lambda,\bar{\lambda},t) \doteq -\frac{1}{2\lambda} \iint_C dz \wedge d\bar{z}\, e^{2i\lambda z + 2i\bar{\lambda}\bar{z}} \theta_z(z,\bar{z},t)\mu_{22}(z,\bar{z};\lambda,t). \tag{7.5.18}$$

Similar to the DS-II equation case one has the involution

$$\sigma_2 \mu(\lambda,\bar{\lambda})\sigma_2^{-1} = \overline{\mu(-\bar{\lambda},-\lambda)}. \tag{7.5.19}$$

To find the structure of the singular part $\mu^{(s)}$ of $\mu$ one should take into account the fact that each column $\mu^{(s)}$ may have its own set of poles and involution (7.5.18). The nontrivial $\mu^{(s)}$ correspond to the choice

$$\mu^{(s)} = \sum_{i=1}^N \frac{C_i}{\lambda - \lambda_i} \begin{pmatrix} \mu_{11i} & 0 \\ \mu_{21i} & 0 \end{pmatrix} + \sum_{i=1}^N \frac{\bar{C}_i}{\lambda + \bar{\lambda}_i} \begin{pmatrix} 0 & -\overline{\mu_{21i}} \\ 0 & \overline{\mu_{11i}} \end{pmatrix}.$$

So the full $\bar{\partial}$-problem is of the form

$$\frac{\partial \mu}{\partial \bar{\lambda}} = \mu \begin{pmatrix} 0 & \overline{\lambda F(\lambda,\bar{\lambda})}e^{-2i\lambda z - 2i\bar{\lambda}\bar{z}} \\ \lambda F(\lambda,\bar{\lambda})e^{2i\lambda z + 2i\bar{\lambda}\bar{z}} & 0 \end{pmatrix}$$

$$+ \sum_{i=1}^N C_i \pi \delta(\lambda - \lambda_i) \begin{pmatrix} \mu_{11i} & 0 \\ \mu_{21i} & 0 \end{pmatrix} + \sum_{i=1}^N \bar{C}_i \pi \delta(\lambda + \bar{\lambda}_i) \begin{pmatrix} 0 & -\overline{\mu_{21i}} \\ 0 & \overline{\mu_{11i}} \end{pmatrix}. \tag{7.5.20}$$

For the first column $\mu^{(1)} = \begin{pmatrix} \mu_{11} \\ \mu_{21} \end{pmatrix}$ one has

$$\frac{\partial \mu^{(1)}(\lambda,\bar{\lambda})}{\partial \bar{\lambda}} = \pi \sum_{i=1}^N \delta(\lambda - \lambda_i)\mu_i^{(1)} - \lambda F(\lambda,\bar{\lambda})e^{2i\lambda z + 2i\bar{\lambda}\bar{z}} \sigma_2 \overline{\mu^{(1)}(\lambda,\bar{\lambda})} \tag{7.5.21}$$

where we normalize $\mu_i^{(1)}$ as $\mu^{(1)} \underset{|z|\to\infty}{\longrightarrow} \frac{-2i\lambda_i}{z}\begin{pmatrix}1\\0\end{pmatrix}$, i. e. $C_i = -2i\lambda_i$. The generalized Cauchy formula gives

$$\mu^{(1)}(\lambda,\bar\lambda) = \begin{pmatrix}1\\0\end{pmatrix} - 2\sum_{i=1}^{N}\frac{i\lambda_i\mu_i^{(1)}(z,\bar z)}{\lambda - \lambda_i}$$

$$-\frac{1}{2\pi}\iint_C \frac{d\lambda' \wedge d\bar\lambda'}{\lambda' - \lambda}\lambda' F(\lambda',\bar\lambda')e^{2i\lambda' z + 2i\bar\lambda'\bar z}\sigma_2\overline{\mu^{(1)}(\lambda',\bar\lambda')}. \qquad (7.5.22)$$

Equation (7.5.22) is the part of the inverse problem equations. To derive the rest of such equations one should proceed in equation (7.5.22) to the limit $\lambda \to \lambda_k$ to obtain the system of equations for $\mu^{(1)}$. To do this we must calculate the limit

$$\lim_{\lambda\to\lambda_i}\left(\mu^{(1)} + \frac{2i\lambda_i\mu_i^{(1)}}{\lambda - \lambda_i}\right). \qquad (7.5.23)$$

Completely similar to the DS-II equation case one can show that the quantity

$$\lim_{\lambda\to\lambda_i}\left(\mu^{(1)} + \frac{2i\lambda_i\mu_i^{(1)}}{\lambda - \lambda_i}\right) + 2\lambda_i z\mu_i^{(1)} \qquad (7.5.24)$$

is also the solution of the homogeneous integral equation for $\mu$ certain given constraints on $\theta$ and $\mu_i^{(1)}$. Therefore the expression (7.5.24) is the linear superposition of the two independent solutions of the homogeneous equation, i. e.

$$\lim_{\lambda\to\lambda_i}\left(\mu^{(1)} + \frac{2i\lambda_i\mu_i^{(1)}}{\lambda - \lambda_i}\right) + 2\lambda_i z\mu_i^{(1)} = \gamma_i\mu_i^{(1)} - i\tilde\gamma_i\sigma_2\bar\mu_i^{(1)}, \qquad (7.5.25)$$

where $\gamma_i$ and $\tilde\gamma_i$ are some constants.

Now, proceeding to the limit $\lambda \to \lambda_k$ in equation (7.5.22) and taking into account the identity (7.5.25), we obtain

$$(-2\lambda_k z + \gamma_k)\mu_k^{(1)} - i\tilde\gamma_k\sigma_2\bar\mu_k^{(1)}\exp(2i\lambda_k z + 2i\bar\lambda_k\bar z)$$

$$= \begin{pmatrix}1\\0\end{pmatrix} + \sum_{j\ne k}^{N}\frac{-2i\lambda_j\mu_j^{(1)}}{\lambda_j - \lambda_k} - \frac{1}{2\pi}\iint_C \frac{d\mu \wedge d\bar\mu}{\mu - \lambda_k}\mu F(\mu,\bar\mu)e^{2i\mu z + 2i\bar\mu\bar z}\sigma_2\overline{\mu^{(1)}(\mu,\bar\mu)}, \qquad (7.5.26)$$

$$k = 1,\ldots,N.$$

At last, the reconstruction formula for the potential $\theta$ is given by

$$\theta_z = 2i\sum_{k=1}^{N}\lambda_k\chi_{21k} - \frac{i}{\pi}\iint_C d\mu \wedge d\bar\mu\,\mu F(\mu,\bar\mu)e^{2i\mu z + 2i\bar\mu\bar z}\overline{\mu_{11}(\mu,\bar\mu)}. \qquad (7.5.27)$$

The formulae (7.5.22), (7.5.26) and (7.5.27) form the set of the inverse problem equations for the linear problem (7.5.5) with $\sigma^2 = -1$. The quantities $\{F(\lambda,\bar{\lambda}), \lambda_i, \gamma_i, \mu_i, (i=1,\ldots,N)\}$ are the inverse problem data.

Now one should find the time evolution of the inverse problem data. Substituting (7.5.21) and (7.5.25) into equation (7.5.3) and considering the limit $|z| \to \infty$, one obtains

$$\frac{\partial F}{\partial t} = \frac{i}{2}\left(\frac{m}{\lambda} + \frac{\bar{m}}{\bar{\lambda}}\right) F,$$

$$\frac{\partial \lambda_i}{\partial t} = 0, \quad \frac{\partial \gamma_i}{\partial t} = 0, \quad (7.5.28)$$

$$\frac{\partial \mu_i}{\partial t} = \frac{i}{2}\left(\frac{m}{\lambda_i} + \frac{\bar{m}}{\bar{\lambda}_i}\right)\mu_i, \quad (i=1,\ldots,N).$$

So

$$F(\lambda,\bar{\lambda},t) = F(\lambda,\bar{\lambda},0)\exp\left(\frac{i}{2}\left(\frac{m}{\lambda} + \frac{\bar{m}}{\bar{\lambda}}\right)t\right),$$

$$\gamma_i(t) = \gamma_i(0), \quad (7.5.29)$$

$$\mu_i(t) = \mu_i(0)\exp\left(\frac{i}{2}\left(\frac{m}{\lambda_i} + \frac{\bar{m}}{\bar{\lambda}_i}\right)t\right), \quad (i=1,\ldots,N).$$

The inverse problem equations (7.5.22), (7.5.26), (7.5.27) and time evolution (7.5.29) reduce the solution of the initial value problem for the 2DISG-II equation to the standard IST set of the linear problems.

As usual, for the pure discrete data ($F(\lambda,\bar{\lambda}) = 0$) the inverse problem equations are reduced to the system of the linear algebraic equations and can be solved explicitly. The corresponding solutions with the real valued $\theta$ are not found.

# Chapter 8

# Extensions of the $\bar{\partial}$-dressing method

The $\bar{\partial}$-dressing method is a powerful method with a broad variety of opportunities. In the previous chapters we considered, in fact, the simplest cases which correspond to the single canonical normalization and choices of the simple operators $\hat{B}_i$ ($i = 1, 2, 3$). In this chapter we will expose several possible extensions and generalizations of the $\bar{\partial}$-dressing method.

## 8.1. $\bar{\partial}$-dressing for the DS-equation with constant asymptotics

As we already mentioned the $\bar{\partial}$-dressing for the DS-equation described in section (5.1) is appropriate for nonconstant asymptotics of $q$ and $r$ at the infinity $x^2 + y^2 \to \infty$. The case

$$q \to q_\infty, \quad r \to r_\infty \tag{8.1.1}$$

as $x^2 + y^2 \to \infty$ where $q_\infty$ and $r_\infty$ are constants requires a different treatment. Our construction here is based mainly on the papers [259, 260].

So we start with the scalar $\bar{\partial}$-problem

$$\frac{\partial \chi(\lambda, \bar{\lambda})}{\partial \bar{\lambda}} = \iint_C d\mu \wedge d\bar{\mu}\, \chi(\mu, \bar{\mu}) R(\mu, \bar{\mu}; \lambda, \bar{\lambda}). \tag{8.1.2}$$

and choose

$$\hat{B}_1 = -r_\infty \lambda, \quad \hat{B}_2 = -q_\infty \lambda^{-1}, \quad \hat{B}_3 = -i(r_\infty^2 \lambda^2 + q_\infty^2 \lambda^{-2}). \tag{8.1.3}$$

Thus, the operators $D_i$ ($i = 1, 2, 3$) are ($x_1 = \xi$, $x_2 = \eta$, $x_3 = t$)

$$D_1 = \partial_\xi - r_\infty \lambda,$$

$$D_2 = \partial_\eta - q_\infty \lambda^{-1}, \tag{8.1.4}$$

$$D_3 = \partial_t - i r_\infty^2 \lambda^2 - i q_\infty^2 \lambda^{-2}.$$

Using these operators, one should construct the operators $L_i$ which obey the conditions (2.3.15) and (2.3.16). It is not difficult to convince yourself that it is impossible to construct the operators $L_i$ which would be linear in $D_1$ and $D_2$ using the only solution of the $\bar{\partial}$-problem (8.1.2) with fixed normalization, say canonical one.

This fact and the form (8.1.4) of the operators $D_i$ indicate that the use of the two solutions of the $\bar{\partial}$-problem (8.1.2) normalized at $\lambda = \infty$ and $\lambda = 0$ may solve the problem.

So let us consider the two solutions $\chi, \tilde{\chi}$ of the $\bar{\partial}$-problem normalized as

$$\chi \to 1 + \frac{1}{\lambda}\chi_{-1} + \frac{1}{\lambda^2}\chi_{-2} + \ldots \quad \text{as} \quad \lambda \to \infty \tag{8.1.5}$$

and

$$\tilde{\chi} = \frac{1}{\lambda} + \tilde{\chi}_0 + \lambda \tilde{\chi}_1 + \ldots \quad \text{as} \quad \lambda \to 0. \tag{8.1.6}$$

These functions are the solutions of the integral equations (see the formula (2.2.3))

$$\chi(\lambda, \bar{\lambda}) = 1 + \frac{1}{2\pi i} \iint_C \frac{d\lambda' \wedge d\bar{\lambda}'}{\lambda' - \lambda} \iint_C d\mu \wedge d\bar{\mu}\, \chi(\mu, \bar{\mu}) R(\mu, \bar{\mu}; \lambda', \bar{\lambda}'), \tag{8.1.7}$$

$$\tilde{\chi}(\lambda, \bar{\lambda}) = \frac{1}{\lambda} + \frac{1}{2\pi i} \iint_C \frac{d\lambda' \wedge d\bar{\lambda}'}{\lambda' - \lambda} \iint_C d\mu \wedge d\bar{\mu}\, \tilde{\chi}(\mu, \bar{\mu}) R(\mu, \bar{\mu}; \lambda', \bar{\lambda}'), \tag{8.1.8}$$

where

$$R(\mu, \bar{\mu}; \lambda, \bar{\lambda}) = e^{F(\mu) - F(\lambda)} R_0(\mu, \bar{\mu}; \lambda, \bar{\lambda}) \tag{8.1.9}$$

and

$$F(\lambda) = -r_\infty \lambda \xi - q_\infty \frac{1}{\lambda}\eta - i(r_\infty^2 \lambda^2 + q_\infty^2 \lambda^{-2})t. \tag{8.1.10}$$

Note that equation (8.1.8) implies that

$$\tilde{\chi} = \frac{1}{\lambda} + \frac{1}{\lambda}\tilde{\chi}_{-1} + \frac{1}{\lambda^2}\tilde{\chi}_{-2} + \ldots \quad \text{as} \quad \lambda \to \infty. \tag{8.1.11}$$

Using the formulae (8.1.9), (8.1.5), (8.1.6) and (8.1.11), one readily concludes that the quantity

$$D_2 \chi = \chi_\eta - \frac{q_\infty}{\lambda}\chi \tag{8.1.12}$$

has no singularity as $\lambda \to \infty$ and has a simple pole as $\lambda \to 0$. Since the function $\tilde{\chi}$ also has a pole at $\lambda = 0$, the combination

$$D_2 \chi + q\tilde{\chi} = \chi_\eta - \frac{q_\infty}{\lambda}\chi + q\tilde{\chi} \tag{8.1.13}$$

has no singularity at $\lambda = 0$ if

$$q(\xi, \eta, t) = q_\infty \chi_0. \tag{8.1.14}$$

Further, $D_2 \chi + q\tilde{\chi} \to 0$ as $\lambda \to \infty$.

Thus, we have the first desired linear equation

$$D_2\chi + q\tilde{\chi} = 0 \tag{8.1.15}$$

where $q(\xi,\eta,t)$ is given by (8.1.14).

Analogously it is easy to see that the quantity

$$D_1\tilde{\chi} = \tilde{\chi}_\xi - r_\infty \lambda \tilde{\chi} \tag{8.1.16}$$

has no singularities both as $\lambda \to \infty$ and $\lambda \to 0$. Then $D_1\tilde{\chi} + r\chi \to 0$ as $\lambda \to \infty$ if

$$r(\xi,\eta,t) = r_\infty + r_\infty \tilde{\chi}_{-1}. \tag{8.1.17}$$

Thus, we have the second linear equation

$$D_1\tilde{\chi} + r\chi = 0 \tag{8.1.18}$$

where the function $r$ is given by (8.1.17).

Equations (8.1.15) and (8.1.18) form the first matrix linear problem

$$\begin{pmatrix} D_2 & 0 \\ 0 & D_1 \end{pmatrix} \begin{pmatrix} \chi \\ \tilde{\chi} \end{pmatrix} + \begin{pmatrix} 0 & q \\ r & 0 \end{pmatrix} \begin{pmatrix} \chi \\ \tilde{\chi} \end{pmatrix} = 0. \tag{8.1.19}$$

Note that, by virtue of (8.1.14) and (8.1.17), $q \to q_\infty$ and $r \to r_\infty$ as $\xi^2 + \eta^2 \to \infty$.

In order to construct the second linear problem one should use the operator $D_3$. The quantity $D_3\chi$ has the second order singularity as $\lambda \to \infty$ and the second order pole as $\lambda \to 0$. It is obvious that to compensate these singularities one should add to $D_3\chi$ the term $-iD_1^2\chi$ and $-iD_2^2\chi$. The quantity

$$D_3\chi - iD_1^2\chi - iD_2^2\chi = \chi_t - i\chi_{\xi\xi} - i\chi_{\eta\eta} + 2i\lambda r_\infty \chi_\xi + 2i\frac{q_\infty}{\lambda}\chi_\eta \tag{8.1.20}$$

has only the first order singularities as $\lambda \to \infty$ and $\lambda \to 0$.

Their compensation can be achieved by adding to (8.1.20) the term $V_1 D_1\chi$ and $V_2\tilde{\chi}$ with the proper $V_1$ and $V_2$. Further, to satisfy the condition (2.3.16) we also add the term $V_3\chi$. All these requirements for the quantity

$$D_3\chi - iD_1^2\chi - iD_2^2\chi + V_1 D_1\chi + V_2\tilde{\chi} + V_3\chi \tag{8.1.21}$$

give

$$V_1 = 0, \quad V_2 = -2iq_\infty \chi_{0\eta}, \quad V_3 = -2ir_\infty \chi_{-1\xi}. \tag{8.1.22}$$

So we have the equation

$$D_3\chi - iD_1^2\chi - iD_2^2\chi - 2iq_\infty \chi_{0\eta}\tilde{\chi} - 2ir_\infty \chi_{-1\xi}\chi = 0. \tag{8.1.23}$$

## Chapter 8

In a similar manner one can find one more equation. It is

$$D_3\tilde{\chi} - iD_1^2\tilde{\chi} - iD_2^2\tilde{\chi} - 2iq_\infty\tilde{\chi}_{0\eta}\tilde{\chi} - 2ir_\infty\tilde{\chi}_{-1\xi}\chi = 0. \qquad (8.1.24)$$

So, we get the second matrix linear problem

$$(D_3 - iD_1^2 - iD_2^2)\begin{pmatrix} \chi \\ \tilde{\chi} \end{pmatrix} - 2i \begin{pmatrix} r_\infty\chi_{-1\xi} & q_\infty\chi_{0\eta} \\ q_\infty\tilde{\chi}_{0\eta} & r_\infty\tilde{\chi}_{-1\xi} \end{pmatrix} \begin{pmatrix} \chi \\ \tilde{\chi} \end{pmatrix} = 0. \qquad (8.1.25)$$

The system (8.1.19), (8.1.25) is compatible by the construction and implies the nonlinear equations for $q$ and $r$.

Evaluating equation (8.1.23) at $\lambda = 0$, one gets

$$\chi_{0t} - i\chi_{0\xi\xi} - i\chi_{0\eta\eta} + 2iq_\infty\chi_{1\eta} - 2iq_\infty\chi_{0\eta}\tilde{\chi}_0 - 2ir_\infty\chi_{-1\xi}\chi_0 = 0. \qquad (8.1.26)$$

Equation (8.1.15) implies

$$q_\infty\chi_1 = \chi_{0\eta} + q\tilde{\chi}_0. \qquad (8.1.27)$$

Further, substituting the expansion (8.1.11) into equation (8.1.29) and taking terms of order $\frac{1}{\lambda}$, one gets

$$\tilde{\chi}_{-1t} - i\tilde{\chi}_{-1\xi\xi} - i\tilde{\chi}_{-1\eta\eta} + 2ir_\infty\tilde{\chi}_{-2\xi} - 2iq_\infty\tilde{\chi}_{0\eta}(1 + \tilde{\chi}_{-1}) - 2ir_\infty\tilde{\chi}_{-1\xi}\chi_{-1} = 0. \qquad (8.1.28)$$

From equation (8.1.18) one has

$$r_\infty\tilde{\chi}_{-2} = \tilde{\chi}_{-1\xi} + r\chi_{-1} \qquad (8.1.29)$$

and

$$\tilde{\chi}_{0\xi} = r_\infty - r\chi_0 \qquad (8.1.30)$$

while equation (8.1.15) implies

$$\chi_{-1\eta} - q_\infty + q(1 + \tilde{\chi}_{-1}) = 0. \qquad (8.1.31)$$

Using (8.1.27), (8.1.29), (8.1.30), one rewrites equations (8.1.26), (8.1.28) in the form

$$iq_t + q_{\xi\xi} - q_{\eta\eta} + 2(r_\infty\chi_{-1\xi} - q_\infty\tilde{\chi}_{0\eta})q = 0,$$

$$ir_t - r_{\xi\xi} + r_{\eta\eta} - 2(r_\infty\chi_{-1\xi} - q_\infty\tilde{\chi}_{0\eta})r = 0. \qquad (8.1.32)$$

Denoting

$$C_1 = -q_\infty\tilde{\chi}_0, \quad C_2 = -r_\infty\chi_{-1} \qquad (8.1.33)$$

and using (8.1.30), (8.1.31), one finally gets the system

$$iq_t + q_{\xi\xi} - q_{\eta\eta} + 2(C_{1\eta} - C_{2\xi})q = 0,$$

$$ir_t - r_{\xi\xi} + r_{\eta\eta} - 2(C_{1\eta} - C_{2\xi})r = 0, \qquad (8.1.34)$$

$$C_{1\xi} = C_{2\eta} = qr - q_\infty r_\infty.$$

This system is apparently the extension of the DS system (5.1.26) to the case $q_\infty \neq 0$, $r_\infty \neq 0$.

Thus, one can apply the $\bar\partial$-dressing method for the construction of the exact explicit solutions of the DS equation with constant asymptotics of $q$ and $r$. Considering, as usual, the degenerate kernel $R$ of equations (8.1.7) and (8.1.8) and using the general formulae of section (2.2), one can find the solutions of the system (8.1.34) with functional parameters, line solitons and rational-exponential solutions. We leave this construction to the readers. Note that for the first time the bounded rational-exponential solutions of the system (8.1.34) have been calculated in [209] by the Hirota method.

In a similar manner using the solutions of the $\bar\partial$-problem with two different normalizations one can treat the Ishimori and 2DISG equations with the constant asymptotics $\vec{S} \to (S_{01}, S_{02}, S_{03})$ and $\theta_\xi \to$ const and also other integrable equations.

## 8.2. $\bar\partial$-dressing with variable normalization

In the previous section we demonstrated that the use of solutions of the $\bar\partial$-problem with two different normalizations allows us to treat a new case. This is the particular example of the general scheme of the $\bar\partial$-dressing which uses the variety of solutions with various normalizations. We will show that such approach provides the interesting and beautiful results, in particular, it allows to construct the integrable system of the general position in the explicit form and to establish the interrelations between the different integrable systems [145]. We will follow here the paper [145].

So, we consider the nonlocal $\bar\partial$-problem with the general normalization $\eta(\lambda)$ (section (2.2))

$$\frac{\partial \chi}{\partial \bar\lambda} = \chi * R + \frac{\partial \eta}{\partial \bar\lambda} \qquad (8.2.1)$$

with the boundary condition

$$\chi(\lambda) - \eta(\lambda) \to 0, \quad \lambda \to \infty \qquad (8.2.2)$$

where $\eta(\lambda)$ is, in general, an arbitrary rational function on $\lambda$.

Equation (8.2.1) is the singular inhomogeneous $\bar\partial$-equation. It can be rewritten in the form of the inhomogeneous $\bar\partial$-equation for the regular function $\varphi \doteq \chi - \eta$:

$$\frac{\partial \varphi}{\partial \bar\lambda} = \varphi * R + \eta * R \qquad (8.2.3)$$

where $\varphi \to 0$ as $\lambda \to \infty$. This equation is uniquely solvable at least for small $R$. In what follows we will assume the unique solvability of the problem (8.2.1) – (8.2.2) for an arbitrary $R$.

We assume that the dependence of the function $R$ on the additional variables $x_1, x_2, x_3$ is defined by equation (2.3.3) or (2.3.6). We confine ourselves to $\hat{B}_i = I_i(\lambda)$ where $I_i(\lambda)$ are matrix functions. In this case

$$R(\lambda', \bar{\lambda}'; \lambda, \bar{\lambda}; x) = \exp(\sum_{i=1}^{3} I_i(\lambda')x_i) R_0(\lambda', \bar{\lambda}'; \lambda, \bar{\lambda}) \exp(-\sum_{i=1}^{3} I_i(\lambda)x_i) \qquad (8.2.4)$$

where $R_0$ is an arbitrary matrix-valued function. In what follows we will consider the case of the rational $I_i(\lambda)$ and in view of this we choose $R_0(\lambda', \bar{\lambda}'; \lambda, \bar{\lambda})$ in such a way that $R_0(\lambda', \bar{\lambda}'; \lambda, \bar{\lambda}, x)$ have no essential singularities at the points where $I_i(\lambda)$ have the poles. For such $R_0$ the function $R(\lambda', \bar{\lambda}'; \lambda, \bar{\lambda}, x)$ decreases at the point $\lambda_1$ (pole of $I_i(\lambda)$) more rapidly than any degree $(\lambda - \lambda_1)^n (\lambda' - \lambda_1)^n$ for all $\lambda$. Equation (8.2.1) also implies that $\partial \varphi / \partial \bar{\lambda}$ at the point $\lambda_1$ tends to zero faster than any degree $(\lambda - \lambda_1)^n$.

Let us differentiate equation (8.2.1) with respect to $x_i$. Taking into account (8.2.2), we obtain

$$\frac{\partial}{\partial \bar{\lambda}} D_i \chi = D_i \chi * R + \left[\frac{\partial}{\partial \bar{\lambda}}, D_i\right] \chi + D_i \frac{\partial \eta}{\partial \bar{\lambda}}. \qquad (8.2.5)$$

The last two terms in (8.2.5) can always be represented in the form $\partial \mu_i / \partial \bar{\lambda}$ where $\mu_i$ are some functions. Hence, the functions $D_i \chi$ are also the solutions of the $\bar{\partial}$-problem (8.2.1) but with different normalizations

$$\mu_i = D_i \chi - \left(\frac{\partial}{\partial \bar{\lambda}}\right)^{-1} D_i \frac{\partial \varphi}{\partial \bar{\lambda}}.$$

Let us show that in the case of the rational functions $I_i(\lambda)$ and $\eta(\lambda)$ the functions $\mu_i$ are the rational ones too. To prove this fact one can use the formulae

$$\frac{\partial}{\partial \bar{\lambda}}(\lambda - \lambda_0)^{-n} = \pi \frac{(-1)^{n-1}}{(n-1)!} \frac{\partial^{n-1}}{\partial \lambda^{n-1}} \delta(\lambda - \lambda_0) \qquad (8.2.6a)$$

and

$$\varphi(\lambda) \frac{\partial^n}{\partial \lambda^n} \delta(\lambda - \lambda_0) = \sum_{k=0}^{n} C_n^k \frac{\partial^k \varphi(\lambda_0)}{\partial \lambda^k} \frac{\partial^{n-k}}{\partial \lambda^{n-k}} \delta(\lambda - \lambda_0). \qquad (8.2.6b)$$

It follows from these formulae that $\varphi(\partial I_i / \partial \bar{\lambda})$ can be expressed in terms of the $\delta$-function and its derivatives at the poles of $I_i(\lambda)$. Moreover, the formulae (8.2.6) imply that $(\partial / \partial \bar{\lambda})^{-1} \varphi(\partial I_i(\lambda)/\partial \bar{\lambda})$ are the rational functions. In view of this, after some transformations we obtain

$$\mu_i = \frac{\partial \eta}{\partial x_i} + \eta I_i(\lambda) + \left(\frac{\partial}{\partial \bar{\lambda}}\right)^{-1} \left(\varphi \frac{\partial I_i(\lambda)}{\partial \bar{\lambda}}\right). \qquad (8.2.7)$$

Hence, $\mu_i$ are rational functions.

So, the functions $D_i\chi$ are the solutions of the $\bar\partial$-equation (8.2.1) with the rational normalizations. Then, the relation

$$D_i\chi - \mu_i = D_i\varphi - \left(\frac{\partial}{\partial\bar\lambda}\right)^{-1}\varphi\frac{\partial I_i}{\partial\bar\lambda}$$

gives

$$\frac{\partial}{\partial\bar\lambda}(D_i\chi - \mu_i) = D_i\varphi. \qquad (8.2.8)$$

It follows from the formula (8.2.8) and the fact that $\partial\varphi/\partial\bar\lambda$ decreases faster than any degree $(\lambda - \lambda_1)^n$ at the poles of $I_i(\lambda)$, that the functions $D_i\chi - \mu_i$ are regular ones. Therefore, $\mu_i$ are indeed the normalizations of the functions $D_i\chi$.

Thus, using the operators $D_i$ one can multiply the solutions of the $\bar\partial$-problem (8.2.1). It is important that the number of solutions which can be constructed in such a way increases with the increase of the order of the operators $D_i$ faster than the number of normalization poles. Just this circumstance allows us to construct the differential equations for the solutions $\chi$ of the $\bar\partial$-problem (8.2.1) [145].

Let us consider directly the generic case when the functions $I_i(\lambda)$ have the arbitrary number of the simple distinct poles, i. e.

$$I_i(\lambda) = \sum_{\alpha=1}^n \frac{A_\alpha^{(i)}}{\lambda - \lambda_\alpha^{(i)}}, \quad (i = 1,2,3). \qquad (8.2.9)$$

For the convenience we introduce the multiindex $I = \begin{pmatrix}(i)\\\alpha\end{pmatrix}$. The presence of the common index $I$ means the summation over $\alpha$ from 1 to $n$.

Let $\chi_I(x, \lambda)$ be the solutions of the $\bar\partial$-equation (8.2.1) normalized to $(\lambda - \lambda_I)^{-1}$. These functions possess the following expansions

$$\chi_I(x,\lambda) \underset{\lambda\to\lambda_I}{\to} (\lambda - \lambda_I)^{-1} + \sum_{n=0}^\infty \chi_{II}^{(n)}(x)(\lambda - \lambda_I)^n,$$

$$\chi_I(x,\lambda) \underset{\lambda\to\lambda_J}{\to} \sum_{n=0}^\infty \chi_{IJ}^{(n)}(x)(\lambda - \lambda_J)^n. \qquad (8.2.10)$$

The functions $D_i\chi_J$ are the solutions of the $\bar\partial$-equation (8.2.1) too. In virtue of (8.2.1), the normalizations of $D_i\chi_J$ contain the first order poles at the points $\lambda_I$ and $\lambda_J$ where $y$ is fixed and $I$ is an arbitrary one. On the other hand, the normalizations of the functions $\chi_I$ and $\chi_J$ contain the simple poles at the points $\lambda_I$ and $\lambda_J$ respectively. By this reason, one can construct the linear combination of the functions $D_i\chi_J$, $\chi_I$ and $\chi_J$ which has the vanishing normalization. This combination is of the form

$$(L\chi)_{iJ} = D_i\chi_J - \frac{A_I}{\lambda_J - \lambda_I}\chi_J - \chi_{JI}A_I\chi_I \qquad (8.2.11)$$

where $\chi_{JT} \doteq \chi_J(x, \lambda_I)$. In virtue of the unique solvability of the $\bar{\partial}$-problem (8.2.1), one, therefore, has [145]

$$(L_i \chi)_J = D_i \chi_J - \frac{A_i}{\lambda_J - \lambda_I} \chi_J - \chi_{JI} A_I \chi_I = 0. \qquad (8.2.12)$$

Together with (8.2.12) the equations with the cycled indices hold.

So, in the generic case (8.2.9) we have the system of the linear spectral problems (8.2.12) which are also linear on $D_i$. The possibility of construction of the linear on $D_i$ auxiliary problems is connected obviously, with the use of the rational normalizations of the functions $\chi_J$. Recall that in the case of the fixed canonical normalization $\eta = 1$ the spectral problems (2.4.31) are quadratic on $D_i$ even in the simplest case $n = 1$. With the increase of $n$ the order of the spectral problems rapidly increases and their construction becomes technically rather complicated problem. The use of the variable rational normalization, as we saw, allows us to construct quickly the spectral problems, even in the generic case (8.2.9).

Linear equations (8.2.12) possess the important property. They are, in the certain sense, the basic ones within the $\bar{\partial}$-dressing scheme under consideration. Let us demonstrate this. We already mentioned that one can multiply the solutions of the $\bar{\partial}$-problem (8.2.1) by the operators $D_i$. Let $\chi_I$ be the solution of (8.2.1) normalized to $(\lambda - \lambda_I)^{-1}$. Let us consider the class of solutions of the problem (8.2.1) of the form

$$(L\chi)(x, \lambda) = \sum_{n_1, n_2, n_3 = 0}^{N} U_{(I) n_1, n_2, n_3}(x) D_i^{n_1} D_j^{n_2} D_k^{n_3} \chi_I(x, \lambda) + \text{c. p.} \qquad (8.2.13)$$

where c. p. means the cyclic permutation of the indices $I$, $J$, $K$. Using the definition (8.2.11) of $L_{iJ}(\chi)$ and excluding the terms $D_i \chi_J$ from the expression (8.2.13), one can show that

$$(L\chi)(x, \lambda) = \sum_{n=0}^{N} U_{I_n}(x) D_i^n \chi_I(x, \lambda)$$
$$+ \sum_{n_1, n_2, n_3 = 0}^{N} U_{(iJ) n_1, n_2, n_3}(x) D_i^{n_1} D_j^{n_2} D_k^{n_3} L_{iJ}(\chi) + \text{ c. p.} \qquad (8.2.14)$$

Further, considering subsequently the poles of the orders $N, N + 1$ and so on, one can verify that, if

$$\sum_{n=0}^{N} U_{I_n}(x) D_i^n \chi_I(x, \lambda) + \text{c. p.} = 0 \qquad (8.2.15)$$

then the l. h. s. of (8.2.15) vanishes identically.

This result and the formula (8.2.14) imply that, if $L\chi = 0$ then $L$ is representable in the form

$$L\chi \equiv \sum_{n_1,n_2,n_3=0}^{N} U_{(iJ)n_1,n_2,n_3}(x) D_i^{n_1} D_j^{n_2} D_k^{n_3} L_{iJ}(\chi) + \text{c. p.} \qquad (8.2.16)$$

Hence any linear equation $L(\chi) = 0$ where the operator $L$ is of the form (8.2.13) is, in fact, the linear superposition of equations $(L_i\chi)_J = 0$ of the type (8.2.12) [145].

So, the spectral problems (8.2.12) are the fundamental objects in the $\bar{\partial}$-dressing method with the variable normalization. Correspondingly the nonlinear system associated with these spectral problems plays the principal role too. This nonlinear system can be constructed in a very simple way. Indeed, expanding the functions $\chi_J(x,\lambda)$ from (8.2.12) into the series near the points $\lambda_k$, we find, in the main order, the system [145]

$$\frac{\partial}{\partial x_i}\chi_{Jk}(x) + \chi_{Jk}(x)\frac{A_I}{\lambda_k - \lambda_I} - \frac{A_I}{\lambda_J - \lambda_k}\chi_{Jk}(x) - \chi_{JI}(x)A_I\chi_{Jk}(x) = 0. \qquad (8.2.17)$$

Together with (8.2.17) one has also the equations which differ from (8.2.17) by the cyclic permutation of the indices $I$, $J$, $K$. The system (8.2.17) is the fundamental integrable system of the $6n^2$ equations for $6n^2$ unknown functions $\chi_{IK}(x)$, associated with the fundamental auxiliary linear system (8.2.12).

The system of the nonlinear equations (8.2.17) occupies the central place within the class of the nonlinear systems integrable by the $\bar{\partial}$-problem (8.2.1). The solutions of the nonlocal $\bar{\partial}$-problem (8.2.1) appropriately normalized give the particular solutions of the fundamental system (8.2.17). These solutions depend on the functional parameter $R_0(\lambda', \bar{\lambda}'; \lambda, \bar{\lambda})$ and the natural boundary condition for $\chi_{IJ}(x)$ is the condition $\chi_{IJ}(x) \underset{|x|\to\infty}{\longrightarrow} (\lambda_J - \lambda_I)^{-1}$. In the case of the degenerate $R_0(\lambda', \bar{\lambda}'; \lambda, \bar{\lambda})$ the solutions of the system (8.2.17) can be constructed in the explicit form, as usual.

The fundamental system (8.2.17) is close in its form to the system of the three resonantly interacting waves with the special matrix structure. This circumstance indicates that the system (8.2.17) should be a Lagrangian one. Indeed, the corresponding Lagrangian is [145]

$$\mathcal{L} = \text{tr}\Bigg[\text{sgn}(ijk)\left\{\frac{1}{2}\chi_{IJ}A_J\frac{\partial\chi_{JI}}{\partial x_k}A_I - \chi_{JI}A_I\frac{\partial\chi_{IJ}}{\partial x_k}A_J\right\}$$
$$- \frac{A_k}{\lambda_I - \lambda_k}\chi_{IJ}A_J\chi_{JI}A_I - \frac{A_k}{\lambda_J - \lambda_k}\chi_{JI}A_I\chi_{IJ}A_J$$
$$+ \frac{1}{3}A_I\chi_{Ik}A_k\chi_{kJ}A_J\chi_{JI} - \frac{1}{3}A_I\chi_{IJ}A_J\chi_{Jk}A_k\chi_{kI}\Bigg] \qquad (8.2.18)$$

where sgn$(ijk)$ is the sign of the transposition of the indices $i$, $j$, $k$ and the summation is assumed over the arbitrary transpositions of the indices 1, 2, 3 and over $\alpha$ for the common indices $I$.

Note that the linear system (8.2.12) gives not only the fundamental system (8.2.17) but also the infinite set of the conservation laws for (8.2.17). To derive these conservation laws it is sufficient to expand (8.2.11) near the point $\lambda_J$. The first two conservation laws are local ones while all the higher conservation laws are nonlocal.

The system (8.2.17) is of fundamental character since it corresponds to the generic functions $I_i(\lambda)$ with the arbitrary number of the simple distinct poles. Any rational functions $I_i(\lambda)$ can be obtained from such generic $I_i(\lambda)$ by the appropriate limiting processes. The corresponding nonlinear integrable systems are obtained as the appropriate limits of the fundamental system (8.2.17) too. The system (8.2.17) is the universal system in this sense.

Let us illustrate now the interrelation between the fundamental system (8.2.17) and the different concrete integrable systems. As a first example we will consider the fundamental system (8.2.17) at $n = 1$, i.e. $J \to j$ and $K \to k$. It is the system of six equations which under the additional reduction gives the system of the three resonantly interacting waves. In the last case the fundamental system turns out to be equivalent to the system (2.4.32) and the interrelation between the functions $\chi_{ij}(x)$ and $\chi_i(x)$ is given by the formula [145]

$$\chi_{ij}(x) = A_i^{-1}\chi_i^{-1}\left(\frac{\partial \chi_j}{\partial x_i} + \chi_j\frac{A_i}{\lambda_j - \lambda_i} - \frac{A_i}{\lambda_j - \lambda_i}A_j\right) \quad (i,j = 1,2,3; i \neq j). \quad (8.2.19)$$

The simplest example of the degeneration of the generic system (8.2.17) is the case when the sets of the poles for the functions $I_1(\lambda)$, $I_2(\lambda)$ and $I_3(\lambda)$ coincide

$$I_i(\lambda) = \sum_{k=1}^{n} \frac{A_{ik}}{\lambda - \lambda_k} \quad (i = 1, 2, 3). \quad (8.2.20)$$

In this case it is convenient to introduce the new functions $Q_{kp}(x) = \chi_{kp} - \frac{\delta_{kp}}{\lambda_p - \lambda_k}$. In terms of these variables the fundamental system (8.2.17) looks like [145]

$$\left(A_{jq}\frac{\partial}{\partial x_i} - A_{iq}\frac{\partial}{\partial x_j}\right)Q_{qr}A_{kr} + (A_{jq}Q_{qr}A_{ip} - A_{iq}Q_{qr}A_{jp})A_{kr}\frac{1}{\lambda_r - \lambda_p}$$

$$-A_{iq}Q_{qp}A_{ip}Q_{qr} + A_{iq}Q_{qp}A_{jp}Q_{pr} + \text{c.p.} = 0 \quad (8.2.21)$$

where the indices $q$ and $p$ are fixed while there is a summation over the other indices. Note that the singularities in (8.2.21) are canceled identically.

The nonlinear integrable system (8.2.21) for the $n^2$ functions $Q_{qr}$ is similar to the resonantly interacting waves system. The Lagrangian of the system (8.2.21) is given by (8.2.18) with the obvious substitution $\chi_{IJ} \to Q_{ij}$.

The degeneration of the generic $I_i(\lambda)$ to the functions $I_i(\lambda)$ with the multiple poles is a much less trivial case. The KP equation examplifies such situation. As we have mentioned in section (4.5) the KP equation can be imbedded into the $\bar{\partial}$-dressing method by the following choice of operators $D_i$:

$$D_1 = \partial_x + \lambda^{-1}, \quad D_2 = \partial_y + \lambda^{-2}, \quad D_3 = \partial_t + \lambda^{-3} \qquad (8.2.22)$$

and the function $\chi(x,\lambda)$ normalized as $\chi(x,\lambda) \to \lambda^{-1}$ when $\lambda \to 0$. These operators $D_i$ contain the multiple poles at $\lambda = 0$. One can obtain these multiple poles as the degeneration of the situation with the simple poles. Indeed, if one considers the operators

$$D_1 = \partial_x + \lambda^{-1},$$

$$D_2 = \partial_y + \frac{1}{2\epsilon}[(\lambda - \epsilon)^{-1} - (\lambda + \epsilon)^{-1}], \qquad (8.2.23)$$

$$D_3 = \partial_t + \frac{1}{2\epsilon^2}[(\lambda + \epsilon)^{-1} + (\lambda - \epsilon)^{-1} - 2\lambda^{-1}]$$

and passing to the limit $\epsilon \to 0$, then one gets the operators (8.2.22).

The disadvantage of the operators (8.2.23) is that the corresponding $I_i(\lambda)$ contain the coinciding poles. To remove these coinciding poles it is convenient to perform the change of independent variables $(x, y, t) \to (x, \xi, \eta)$ defined by

$$D_1' = D_1 = \partial_x + \lambda^{-1},$$

$$D_2' = \frac{1}{2}(\epsilon^2 D_3 + \epsilon D_2 + D_1) \doteq \partial_\xi + (\lambda - \epsilon)^{-1}, \qquad (8.2.24)$$

$$D_3' = \frac{1}{2}(\epsilon^2 D_3 - \epsilon D_2 + D_1) \doteq \partial_\eta + (\lambda + \epsilon)^{-1}.$$

The functions $I_1(\lambda)$ which correspond to the operators (8.2.24) only have the simple poles at points $0, \epsilon, -\epsilon$. Let us denote these poles by the indices 1, 2, 3.

The case (8.2.24) is the very special case of the generic situation (8.2.9) and the Lagrangian (8.2.18) takes the form

$$\mathcal{L} = \chi_{12}\partial_\eta \chi_{21} + \chi_{23}\partial_\xi \chi_{32} + \chi_{31}\partial_\xi \chi_{13} + \frac{1}{2\epsilon}\chi_{12}\chi_{21}$$

$$+ \frac{1}{2\epsilon}\chi_{13}\chi_{31} + \frac{2}{\epsilon}\chi_{23}\chi_{32} - \chi_{13}\chi_{32}\chi_{21} - \chi_{12}\chi_{23}\chi_{31}. \qquad (8.2.25)$$

Now let us proceed to the limit $\epsilon \to 0$. We will demonstrate, following [145], that the Lagrangian (8.2.25) is reduced as $\epsilon \to 0$ to the KP equation Lagrangian.

Firstly, we should use the fundamental equations (8.2.17) in the case (8.2.24). These equations give

$$\partial_\xi \chi_{13} + \frac{1}{2\epsilon}\chi_{13} - \chi_{12}\chi_{23} = 0,$$

*Chapter 8* 251

$$\partial_\eta \chi_{12} - \frac{1}{2\epsilon}\chi_{12} + \chi_{13}\chi_{32} = 0. \qquad (8.2.26)$$

They allow us to exclude the functions $\chi_{23}$ and $\chi_{32}$ from the Lagrangian (8.2.25). As a result, the Lagrangian (8.2.25) becomes

$$\mathcal{L} = \chi_{12}^{-1}\left(\partial_\xi \chi_{13} + \frac{1}{2\epsilon}\chi_{13}\right)\partial_x\left\{\chi_{13}^{-1}\left(\partial_\eta \chi_{12} - \frac{1}{2\epsilon}\chi_{12}\right)\right\}$$
$$+ \frac{2}{\epsilon}(\chi_{12}\chi_{13})^{-1}\partial_\xi \chi_{13}\partial_\eta \chi_{12} \qquad (8.2.27)$$

up to the total derivative. Further, we introduce the functions $\varphi_{ij} \doteq \chi_{ij} - \frac{1}{\lambda_j - \lambda_i}$. In terms of $\varphi_{ij}$ the Lagrangian (8.2.27) looks like

$$\mathcal{L} = \frac{2}{\epsilon}(\partial_\eta \ln(1 + \epsilon\varphi_{12}))\partial_\xi \ln(1 - \epsilon\varphi_{13})$$
$$+ \frac{\varphi_{12} + \varphi_{13} + \epsilon\partial_\xi \varphi_{13}}{2(1 + \epsilon\varphi_{12})}\partial_x\left\{\frac{-\frac{1}{2}(\varphi_{12} + \varphi_{13}) + \epsilon\partial_\eta \varphi_{12}}{-1 + \epsilon\varphi_{13}}\right\}. \qquad (8.2.28)$$

Now, we are proceeding to the limit $\epsilon \to 0$. Using the expansions

$$\varphi_{12} = \varphi_{11} + \varphi_{11}^{(1)}\epsilon + \varphi_{11}^{(2)}\epsilon^2 + \varphi_{11}^{(3)}\epsilon^3 + o(\epsilon^3),$$

$$\varphi_{13} = \varphi_{11} - \varphi_{11}^{(1)}\epsilon + \varphi_{11}^{(2)}\epsilon^2 - \varphi_{11}^{(3)}\epsilon^3 + o(\epsilon^3),$$

denoting $v \doteq \chi_{11}(x, y, t)$ and then excluding $\varphi_{11}^{(1)}$ with the use of the formula

$$\partial_x \varphi_{11}^{(1)} = \frac{1}{2}v_y + vv_x - \frac{1}{2}v_{xx}$$

which follows from (8.2.17), we obtain the expansion of the Lagrangian (8.2.28) into the series on $\epsilon$:

$$\mathcal{L} = 2\epsilon^3\left(-v_t v_x - \frac{1}{4}v_{xx}^2 - v_x^3 + \frac{3}{4}v_y^2\right) + o(\epsilon^3). \qquad (8.2.29)$$

The first nontrivial term in (8.2.29) is, in fact, the Lagrangian of the KP equation. Indeed, it is easy to verify the corresponding Euler-Lagrange equation

$$\left(v_t - \frac{1}{4}v_{xxx} + \frac{3}{2}v_x^2\right)_x - \frac{3}{4}v_{yy} = 0$$

after the change $u(x, y, t) \doteq v_x$ and the obvious rescaling of the independent variables coincides with the KP equation.

Thus, we have justified that Lagrangian of the KP equation is obtainable as the special limit of the fundamental Lagrangian (8.2.18). Similarly the other nonlinear integrable equations can be reproduced as the limits of the fundamental nonlinear integrable system (8.2.17) to [145].

We see that as a whole the $\bar{\partial}$-dressing method with the variable normalization is a very effective one for the study of the general structure of the three-dimensional integrable systems and their deep interrelations.

## 8.3. Further generalizations

The $\bar{\partial}$-dressing method as it was formulated in section (2.3) is a very general method. The structure and properties of the nonlinear integrable equations are completely determined by the operators $\hat{B}_i$. In general, these operators may be the integral operators in the complex spectral space $\lambda$. A main requirement to the operators $\hat{B}_i$ is that one should be able to effectively construct the corresponding operators $L_i$ which obey the conditions (2.3.15), (2.3.16) and, consequently, linear problems (2.3.18).

In the previous chapters we considered the two cases:

1. The operators $\hat{B}_i$ are the operators of multiplication by matrix valued functions, i. e.
$$\hat{B}_i = I_i(\lambda) \tag{8.3.1}$$
where $I_i(\lambda)$ are, in general, arbitrary rational functions.

2. The operators $\hat{B}_i$ are the integral operators of convolutive type
$$(\hat{B}_i f)(\lambda) = \iint_C d\mu \wedge d\bar{\mu}\, B_i(\lambda - \mu) f(\mu) \tag{8.3.2}$$
or close type. The Fourier transform of these operators have been the differential operators associated with the background integrable system.

The operators (8.3.1) and (8.3.2) are, of course, very special integral operators.

An analysis of the case of more or less general integral operators $\hat{B}_i$ is a difficult problem. Here we will discuss one more special class of the integral operators $\hat{B}_i$, the most known one, namely, the class of differential operators
$$\hat{B}_i \cdot = \sum_{k=0}^{n_i} b_{(i)k}(x_i, \lambda) \frac{\partial^k}{\partial \lambda^k}, \tag{8.3.3}$$
where $b_{(i)k}(x_1, x_2, x_3, \lambda)$ are matrix-valued functions.

For the choice (8.3.3) equations (2.3.7) or, equivalently, equations
$$[\partial_{x_i} - \hat{B}_i,\ \partial_{x_k} - \hat{B}_k] = 0 \tag{8.3.4}$$
represent the typical commutativity operator form of the integrable equations discussed in section (1.2) (see (1.2.1) with the change $x \to \lambda$). So, for characterization of the operators $\hat{B}_i$ (8.3.3) one can use all the known results for the 2+1-dimensional integrable equations. This provides, in principle, an essential extension of the $\bar{\partial}$-dressing method.

The idea to use differential operators in the spectral space as $\hat{B}_i$ in the framework of the $\bar{\partial}$-dressing method was originated by V. Zakharov. He also found some simple examples [144, 147]. Transformations (2.3.6) of the kernel $R$ of the $\bar{\partial}$-problem with the

*Chapter 8* 253

special differential operators $\hat{B}_i$ associated with the nonisospectral symmetries has been considered in [261].

A general, quite obvious, requirement for the operators $\hat{B}_i$ is that they should have some sets of singularities in $\lambda$. Otherwise it will be impossible to construct a finite systems for finite number of the dependent variables.

According to this, the operators $\hat{B}_i$ (8.3.3) with constant coefficients $b_{(i)k}$ are not a proper choice. Indeed, in this case for the canonically normalized $\chi$, i. e.

$$\chi \to 1 + \frac{1}{\lambda}\chi_{-1} + \frac{1}{\lambda^2}\chi_{-2} + \dots \quad (\lambda \to \infty) \tag{8.3.5}$$

one has the linear problems

$$D_i\chi - \chi b_{(i)0} = 0 \quad (i = 1, 2, 3) \tag{8.3.6}$$

with constant coefficients.

By the same reason the operators $\hat{B}_i$ (8.3.3) with variable $b_{(i)k}(x,\lambda)$ but such that

$$b_{(i)k}(x,\lambda) \underset{\lambda \to \infty}{\to} \text{const} \tag{8.3.7}$$

are not suitable too.

For the case of canonical normalization the functions $b_{(i)k}$ should have singularities as $\lambda \to \infty$. We will consider here some examples.

Let us start with the case when $\partial \hat{B}_i / \partial x_k \equiv 0$. Hence

$$[\hat{B}_i(x,\lambda), \hat{B}_k(x,\lambda)] = 0 \tag{8.3.8}$$

where $\hat{B}_i$ are of the form (8.3.3). Equation (8.3.8) with $b_{(i)n} = \text{const}$ is the operator representation of the stationary KdV and, in general, stationary Gelfand—Dikii equations. Such equations have been studied in detail but they have no solutions polynomial in $\lambda$.

The nontrivial for our case corresponds to $b_{(i)n_i} \neq \text{const}$. It is not difficult to convince oneself that the simplest nontrivial scalar choice is [147]

$$B_1^* = \lambda^2 \frac{\partial}{\partial \lambda}, \quad B_2 = B_1^{*2}, \quad B_3 = B_1^{*3}, \dots \tag{8.3.9}$$

One can readily show that the quantity $D_1\chi$ has no singularity as $\lambda \to \infty$. Then

$$D_1\chi + V\chi = \partial_x \chi + \lambda^2 \frac{\partial \chi}{\partial \lambda} + V(x_1, \lambda)\chi \to 0 \tag{8.3.10}$$

as $\lambda \to \infty$ if $V = \chi_{-1}$. Hence, one has the linear problem

$$D_1\chi - \chi_{-1}(x)\chi = 0. \tag{8.3.11}$$

Further, the quantity
$$D_2\chi = \partial_y\chi + \left(\lambda^2\frac{\partial}{\partial\lambda}\right)^2\chi \tag{8.3.12}$$
has the first order singularity at infinity. It is eliminated by the term $-D_1^2\chi$. The condition (2.3.16) is satisfied for
$$D_2\chi - D_1^2\chi + \tilde{V}\chi \tag{8.3.13}$$
if $\tilde{V} = -2\chi_{-1x}$. So the second problem is of the form
$$D_2\chi - D_1^2\chi - 2\chi_{-1x}\chi = 0. \tag{8.3.14}$$

The problems (8.3.11) and (8.3.14) form a complete set. They provide the nonlinear equation for $\chi_{-1}$. Substituting the expansion (8.3.5) into equation (8.3.14), considering the terms of order $\lambda^{-1}$ and using (8.3.11), one gets
$$\chi_{-1y} + \chi_{-1xx} - 2\chi_{-1}\chi_{-1x} = 0. \tag{8.3.15}$$
In terms of $u \equiv -\chi_{-1}$ one has
$$u_y + u_{xx} + 2uu_x = 0 \tag{8.3.16}$$
that is the well-known Burgers equation.

If one starts with the operator $D_3\chi$, then one shall arrive at the following linear problem
$$D_3\chi - D_1^3\chi + 3\chi_{-1x}D_1\chi - 3(\chi_{-1xx} + \chi_{-1}\chi_{-1x})\chi = 0. \tag{8.3.17}$$
Considering the $\frac{1}{\lambda}$ order terms on the l. h. s. of (8.3.17) and eliminating $\chi_{-2}$, one finds
$$\chi_{-1t} - \chi_{-1xxx} + 3(\chi_{-1x})^2 - 3\chi_{-1}\chi_{-1xx} - 3\chi_{-1}^2\chi_{-1x} = 0 \tag{8.3.18}$$
that is the first higher Burgers equation in terms of $u = -\chi_{-1}$.

So the choice (8.3.9) gives rise to the Burgers equations. This example has been presented for the first time in [147]. It has been mentioned in [147] that the common solution (more precisely $u_{KP} = -2\chi_{-1x}$) of both the Burgers and higher Burgers equations (8.3.17) and (8.3.19) is the solution of the KP equation up to the obvious rescaling of variables. One can check this fact straightforwardly.

A more simple way to convince yourself in this fact is to compare the linear problems (8.3.14) and (8.3.17) (with the substitution $\chi_{-1xx} = -\chi_{-y} + 2\chi_{-1}\chi_{-1x}$, due to (8.3.15)) with the linear problems (3.1.8) and (3.1.21). They coincide up to the rescaling of independent variables. The operators $D_i$ ($i = 1, 2, 3$) in these problems are different, but this does not affect the corresponding nonlinear integrable equation.

A different situation arises if one considers, for instance, the operators $\hat{B}_i$ of the form

$$\hat{B}_1^* = \alpha\lambda + \beta\frac{\partial}{\partial\lambda}, \quad \hat{B}_2^* = (\hat{B}_1^*)^2, \quad B_3^* = 4(\hat{B}_1^*)^3 \tag{8.3.19}$$

where $\alpha$ and $\beta$ are arbitrary constants. The corresponding operators $D_i$ have the first, second and third order singularities as $\lambda \to \infty$ as for the KP equation. This will lead to the same linear problems as for the KP equation. Indeed, the first problem is

$$D_2\chi - D_1^2\chi + u\chi = 0 \tag{8.3.20}$$

where

$$u = 2\alpha\chi_{-1x}. \tag{8.3.21}$$

One can show that the second problem looks like

$$D_3\chi - 4D_1^3\chi + 12\alpha\chi_{-1}D_1\chi + 12(\alpha\chi_{-1xx} + \alpha^2\chi_{-2x} - \alpha^2\chi_{-1}\chi_{-1x})\chi = 0 \tag{8.3.22}$$

or, equivalently,

$$D_3\chi - 4D_1^3\chi + 6uD_1\chi + (3u_x - 3w)\chi = 0 \tag{8.3.23}$$

where $w_x = u_y$.

We see that the problems (8.3.20), (8.3.23) are exactly the same as those (3.1.8), (3.1.21) constructed in section (3.1) in the case $\beta = 0$. Consequently, the corresponding integrable equation is again the KP equation. But the choice (8.3.19) allows us to construct a new class of solutions of the KP equation. Indeed, the $x, y$ and $t$ dependence of the kernel $R$ is defined now by equations (2.3.6):

$$\frac{\partial R(\mu, \lambda; x, y, t)}{\partial x} = \alpha(\mu - \lambda)R - \beta\frac{\partial}{\partial\mu}R - \beta\frac{\partial R}{\partial\lambda},$$

$$\frac{\partial R(\mu, \lambda; x, y, t)}{\partial y} = \left(\alpha\mu - \beta\frac{\partial}{\partial\mu}\right)^2 R - \left(\alpha\lambda + \beta\frac{\partial}{\partial\lambda}\right)^2 R, \tag{8.3.24}$$

$$\frac{\partial R(\mu, \lambda; x, y, t)}{\partial t} = \left(\alpha\mu - \beta\frac{\partial}{\partial\mu}\right)^3 R - \left(\alpha\lambda + \beta\frac{\partial}{\partial\lambda}\right)^3 R.$$

So, for instance, the solutions with functional parameters are defined by arbitrary solutions of the system of equations

$$\frac{\partial f(\mu)}{\partial x} - \left(\alpha\mu - \beta\frac{\partial}{\partial\mu}\right)f(\mu) = 0,$$

$$\frac{\partial f(\mu)}{\partial y} - \left(\alpha\mu - \beta\frac{\partial}{\partial\mu}\right)^2 f(\mu) = 0, \tag{8.3.25}$$

$$\frac{\partial f(\mu)}{\partial t} - \left(\alpha\mu - \beta\frac{\partial}{\partial \mu}\right)^3 f(\mu) = 0.$$

Similar situation takes place also for the operators $\hat{B}_i$ of the form

$$\hat{B}_1^* = \alpha\lambda + \lambda^{n+k}\frac{\partial}{\partial \lambda^n}, \quad \hat{B}_2^* = (\hat{B}_1^*)^2, \ldots \qquad (8.3.26)$$

with $k = 1, 2$ and $n$ is an arbitrary integer.

Now let us include the dependence on $x_i$ in the operators $\hat{B}_i$. So equations (8.3.4) are equivalent to the nonlinear integrable equations. For effectiveness of the $\bar{\partial}$-dressing the coefficients $b_{(i)k}$ should have singularities. Thus, one should use the solutions of the nonlinear integrable equations with singularities. The first possibility is to consider the solutions which grow polynomially as $|\lambda| \to \infty$. The other choice is the solutions of the nonlinear integrable equations with pole singularities. Such solutions are well studied by the IST method (see e. g. [38, 41]). For the functions $b_{(i)k}(\lambda, x)$ of the form

$$b_{(i)k}(\lambda, x) = \sum_{l=1}^{N} \frac{1}{\lambda - \lambda_i(x)} \qquad (8.3.27)$$

one has the operators $D_i$ with the sets of moving poles. The solutions of the $\bar{\partial}$-problem normalized to $1/(\lambda - \lambda_i(x))$ may be useful for the $\bar{\partial}$-dressing in this case. The construction of the integrable equations and their solutions associated with such operators $\hat{B}_i$ would be of great interest. Such and other possibilities of extension of the $\bar{\partial}$-dressing method have been discussed in [147].

## 8.4. Symmetries of the $\bar{\partial}$-problem and $\bar{\partial}$-dressing

A main achievement of the $\bar{\partial}$-dressing method, as it was described above, is the construction of the nonlinear equations for certain particular values of the function $\chi$ and calculation of their exact solutions. These particular values are the coefficients in the linear problems. The function $\chi$ plays an auxiliary role. It is not an observable quantity in quantum mechanical sense.

Here we will consider another approach in which $\chi$ itself is of a principal meaning. In fact, one can represent the linear problems $L_i\chi = 0$ as the nonlinear equations for $\chi$. For instance, for the KP equation the linear problems are (3.1.8) and (3.1.21), i. e.

$$\sigma\chi_y + \chi_{xx} + 2i\lambda\chi_x + u\chi = 0, \qquad (8.4.1)$$

$$\chi_t + \chi_{xxx} + 12i\lambda\chi_{xx} - 12\lambda^2\chi_x + 6u\chi_x + 6i\lambda u\chi + (3u_x - 3\sigma w)\chi = 0. \qquad (8.4.2)$$

Chapter 8

Then, using (3.1.9) and (3.2.1), one gets

$$\sigma\chi_y(\lambda,\bar{\lambda};x,y,t) + \chi_{xx} + 2i\lambda\chi_x + \frac{1}{\pi}\iint_C d\mu \wedge d\bar{\mu}\,\frac{\partial\chi_x(\mu,\bar{\mu})}{\partial\bar{\mu}} = 0, \tag{8.4.3}$$

$$\chi_t + \chi_{xxx} + 12i\lambda\chi_{xx} - 12\lambda^2\chi_x + \frac{6}{\pi}(\chi_x + i\lambda\chi)\iint_C d\mu \wedge d\bar{\mu}\,\frac{\partial\chi(\mu,\bar{\mu})}{\partial\bar{\mu}}$$

$$+\frac{3}{\pi}\chi(\lambda)(\sigma\partial_y + \partial_x^2)\iint_C d\mu \wedge d\bar{\mu}\,\frac{\partial\chi(\mu,\bar{\mu})}{\partial\bar{\mu}} = 0. \tag{8.4.4}$$

So, we have the two nonlinear integral equations for the function $\chi$ in the space of variables $\lambda,\bar{\lambda};x,y,t$. Further, eliminating $u$ from (8.4.1), (8.4.2) or $\iint_C d\mu \wedge d\bar{\mu}\,\frac{\partial\chi(\mu,\bar{\mu})}{\partial\bar{\mu}}$ from (8.4.3), (8.4.4), one obtains the single nonlinear equation for $\chi(x,y,t,\lambda,\bar{\lambda})$. This equation is obtained from equation (4.1.26) by the change of variable $\chi_0 \to \chi \exp\left(i\lambda x - \frac{1}{\sigma}\lambda^2 y - 4i\lambda^3 t\right)$ [104]. In such a way one can construct the nonlinear equations for $\chi$ for all integrable equations [104].

In this section we will present a different viewpoint. We again start with the $\bar{\partial}$-problem

$$\frac{\partial\chi(\lambda,\bar{\lambda})}{\partial\bar{\lambda}} = (\chi * R)(\lambda,\bar{\lambda}) \tag{8.4.5}$$

but now we ask a different question. We put one of the most important question for differential equations, i. e. the question about symmetries of (8.4.5).

So we will looking for the symmetries of the $\bar{\partial}$-problem (8.4.5). It is convenient to consider together with (8.4.5) the adjoint $\bar{\partial}$-problem

$$\frac{\partial\chi^*(\lambda,\bar{\lambda})}{\partial\bar{\lambda}} = -(R * \chi^*)(\lambda,\bar{\lambda}) = -\iint_C d\mu \wedge d\bar{\mu}\, R(\lambda,\bar{\lambda};\mu,\bar{\mu})\chi^*(\mu,\bar{\mu}). \tag{8.4.6}$$

We assume that the functions $\chi$ and $\chi^*$ are canonically normalized and the problems (8.4.5) and (8.4.6) are uniquely solvable.

Since equation (8.4.5) is an integral one the analysis of its symmetries is a rather nontrivial problem. As usual, it is convenient to consider the infinitesimal symmetry transformations

$$\chi \to \chi' = \chi + \epsilon\frac{\partial\chi}{\partial\tau},$$

$$R \to R' \to R + \epsilon\frac{\partial R}{\partial\tau}, \tag{8.4.7}$$

where $\tau$ is a symmetry parameter. The conditions of invariance of (8.4.5) and (8.4.6) under the transformations (8.4.7) look like

$$\frac{\partial}{\partial\bar{\lambda}}\frac{\partial\chi}{\partial\tau} = \frac{\partial\chi}{\partial\tau} * R + \chi * \frac{\partial R}{\partial\tau}, \tag{8.4.8}$$

$$\frac{\partial}{\partial\bar{\lambda}}\frac{\partial\chi^*}{\partial\tau} = -\frac{\partial R^*}{\partial\tau} * \chi^* - R^* * \frac{\partial\chi^*}{\partial\tau}. \tag{8.4.9}$$

Here we will consider the symmetry transformations of $R$ of the form

$$\frac{\partial}{\partial\tau}R(\lambda',\bar{\lambda}';\lambda,\bar{\lambda}) = \hat{B}(\lambda')R - \hat{B}^*(\lambda)R \tag{8.4.10}$$

where

$$B(\lambda) = \sum_{\alpha,\beta=0}^{\alpha+\beta=n} u_{\alpha\beta}(\tau,\lambda,\bar{\lambda})\frac{\partial^\alpha}{\partial\lambda^\alpha}\frac{\partial^\beta}{\partial\bar{\lambda}^\beta} \tag{8.4.11}$$

where $u_{\alpha\beta}(\tau,\lambda,\bar{\lambda})$ are matrix-valued functions, $\hat{B}^*$ is the operator formally adjoint to $\hat{B}$:

$$\hat{B}^*(\lambda)\cdot = \sum_{\alpha,\beta=0}^{\alpha+\beta=n} (-1)^\alpha(-1)^\beta \frac{\partial^\alpha}{\partial\lambda^\alpha}\frac{\partial^\beta}{\partial\bar{\lambda}^\beta}(\cdot u_{\alpha\beta}). \tag{8.4.12}$$

Substituting (8.4.10) into (8.4.8), combining the result with (8.4.5) and (8.4.6) and assuming that $R(\lambda',\bar{\lambda}';\lambda,\bar{\lambda})\underset{\lambda\to\infty}{\longrightarrow} 0$, one obtains

$$\iint_C d\lambda \wedge d\bar{\lambda} \left\{ \left( \frac{\partial\chi(\lambda,\bar{\lambda})}{\partial\tau} + B^*(\lambda)\hat{\chi}(\lambda,\bar{\lambda}) \right) \chi^*(\lambda,\bar{\lambda}) \right\}$$

$$= \iint_C d\lambda \wedge d\bar{\lambda} \frac{\partial B^*(\lambda)}{\partial\bar{\lambda}} \chi(\lambda,\bar{\lambda})\chi^*(\lambda). \tag{8.4.13}$$

The relation (8.4.13) gives rise to the following [262]

$$\frac{\partial\chi(\lambda,\bar{\lambda})}{\partial\tau} = -B^*(\lambda)\chi(\lambda,\bar{\lambda}) + \text{Anal}(B^*(\lambda)\chi(\lambda,\bar{\lambda})\cdot\chi^*(\lambda,\bar{\lambda}))\cdot\chi^{*-1}(\lambda,\bar{\lambda})$$

$$+ \frac{1}{2\pi i}\iint_C \frac{d\lambda' \wedge d\bar{\lambda}'}{\lambda' - \lambda}\left( \frac{\partial B^*(\lambda')}{\partial\bar{\lambda}'}\chi(\lambda',\bar{\lambda}')\cdot\chi^*(\lambda',\bar{\lambda}') + B_1(\lambda',\bar{\lambda}') \right)\chi^{*-1}(\lambda,\bar{\lambda}) \tag{8.4.14}$$

where $B_1(\lambda',\bar{\lambda}')$ is the matrix-valued function which obey the constraint $\iint_C d\lambda \wedge d\bar{\lambda} B_1(\lambda,\bar{\lambda}) = 0$ and Anal $\Phi(\lambda,\bar{\lambda})$ means the analytic part of function $\Phi$ $(\partial/\partial\bar{\lambda})$Anal $\Phi \doteq 0$.

In a similar manner one obtains [262]

$$\frac{\partial\chi^*(\lambda,\bar{\lambda})}{\partial\tau} = B(\lambda)\chi^*(\lambda,\bar{\lambda}) + \chi^{*-1}(\lambda,\bar{\lambda})\,\text{Anal}\,(\chi(\lambda,\bar{\lambda})B(\lambda)\chi^*(\lambda,\bar{\lambda}))$$

$$-\chi^{*-1}(\lambda,\bar{\lambda})\frac{1}{2\pi i}\iint_C \frac{d\lambda' \wedge d\bar{\lambda}'}{\lambda' - \lambda}\left( \chi(\lambda',\bar{\lambda}')\frac{\partial B(\lambda')}{\partial\bar{\lambda}'}\chi^*(\lambda',\bar{\lambda}') + B_2(\lambda',\bar{\lambda}') \right) \tag{8.4.15}$$

where $B_2(\lambda,\bar{\lambda})$ is the matrix-valued function such that $\iint_C d\lambda \wedge d\bar{\lambda} B_2(\lambda,\bar{\lambda}) = 0$.

The formulae (8.4.10), (8.4.14) and (8.4.15) give us $\partial R/\partial\tau$, $\partial\chi/\partial\tau$ and $\partial\chi^*/\partial\tau$ which, by construction, satisfy equations (8.4.8) and (8.4.9).

Thus, the formulae (8.4.10), (8.4.14) and (8.4.15) define the class of symmetries of the linear problems (8.4.5) and (8.4.6) [262]. This class of symmetry transformations is characterized by an arbitrary differential operator $B(\lambda)$, and by almost arbitrary "constants of integration" $\mathrm{Anal}(B^*\chi \cdot \chi^*)$ and $\mathrm{Anal}(\chi B \chi^*)$. These symmetry transformations are linear for the "inverse problem data" $R$ and, in general, nonlinear and nonlocal for the "wavefunctions" $\chi$ and $\chi^*$. Emphasize that the symmetry transformations (8.4.14) and (8.4.15) are the joint transformations of the function $\chi$ and adjoint function $\chi^*$. The disadvantage of the formulae (8.4.14) and (8.4.15) is that they contain the matrices $B_1$, $B_2$ which are implicit functionals on $\chi$ and $\chi^*$.

The nonlocal $\bar{\partial}$-problem, as we mentioned above, is reduced to the nonlocal Riemann—Hilbert problem if

$$R(\lambda', \bar{\lambda}'; \lambda, \bar{\lambda}) = \delta(\Gamma(\lambda'))R_0(\lambda, \lambda')\delta(\Gamma(\lambda))$$

where the equation $\Gamma(\lambda) = 0$ defines the curve on the complex plane $\lambda$. The corresponding symmetry transformation of the nonlocal RH problem in the particular case $n = 0$, $(B = Y(\lambda))$, is of the form

$$\frac{\partial \chi(\lambda)}{\partial \tau} = -\chi(\lambda)Y(\lambda) + u(\lambda)\chi^{*-1}(\lambda),$$

$$\frac{\partial \chi^*(\lambda)}{\partial \tau} = Y(\lambda)\chi^*(\lambda) - \chi^{-1}(\lambda)u(\lambda) \qquad (8.4.16)$$

where $u(\lambda)$ is an arbitrary rational function on $\lambda$. Under the further reduction to the local RH problem $R_0(\lambda, \lambda') = \delta(\lambda - \lambda')R_0(\lambda)$ one has $\chi^* = \chi^{-1}$ and the symmetry transformation in terms of $\psi = \chi e^{\tau Y(\lambda)}$ converts into the linear equation

$$\frac{\partial \psi(\lambda, \tau)}{\partial \tau} = u(\lambda, \tau)\psi \qquad (8.4.17)$$

which is nothing but the spectral problem (1.1.38).

Another particular case of the $\bar{\partial}$-problem (8.4.5) is the local $\bar{\partial}$-problem which corresponds to the kernel $R(\mu, \bar{\mu}; \lambda, \bar{\lambda}) = \delta(\mu - \lambda)R_0(\lambda)$. In this case $B_1 = B_2 \equiv 0$ and one can identify $\chi^* = \chi^{-1}$. So we have the symmetry transformations

$$\frac{\partial \chi}{\partial \tau} = -B^*(\lambda)\chi(\lambda) + \mathrm{Anal}(B^*(\lambda)\chi(\lambda) \cdot \chi^{-1}(\lambda))\chi(\lambda, \bar{\lambda})$$

$$+ \frac{1}{2\pi i} \iint_C \frac{d\lambda' \wedge d\bar{\lambda}'}{\lambda' - \lambda} \left( \frac{\partial B^*(\lambda)}{\partial \bar{\lambda}'} \chi(\lambda', \bar{\lambda}') \cdot \chi^{-1}(\lambda', \bar{\lambda}') \right) \chi(\lambda, \bar{\lambda}). \qquad (8.4.18)$$

In the particular case $n = 0$ (i. e. when $B(\lambda)$ is the multiplication operator) the formula (8.4.18) has been derived in [263] and [264] within the different approaches. In [263] the analog of equation (3.4.18) for $\chi$ has been used for study of the forced integrable equations. In [264] the same equation has been independently treated as the nonlinear integrable equation for $\chi$.

Equations (8.4.14) and (8.4.15) can be treated in the same manner as the nonlinear nonlocal 2+1 dimensional equations solvable with the help of the $\bar{\partial}$-problem (8.4.5). A simple example corresponds to the choice $B = -(\partial^2/\partial\lambda\partial\bar{\lambda})$, and it is of the form [262]

$$\frac{\partial \chi}{\partial \tau} + \frac{\partial^2 \chi}{\partial \lambda \partial \bar{\lambda}} - \frac{1}{2\pi i} \iint_C \frac{d\lambda' \wedge d\bar{\lambda}'}{\lambda' - \lambda} B_1(\lambda', \bar{\lambda}') \chi^{*-1}(\lambda) = 0,$$

$$\frac{\partial \chi^*}{\partial \tau} - \frac{\partial^2 \chi^*}{\partial \lambda \partial \bar{\lambda}} + \chi^{*-1}(\lambda) \frac{1}{2\pi i} \iint_C \frac{d\lambda' \wedge d\bar{\lambda}'}{\lambda' - \lambda} B_2(\lambda', \bar{\lambda}') = 0. \qquad (8.4.19)$$

Now let us consider the infinite set of symmetries of the problem (8.4.5) of the form (8.4.10), (8.4.14), (8.4.15) which parametrized by $\tau_1, \tau_2, \tau_3, \ldots$. This infinite-dimensional symmetry algebra may have different structures. Here we will consider the abelian infinite-dimensional symmetry algebras of the problem (8.4.5). The commutativity of the symmetry transformations (8.4.10) implies

$$\left[\frac{\partial}{\partial \tau_i} + B_i(\lambda), \frac{\partial}{\partial \tau_k} + B_k(\lambda)\right] = 0 \quad (i, k = 1, 2, 3, \ldots) \qquad (8.4.20)$$

that is equivalent to certain nonlinear differential equations for the coefficients $u_{i\alpha\beta}(\tau_1, \ldots; \lambda, \bar{\lambda})$.

The commutativity of the flows

$$\frac{\partial \chi}{\partial \tau_i} = -B_i^*(\lambda) \chi + \Delta_i \cdot \chi^{*-1}(\lambda),$$

$$\frac{\partial \chi^*}{\partial \tau_i} = B_i(\lambda) \chi^* + \chi^{-1} \Delta_i^* \quad (i = 1, 2, \ldots) \qquad (8.4.21)$$

where $\Delta_i$ and $\Delta_i^*$ are the nonlinear in $\chi$, $\chi^*$ expressions defined by (8.4.14) and (8.4.15), in general case, gives rise to more complicated nonlinear equations. Equations (8.4.21) contain the wave functions $\chi$, $\chi^*$ and potentials (Anal($B^*\chi\chi^*$), Anal($\chi B\chi^*$)). The elimination of the wave functions from (8.4.21) gives rise to the nonlinear integrable equations for potentials. An alternative procedure of the elimination of potentials ("constants of integrations") leads to the nonlinear integrable equations for the wavefunctions $\chi$, $\chi^*$.

Thus, the nonlinear integrable equations associated with the $\bar{\partial}$-problem (8.4.5) express, in fact, the symmetry property of the problem (8.4.5). The following is the simple illustrative example. Let us consider the 2×2 local $\bar{\partial}$-problem and its infinite-dimensional symmetry algebra with $B_i = \lambda^i \sigma_3$ $\left(\sigma_3 = \begin{pmatrix} 1 & 0 \\ 0 & -1 \end{pmatrix}\right)$. In this case the symmetry transformations of wavefunctions in terms of $\psi = \hat{\chi} \exp(\sum_{i=1}^{\infty} \sigma_3 \tau_i \lambda^i)$ are of the form

$$\frac{\partial \psi}{\partial \tau_i} = \sum_{\alpha=0}^{i} \lambda^{i-\alpha} P_\alpha(\tau_1, \tau_2, \ldots) \psi \quad (i = 1, 2, 3, \ldots) \tag{8.4.22}$$

where $P_\alpha$ are the $2 \times 2$ matrix-valued functions on $\tau_1, \tau_2, \ldots$ and $P_0 = \sigma_3$. The commutativity of the transformations (8.4.22) gives rise to the well-known AKNS hierarchy [265]. In fact, the infinite system (8.4.22) completely coincides with that discussed in [252] within the Lie-algebraic approach. This coincidence demonstrates also that the infinite-dimensional algebra considered in (8.4.22) is, in fact, the symmetry algebra (in a sense we discussed here) of the $2 \times 2$ local $\bar{\partial}$-problem. Within such a treatment the AKNS hierarchy becomes completely analogous to the Kadomtsev—Petviashvili (KP) hierarchy in Sato approach (see e. g. [85]). The difference is that in the previous case the infinitesimal transformations $\partial \psi / \partial \tau_n = (L^n)_+ \psi$ are the symmetry transformations of the pseudodifferential linear problem $L\psi = \lambda \psi$ where $L = \partial_x + \sum_{i=1}^{\infty} u_i \partial_x^{-i}$.

Emphasize also that the symmetries of the $\bar{\partial}$-problem are also the symmetries of the corresponding integrable equations for the potentials and wave functions. Hence, the transformations (8.4.14), (8.4.15) provides us the class of hidden symmetries of these integrable equations. Some of such interrelations have been discussed in [264].

A compact formula for the symmetry transformations for the nonlocal $\bar{\partial}$-problem can be derived with the use of its solutions with variable normalization [266].

Let us consider, following [266], the infinite family of functions $\chi(\lambda, \bar{\lambda}, a)$ which solve the $\bar{\partial}$-problems

$$\frac{\partial \chi}{\partial \bar{\lambda}} = \delta(\lambda - a) + (\chi * R)(\lambda, \bar{\lambda}) \tag{8.4.23}$$

where $a$ is an arbitrary complex parameter and, hence, normalized as $\chi \to (\pi(\lambda - a))^{-1}$ as $\lambda \to a$.

The symmetry transformation of $R$ is chosen to be

$$\frac{\partial R(\mu, \bar{\mu}; \lambda, \bar{\lambda})}{\partial \tau} = \Omega(\mu) R - R \Omega(\lambda) \tag{8.4.24}$$

where we confine ourselves for simplicity to operators of multiplication by $\Omega(\lambda)$ as $\hat{B}(\lambda)$ in (8.4.10). Differentiating (8.4.23) with respect to $\tau$ and using (8.4.24), one gets

$$\frac{\partial}{\partial \bar{\lambda}} \frac{\partial \chi}{\partial \tau} = \left(\frac{\partial \chi}{\partial \tau} + \chi \Omega\right) * R - (\chi * R) \Omega. \tag{8.4.25}$$

Then the substitution of

$$\chi * R = \frac{\partial \chi}{\partial \bar{\lambda}} - \delta(\lambda - a) \tag{8.4.26}$$

into (8.4.25) gives

$$\frac{\partial}{\partial \bar{\lambda}}\left(\frac{\partial \chi}{\partial \tau}+\chi\Omega\right) = \left(\frac{\partial \chi}{\partial \tau}+\chi\Omega\right)*R + \delta(\lambda - a)\Omega(a) + \chi(a,\lambda)\frac{\partial \Omega}{\partial \bar{\lambda}}. \qquad (8.4.27)$$

Comparing now the $\bar{\partial}$-problems (8.4.27) and (8.4.23), one concludes [266]

$$\frac{\partial \chi(\lambda,\bar{\lambda},a)}{\partial \tau} = -\chi(\lambda,\bar{\lambda},a)\Omega(\lambda) + \Omega(a)\chi(\lambda,\bar{\lambda},a)$$

$$+ \iint_C d\mu \wedge d\bar{\mu}\, \chi(\mu,\bar{\mu},a)\frac{\partial \Omega(\mu)}{\partial \bar{\mu}}\chi(\lambda,\bar{\lambda};\mu). \qquad (8.4.28)$$

Thus, we have the simple and explicit (in contrast to (8.4.14), (8.4.15)) symmetry transformation for $\chi$. But this form of symmetry transformation includes the infinite set of function $\chi(\lambda,\bar{\lambda},a)$ for all $a$. So in both cases (8.4.14), (8.4.15) and (8.4.28) we have no, in fact, closed form of symmetry transformations for $\chi$. It seems an essential feature of the $\bar{\partial}$-problem which is formulated as the equation for function $\chi(\lambda,\bar{\lambda})$. The closed form of the symmetry transformations can be obtained within the operator formulation. For instance, in the Sato approach (see e. g. [85]), the symmetry transformations are of the form

$$\frac{\partial W}{\partial \tau} = (WAW^{-1})_- W \qquad (8.4.29)$$

where $W = 1 + w_1 \partial_x^{-1} + w_2 \partial_x^{-2} + \ldots$ is the pseudodifferential operator, $A$ is the differential operator and $-$ denotes the protection onto the pure integral part.

## 8.5. 2D Harry Dym equation and the $\bar{\partial}$-problem with essential singularity

The $\bar{\partial}$-dressing method with normalizations to rational functions allows to construct and solve a very wide class of the 2+1-dimensional integrable equations. However it does not cover all the integrable equations. An example of such equation is given by the 2+1-dimensional Harry Dym equation

$$u_t + u^3 u_{xxx} + 3\frac{\sigma^2}{u}\left(u^2 \partial_x^{-1}\left(\frac{u_y}{u^2}\right)\right)_y = 0 \qquad (8.5.1)$$

where $u(x,y,t)$ is the scalar function. Equation (8.5.1) has been introduced in [199]. It is the generalization of the well-known 1+1-dimensional Harry Dym equation (see e. g. [44]). Equation (8.5.1) is equivalent to the compatibility condition for the linear system [199]

$$\sigma\psi_y + u^2\psi_{xx} = 0, \qquad (8.5.2)$$

Chapter 8

$$\psi_t + 4u^3\psi_{xxx} + 6u^2\left(u_x - \sigma\partial_x^{-1}\left(\frac{u_y}{u^2}\right)\right)\psi_{xx} = 0. \qquad (8.5.3)$$

The difference of the system (8.5.2), (8.5.3) from all considered above is that it contains the highest order derivatives with the variable coefficients. It is well known that this circumstance leads to essentially new properties of the corresponding equation.

In our case it gives rise to the essential singularity in the spectral problems associated with (8.5.2), (8.5.3). Indeed, let us convert the linear system (8.5.2), (8.5.3) into the spectral problem via consideration of the class of its solutions of the type

$$\psi = \mu(x,y,t,\lambda)e^{i\frac{x}{\lambda} + \frac{1}{\sigma\lambda^2}y} \qquad (8.5.4)$$

where $\lambda$ is an arbitrary parameter and $\mu \to 1$ as $\lambda \to \infty$. We assume that $u(x,y,t) \to 1$ as $x^2 + y^2 \to \infty$. The function $\mu$ obeys the system of equations

$$\sigma\mu_y + \frac{1}{\lambda^2}\mu + u^2\left(\mu_{xx} + \frac{2i}{\lambda}\mu_x - \frac{1}{\lambda^2}\mu\right) = 0, \qquad (8.5.5)$$

$$\mu_t + 4u^3\left(\mu_{xxx} + \frac{3i}{\lambda}\mu_{xx} - \frac{3}{\lambda^2}\mu_x - \frac{i}{\lambda^3}\mu\right)$$

$$-6u^2\left(u_x - \sigma\partial_x^{-1}\left(\frac{u_y}{u^2}\right)\right)\left(\mu_{xx} + \frac{2i}{\lambda}\mu_x - \frac{\mu}{\lambda^2}\right) = 0. \qquad (8.5.6)$$

Let us analyze the behaviour of $\mu$ near the origin $\lambda = 0$. It is quite clear that $\mu(x,y,t,\lambda = 0)$ cannot be bounded. Then, let us assume that

$$\mu = \sum_{n=-N}^{\infty} \lambda^n \mu_n^{(0)}(x,y,t) \qquad (8.5.7)$$

where $N$ is some positive integer. Substituting the Laurent series (8.5.7) into (8.5.5), one gets the following recurrent relations

$$(1-u^2)\mu_{-N}^{(0)} = 0,$$

$$(1-u^2)\mu_{-N+1}^{(0)} + 2iu^2\mu_{-Nx}^{(0)} = 0 \qquad (8.5.8)$$

and

$$\sigma\mu_{ny}^{(0)} + u^2\mu_{nxx}^{(0)} + 2iu^2\mu_{n+1x}^{(0)} + (1-u^2)\mu_{n+2}^{(0)} = 0 \quad (n = -N+2,\ldots,\infty). \qquad (8.5.9)$$

So if $N$ is finite then $\mu_{-N}^{(0)} = 0$ and, consequently, all $\mu_n^{(0)} = 0$.

Thus, the nontrivial solution $\mu(x,y,t,\lambda)$ of the system (8.5.5), (8.5.6) should have an essential singularity at the origin $\lambda = 0$.

To analyze the structure of this essential singularity it is convenient to introduce the function $\rho$ by

$$\rho = (\log\mu)_x. \qquad (8.5.10)$$

It obeys the equation

$$\sigma u^{-2}\partial_x^{-1}\rho_y + \rho_x + \rho^2 + \frac{2i}{\lambda}\rho + \frac{1}{\lambda^2}(u^{-2}-1) = 0. \qquad (8.5.11)$$

It is not difficult to see that $\rho$ has the pole of the first order at $\lambda = 0$, i.e.

$$\rho(x,y,t,\lambda) = \frac{1}{\lambda}\rho_{-1}(x,y,t) + \rho_0(x,y,t) + \lambda\rho_1 + \cdots \qquad (8.5.12)$$

as $\lambda \to 0$. Equation (8.5.11) implies that

$$\rho_{-1} = i(u^{-1}-1), \quad \rho_0 = \frac{1}{2}(\log u)_x - \frac{\sigma}{2u}\varepsilon_y \qquad (8.5.13)$$

where the function $\varepsilon$ is defined as

$$\varepsilon_x = u^{-1} - 1. \qquad (8.5.14)$$

Using the formulae (8.5.10) and (8.5.12)-(8.5.14), one finally obtains

$$\mu = e^{\frac{i}{\lambda}\varepsilon}\{u^{\frac{1}{2}}\exp(-\frac{\sigma}{2}\partial_x^{-1}(\frac{\varepsilon_y}{u})) + O(\lambda)\} \qquad (8.5.15)$$

as $\lambda \to 0$. This explicit expression is very useful for the further analysis of the spectral problem (8.5.5). In particular, proceeding to the function $\tilde{\mu}$ defined by

$$\mu = \tilde{\mu}e^{+\frac{i}{\lambda}\varepsilon(x,y,t)}, \qquad (8.5.16)$$

one gets the linear problem without essential singularity. This allows us to derive the corresponding inverse problem equations.

The problem with essential singularity aises also if one tries to embed equation (8.5.1) into the framework of the $\bar{\partial}$-dressing method. Let us consider the scalar $\bar{\partial}$-problem

$$\frac{\partial \chi(\lambda,\bar{\lambda})}{\partial \bar{\lambda}} = (\chi * R)(\lambda,\bar{\lambda}) \qquad (8.5.17)$$

with the canonical normalisation $\chi \to 1$ as $\lambda \to \infty$ and choose the operators $D_1, D_2$ and $D_3$ again as

$$D_1 = \partial_x + \frac{i}{\lambda}, \quad D_2 = \partial_y + \frac{1}{\sigma\lambda^2}, \quad D_3 = \partial_t + \frac{4i}{\lambda^3}. \qquad (8.5.18)$$

So, the $x,y,t$-dependence of the kernel $R$ is the same as for the mKP equation, i.e.

$$R(\mu,\bar{\mu},\lambda,\bar{\lambda},x,y,t) = exp(i(\frac{1}{\mu}-\frac{1}{\lambda})x + \frac{1}{\sigma}(\frac{1}{\mu^2}-\frac{1}{\lambda^2})y + 4i(\frac{1}{\mu^3}-\frac{1}{\lambda^3})t) \; R_0(\mu,\bar{\mu},\lambda,\bar{\lambda}). \qquad (8.5.19)$$

Using the operators $D_1, D_2, D_3$ (8.5.18), one should construct the operators $L_i$ which obey the conditions (2.3.15) and (2.3.16). It is clear that for $\chi$ with the pole singularity

## Chapter 8

at $\lambda = 0$ one shall arrive at the mKP type linear problems. The only possibility to obtain the linear problem of the type

$$\sigma D_2 \chi + u^2(x,y,t) D_1^2 \chi = 0 \tag{8.5.20}$$

is to consider the function $\chi$ which has an essential singularity at the origin. Indeed, only in the case the equation

$$\sigma \chi_y + \frac{1}{\lambda^2}\chi + u^2(\chi_{xx} + \frac{2i}{\lambda}\chi_x - \frac{1}{\lambda^2}\chi) = 0 \tag{8.5.21}$$

admits the function $u(x,y,t)$ which is not identically equal to unity. Similarly for the second problem

$$D_3 \chi + 4u^3 D_1^3 \chi + V D_1^2 \chi = 0 \tag{8.5.22}$$

where

$$V = 6u^2(u_x - \partial_x^{-1}(\frac{u_y}{u^2})).$$

The compatibility condition for the system (8.5.20), (8.5.22) is equivalent to equation (8.5.1). However to construct its solutions one should now solve the $\bar{\partial}$-problem (8.5.17) with essential singularity. And this is the ill-posed problem.

The difficulty with the essential singularity can be overcome. We consider again the $\bar{\partial}$-problem (8.5.17) with the canonical normalisation. But now instead of the naive operators (8.5.18) we choose

$$D_1 = \partial_x + \frac{i}{\lambda} + \frac{i}{\lambda}f_x, \quad D_2 = \partial_y + \frac{1}{\sigma\lambda^2} + \frac{i}{\lambda}f_y, \quad D_3 = \partial_t + \frac{4i}{\lambda^3} + \frac{i}{\lambda}f_t \tag{8.5.23}$$

where $f(x,y,t)$ is some function. Such operators $D_i$ are of the type discussed in section 8.3.

Again we start with the quantity

$$\sigma D_2 \chi + u^2 D_1^2 \chi = \sigma \chi_y + \frac{1}{\lambda^2}\chi + \frac{i\sigma}{\lambda}f_y \chi +$$

$$u^2(\chi_{xx} + \frac{2i}{\lambda}(1 + f_x)\chi_x + \frac{i}{\lambda}f_{xx}\chi - \frac{1}{\lambda^2}(1 + f_x)^2 \chi) = 0. \tag{8.5.24}$$

It has no second order pole at $\lambda = 0$ if

$$u = (1 + f_x)^{-1}. \tag{8.5.25}$$

The requirement of absence of the first order singularity gives

$$2\frac{\chi_{0x}}{\chi_0} = -\sigma f_y(1 + f_x) - f_{xx}(1 + f_x)^{-1} \tag{8.5.26}$$

where we assume that
$$\chi = \chi_0(x,y,t) + \lambda\chi_1 + \lambda^2\chi_2 + \cdots \tag{8.5.27}$$

as $\lambda \to 0$. It is easy to see that such a quantity $\sigma D_2\chi + u^2 D_1^2\chi$ obeys both conditions (2.3.15) and (2.3.16). So the first linear problem is of the form
$$\sigma D_2\chi + u^2 D_1^2\chi = 0 \tag{8.5.28}$$

where $u$ is given by (8.5.25). From (8.5.28) it follows also that
$$\sigma f_y \chi_1 + u^2 f_{xx}\chi_1 + 2u^2(1+f_x)\chi_{1x} = i(\sigma\chi_{0y} + u^2\chi_{0xx}). \tag{8.5.29}$$

Emphasize that, in virtue of (8.5.25), (8.5.26), the solution of the problem (8.5.28) bounded at the origin $\lambda = 0$ (see (8.5.27)) do exist, in contrast to the case (8.5.18).

To construct the second linear problem we consider the quantity $D_3\chi + VD_1^3\chi$. It is easy to see that the third order singularity in this quantity is absent if
$$V = 4(1+f_x)^{-3} = 4u^3. \tag{8.5.30}$$

To compensate the second order pole we add the term $WD_1^2\chi$. The residue in the second order pole of the quantity

$$D_3\chi + 4u^3 D_1^3\chi + WD_1^2\chi\chi_t + \frac{i}{\lambda}f_t\chi + 4(1+f_x)^{-3}\chi_{xxx} + \frac{4i}{\lambda}(1+f_x)^{-3}\{3(1+f_x)\chi_{xx}$$

$$+ f_{xx}\chi_x + f_{xxx}\chi\} - \frac{12(1+f_x)^{-2}}{\lambda^2}(f_{xx}\chi + (1+f_x)\chi_x) \tag{8.5.31}$$

$$+ W(\chi_{xx} + \frac{2i}{\lambda}(1+f_x)\chi_x + \frac{i}{\lambda}f_{xx}\chi - \frac{1}{\lambda^2}(1+f_x)^2)\chi$$

is equal to zero if
$$W = \frac{12}{(1+f_x)^4}(\frac{\chi_{0x}}{\chi_0}(1+f_x) - f_{xx}). \tag{8.5.32}$$

Using (8.5.26), one gets
$$W = \frac{6}{(1+f_x)^2}(\sigma f_y + (\frac{1}{1+f_x})_x) = 6u^2(u_x - \sigma\partial_x^{-1}(\frac{u_y}{u})). \tag{8.5.33}$$

Further, the requirement of absence of the first order singularity in (8.5.31) with the use of relations (8.5.29), (8.5.33) gives
$$f_{xt} - \frac{1}{(1+f_x)}(\frac{1}{1+f_x})_{xxx} - 3\sigma^2(1+f_x)^3(\frac{f_y}{(1+f_x)^2})_y = 0. \tag{8.5.34}$$

Finally the quantity (8.5.31) with $V, W$ given by (8.5.30), (8.5.33) and the function $f$, obeying equation (8.5.34), satisfies the condition (2.3.16).

Chapter 8

So our second linear problem looks like

$$D_3\chi + 4u^3 D_1^3 \chi + W D_1^2 \chi = 0 \qquad (8.5.35)$$

where $u$ and $W$ are given by (8.5.25), (8.5.33) and the function $f$ obeys equation (8.5.34).

The compatibility condition for the linear system (8.5.28), (8.5.35) implies equation (8.5.1) for $u$ or equation (8.5.34) for function $f$. In virtue of (8.5.25) they are equivalent to each other.

Proceeding to the function $\psi$ defined by

$$\psi = \chi e^{\frac{i}{\lambda}(x+f) + \frac{1}{\sigma\lambda^2}y + \frac{4i}{\lambda^3}t} \qquad (8.5.36)$$

one converts the system (8.5.25), (8.5.33) into the system (8.5.2), (8.5.3). Note also that $f = \varepsilon$ where $\varepsilon$ is defined in (8.5.14).

The linear system (8.5.28), (8.5.35) is of the same form as the system (8.5.20), (8.5.22) with the different operators $D_1, D_2, D_3$ only. But now, in contrast to the case (8.5.18), the function $\chi$ has no essential singularity. So, for the choice (8.5.23) there is no problem with the solution of the $\bar{\partial}$-problem (8.5.17).

However there is a penalty for such a good property of the function $\chi$. Indeed, for the choice (8.5.23) the dependence of the kernel $R$ on $x, y, t$ is given by

$$R(\mu, \bar{\mu}; \lambda, \bar{\lambda}; x, y, t) = R_0(\mu, \bar{\mu}; \lambda, \bar{\lambda}) \times$$

$$exp(i(\frac{1}{\mu} - \frac{1}{\lambda})(x + f(x,y,t)) + \frac{1}{\sigma}(\frac{1}{\mu^2} - \frac{1}{\lambda^2})y + 4i(\frac{1}{\mu^3} - \frac{1}{\lambda^3})t). \qquad (8.5.37)$$

The $\bar{\partial}$-dressing with the kernel (8.5.37) then implies that the function $f$ obeys equation (8.5.34) which is equivalent to the Harry Dym equation (8.5.1). So the data $R$ in the $\bar{\partial}$-problem (8.5.17) depends on the solution of equation (8.5.1) which we would like to solve using (8.5.1).

As a result, the procedure of construction of exact solutions described above will lead us to the solutions of the Harry Dym equation (8.5.1) in implicit form. The exact formulae for the function $\chi$ are the same as for the mKP equation with the substitution $x \to x + f$. But, in virtue of (8.5.25), (8.5.26) the reconstruction of $u$ or $f$ via $\chi_0$ is rather complicated. At the one-dimensional limit $u_y = 0$ the reconstruction formula is simplified to $u = \chi_0^2$.

So for the Harry Dym equation (8.5.1) the essential singularity of the eigenfunction and implicit character of exact solutions are dual to each other.

The derivation presented above indicates the close connection between Harry Dym equation (8.5.1) and the mKP equation. Indeed, the definitions (8.5.23) and (8.5.25) imply that

$$u D_1 = u \partial_x + \frac{i}{\lambda}. \qquad (8.5.38)$$

Introducing now a new independent variable $\tilde{x}$ via

$$\partial_{\tilde{x}} = u(x,y,t)\partial_x, \quad (8.5.39)$$

one gets the operator

$$\widetilde{D}_1 = uD_1 = \partial_{\tilde{x}} + \frac{i}{\lambda}. \quad (8.5.40)$$

The commutativity of this operator $\widetilde{D}_1$ with $D_2, D_3$ of the form (8.5.23) implies $f =$ const. So after the change (8.5.39), (8.5.40) we obtain the operators of the form (8.5.18) which give rise to the mKP equation in the case of the $\bar{\partial}$-problem without essential singularity.

The interrelation between the mKP and 2DHD equation can be established directly on the level of the linear problems. Indeed, using the trivial identity

$$u^2\partial_x^2 = (u\partial_x)^2 - u_x u\partial_x, \quad (8.5.41)$$

one can rewrite the problem (8.5.2) as

$$\sigma\psi_y + (u\partial_x)^2\psi - u_x(u\partial_x)\psi = 0. \quad (8.5.42)$$

Introducing now a new independent variable $\tilde{x}$ by (8.5.39) one represents (8.5.42) as

$$\sigma\psi_y + \psi_{\tilde{x}\tilde{x}} + v(\tilde{x},y,t)\psi_{\tilde{x}} = 0 \quad (8.5.43)$$

where

$$v(\tilde{x},y,t) \equiv -u_x(x,y,t). \quad (8.5.44)$$

The problem (8.5.43) is just the first linear problem for the mKP equation. The same trick converts the problem (8.5.3) into the second linear problem for the mKP equation.

Thus, the formulae (8.5.39) and (8.5.44) establish the relation between the mKP and 2DHD equations. Using the results of section (4.5), one interrelates the mKP, KP and 2DHD equations. The connection between the KP equation and 2DHD equation has been discussed in [267].

Note in conclusion that the interrelation between the integrable equations which correspond to the same $x, y, t$ dependence of the kernel $R$ but different normalizations (like the mKP, KP and 2DHD equations) is an important problem of the $\bar{\partial}$-dressing method.

## 8.6. Towards the multidimensional systems

The construction of the multidimensional integrable nonlinear systems with four and more independent variables was and still is the main goal of the IST method. But unfortunately there is essentially no examples of nontrivial integrable equations in more than 2+1-dimensions is known yet apart from equations with many "time" variables [44] and celebrated self-dual Yang—Mills equation [268]. There are several obstacles of algebraic [269, 270] and analytic [271–274] character which prevent the straightforward extension of the IST method to the true multidimension.

In spite of this many attempts have been done to use the IST method for the study of the multidimensional equations [254, 255, 275–295]. Here we will discuss some of them which have been done in essence within the framework of $\bar{\partial}$-dressing method.

First, note that in addition to three-dimensional system (2.4.32) one can construct in completely analogous way the multidimensional system considering an arbitrary number of the variables $x_1, \ldots, x_n$. The integrable system is again of the form (2.4.32) but now $i = 1, 2, \ldots, n$. The $\bar{\partial}$-dressing method allows to construct wide classes of exact solutions but these solutions will be effectively three-dimensional [109].

Further, considering the hierarchies of the 2+1-dimensional integrable equations for the times $t_1, t_2, t_3, \ldots$, one can easily combine them into multidimensional nonlinear equation. For instance, taking the common solution $u(x, y, t_1, t_2)$ of the KP equation in $(x, y, t_1)$ and higher KP equation in $(x, y, t_2)$, one can combine them into four-dimensional nonlinear equation in $(x, y, t_1, t_2)$. But effectively this equation describes the completely separated evolutions in the triples of variables $(x, y, t_1)$ and $(x, y, t_2)$. Note that in less trivial way similar trick allows to construct the solutions of the 2+1-dimensional integrable equations in $(x, t_1, t_2)$ using the common solutions of two 1+1-dimensional equations in $(x, t_1)$ and $(x, t_2)$ [296 – 302]. For instance, the common solutions $q(x, t_1, t_2)$, $r(x, t_1, t_2)$ of the systems (1.1.17) and (1.1.18) provides via $u(x, t_1, t_2) = q(x, t_1, t_2)r(x, t_1, t_2)$ the solutions of the KP equation [296 – 304].

The construction of the similar nature has been used in [287] to obtain the 2+2 dimensional nonlinear integrable system. In fact, the paper [287] was based on the two separated $2 \times 2$ matrix $\bar{\partial}$-problems for lines with the special dependence on the variables $x, y, t_1, t_2$. In each triple $(x, y, t_1)$ and $(x, y, t_2)$ one has completely disconnected nonlinear integrable equations. Then the consideration of the bilinear combinations of their solutions gives rise to the solutions of the nonlinear equation in the variables $x, y, t_1, t_2$. The exponentially localized solution of this equation has been found also in [287].

This method of glueing of the multidimensional nonlinear equations from the lower dimensional integrable equations may be useful for some problems. The associated analytic and algebraic problems have been discussed in [293, 294, 305].

One more class of nonlinear integrable systems in multidimensions is given by the nonlinear equations with the constraints [157, 288, 295]. The systems of this type can be constructed as follows. Let we have the triple of linear problems

$$L_i \chi = 0 \quad (i = 1, 2, 3). \tag{8.6.1}$$

This system defines the tree-dimensional integrable system in $(x_1, x_2, x_3)$. Then we introduce one more linear equation

$$L_\alpha \chi = 0 \tag{8.6.2}$$

compatible with the system (8.6.1). All equations (8.6.1) and (8.6.2) define the nonlinear system in the four independent variables $x_1, x_2, x_3$ and say $\tau$. If the operator $L_\alpha$ is of the form $L_\alpha = \frac{\partial}{\partial \tau} + \tilde{L}_\alpha$ then the integrable system under consideration looks like

$$\Omega(u) = 0, \tag{8.6.3}$$

$$\frac{\partial u}{\partial \tau} = f(u) \tag{8.6.4}$$

where equation (8.6.3) is equivalent to the compatibility condition for the system (8.6.1). One can treat the system (8.6.3) and (8.6.4) as the four dimensional equation (8.6.4) constrainted by the condition (8.6.3). Equation (8.6.4) is integrable only together with the constraint (8.6.3). General theory of such systems has been developed in [295].

So this approach provides us the multidimensional integrable systems subject to the integrable constraints [157, 288, 295]. One can construct such a system (8.6.3), (8.6.4) that equation (8.6.4) cannot be reduced to the 2+1-dimensional one, in contrast to the KP and higher KP equations case mentioned above. Thus equation (8.6.4) may be a real multidimensional equation. Several examples of the multidimensional integrable systems with constraints have been considered in [157, 288, 295]. The localized solutions of some of them have been found in [288].

Note however, that such approach, in fact, leads to the construction of the symmetries of the integrable equation (8.6.3). Indeed, the invariance of equation (8.6.3) under the flow (8.6.4) is nothing but the requirement of symmetry. So equation (8.6.4), in fact, describes the symmetry of equation (8.6.3) more general than that in the KP equation case. Nevertheless, the multidimensional systems (8.6.3), (8.6.4) with constraints may be useful for some multidimensional problems.

At last, one more possibility for creating the multidimensional systems is connected with the consideration of the $\bar{\partial}$-problem of the Riemann surfaces. This may lead to multidimensional integrable systems of a very special structure. The $\bar{\partial}$-problem on the torus and the corresponding 2+1-dimensional integrable generalization of the Landau–Lifshitz equation (1.1.25) have been considered in [306].

Study of the possible extensions of the $\bar{\partial}$-dressing method mentioned in this chapter and also an analysis of the dressing method based on the tensor and operator $\bar{\partial}$-problem or other generating problems would be of great importance for the further development of the multidimensional IST method.

# References

1. *C.S. Gardner, J.M. Greene, M.D. Kruskal and R.M. Miura*, Method for solving the Korteweg-de Vries equation, Phys. Rev. Lett., **19**, 1095 (1967).

2. *P.D. Lax*, Integrals of nonlinear equations of evolution and solitary waves, Comm. Pure Appl. Math., **7**, 159 (1968).

3. *V.E. Zakharov and A.B. Shabat*, Exact theory of two-dimensional self-focusing and one-dimensional self-modulation of waves in nonlinear media, Zh. Eksp. Teor. Fiz., **61**, 118 (1971); Soviet Phys., JETP, **34**, 62 (1972).

4. *M. Wadati*, The modified Korteweg-de Vries equation, J. Phys. Soc. Japan, **32**, 168 (1972).

5. *V.E. Zakharov*, On stochastization problem for the one-dimensional nonlinear oscillators chains, ZETF, **65**, 219 (1973).

6. *V.E. Zakharov and S.V. Manakov*, Exact theory of resonant interaction of the waves in nonlinear media, Pis'ma ZETF, **18**, 413 (1973).

7. *M.J. Ablowitz, D.J. Kaup, A.C. Newell and H. Segur*, Method for solving the sine-Gordon equation, Phys. Rev. Lett., **30**, 1262 (1973).

8. *V.E. Zakharov, L.A. Takhtajan and L.D. Faddeev*, A complete description of the solutions of the sine-Gordon equation, DAN SSSR, **219**, 1334 (1974); Soviet Phys. Dokl., **19**, 824 (1975).

9. *A.C. Scott, F.Y.F. Chu and D.W. McLaughlin*, The soliton—a new concept in applied science, Proc. IEEE, **61**, 1443 (1973).

10. *V.E. Zakharov and A.B. Shabat*, A scheme for integrating the nonlinear equations of mathematical physics by the method of the inverse scattering problem. I. Funk. Anal. Pril., **8**, 43 (1974); Func. Anal. Appl., **8**, 228 (1974).

11. *M.J. Ablowitz, D.J. Kaup, A.C. Newell and H. Segur*, The inverse scattering transform-Fourier analysis for nonlinear problems, Stud. Appl. Math., **53**, 249 (1974).

12. M. Lakshmanan, Continuum spin system as an exactly solvable dynamical system, Phys. Lett., **61A**, 53 (1977).

13. L.A. Takhtajan, Integration of the continuous Heisenberg spin chain through inverse scattering method, Phys. Lett., **64A**, 235 (1977).

14. V.E. Zakharov and L.A. Takhtajan, Equivalence of nonlinear Schrödinger equation and equation of Heisenberg ferromagnet, Teor. Mat. Fiz., **38**, 26 (1979).

15. V.E. Zakharov and A.B. Shabat, Integration of the nonlinear equations of mathematical physics by the method of the inverse scattering problem. II. Funk. Anal. Pril., **13**, N3, 13 (1979); Func. Anal. Appl., **13**, 166 (1979).

16. V.E. Zakharov and A.V. Mikhailov, Relativistically invariant two-dimensional models of field theory which are integrable by means of the inverse scattering problem method, ZETF, **74**, 1953 (1978); Sov. Phys., JETP, **47**, 1017 (1978).

17. E.K. Sklyanin, On complete integrability of the Landau–Lifshitz equation, Preprint LOMI, E-3-79, Leningrad, 1979.

18. A.E. Borovik and V.N. Robuk, Linear pseudopotentials and conservation laws for the Landau–Lifshitz equation describing nonlinear dynamics of the ferromagnet with a single-axis anisotropy, Teor. Mat. Fiz., **46**, 371 (1981).

19. V.A. Belinsky and V.E. Zakharov, Integration of the Einstein equations by means of the inverse scattering problem technique and construction of exact soliton solutions, ZETF, **75**, 1955 (1978); Sov. Phys., JETP, **48**, 985 (1978).

20. D. Maison, Are the stationary axially symmetric Einstein equations completely integrable? Phys. Rev. Lett., **41**, 521 (1978).

21. S.P. Burtzev, V.E. Zakharov and A.V. Mikhailov, The method of inverse problem with variable spectral parameter, Teor. Mat. Fiz., **70**, 323 (1987).

22. M.D. Kruskal, The Korteweg-de Vries equation and related evolution equation, in: Nonlinear Wave Motion (A.C. Newell, Ed.), AMS Lectures in Appl. Math., **15**, p.61, American Mathematical Society, Providence, RI, 1974.

23. V.E. Zakharov, The inverse scattering transform method, chapter V in "Theory of the elastic media with microstructure" by I.A. Kunin, Nauka, Moscow, 1975.

24. *R.M. Miura*, The Korteweg-de Vries equation: a survey of results, SIAM Review, **18**, 412 (1976).

25. Bäcklund transformations, the inverse scattering method, solitons and their applications (R.M. Miura, Ed.), Lecture Notes in Math., **515**, Springer, Berlin, 1976.

26. *R. Hermann*, The geometry of nonlinear differential equations, Bäcklund transformations and solitons, part A, Math. Sci. Press, Brooklin, MA, 1976.

27. *R.K. Bullough*, Solitons, in: Interaction of Radiation with Condensed Matter, vol.I, IAEA, Vienna (1977), p.381.

28. *C. Cercignani*, Solitons, theory and applications, Riv. Nuovo Cimento, **7**, 429 (1977).

29. *M.J. Ablowitz*, Lectures on the inverse scattering transform, Stud. Appl. Math., **58**, 17 (1978).

30. Nonlinear evolution equations solvable by the spectral transform (F. Calogero, Ed.), Pitman, London, 1978.

31. Proceedings of a conference on the theory and applications of solitons (H. Flaschka and D.W. McLaughlin, Eds.), Rocky Mountain Journ. Math., **8**, N1, 2 (1978).

32. Solitons in Action (K. Longren and A. Scott, Eds.), Academic Press, New York, 1978.

33. Solitons and Condensed Matter Physics (A.R. Bishop and T. Schneider, Eds.), Springer Series in Solid-State Sciences, **8**, Springer, Berlin, 1978.

34. *V.E. Zakharov and S.V. Manakov*, Soliton theory, Soviet Sci. Reviews, **A1**, 133 (1979), Plenum, London.

35. Solitons in Physics (H. Wilhelmsson, Ed.), Topical issue of Physica Scripta, Phys. Ser., **20**, (1979).

36. Solitons (R.K. Bullough and P.J. Caudrey, Eds.), Topics in Current Physics, **17**, Springer, Berlin, 1980.

37. Nonlinear evolution equations and dynamical systems (M. Boiti, F. Pempinelli and G. Soliani, Eds.), Lecture Notes in Physics, **120**, Springer, Berlin, 1980.

38. *V.E. Zakharov, S.V. Manakov, S.P. Novikov and L.P. Pitaevsky*, Theory of solitons. The inverse problem method, Nauka, Moscow, 1980, (Russian); Plenum Press, 1984.

39. *G.L. Lamb, Jr.*, Elements of soliton theory, Wiley, New York, 1980.

40. Soliton theory, Proceedings of the Soviet-American Symposium on Soliton theory (Kiev, September 1979) (S.V. Manakov and V.E. Zakharov, Eds.); Physica, **3D**, N1, 2 (1981).

41. *M.J. Ablowitz and H. Segur*, Solitons and inverse scattering transform, SIAM, Philadelphia, 1981.

42. *G. Eilenberger*, Solitons, Mathematical Methods for Physicists, Springer Series in Solid State Science, **19**, Springer, Berlin, 1981.

43. *W. Eskhaus and A. van Harten*, The inverse scattering transformation and theory of solitons: An introduction, North-Holland, Math. Studies, **50**, North-Holland, Amsterdam, 1981.

44. *F. Calogero and A. Degasperis*, Spectral transform and solitons: Tools to solve and investigate nonlinear evolution equations, vol.1, North-Holland P. C., Amsterdam, 1982.

45. *R.K. Dodd, J.C. Eilbeck, J.D. Gibbon and H.C. Morris*, Solitons and nonlinear waves, Academic Press, New York, 1982.

46. *C. Rogers and W.F. Shadwick*, Bäcklund transformations and their applications, Academic Press, New York, 1982.

47. Nonlinear Phenomena, Proc. of the CIFMO School and Workshop held at Oaxtepec, Mexico, 1982 (K.B. Wolf, Ed.); Lecture Notes in Physics, **189**, Springer, Berlin, 1983.

48. Nonlinear integrable system — Classical theory and quantum theory, Proc. of RIMS Symposium, May 1981 (M. Jimbo and T. Miwa, Eds.), World Scientific, Singapore, 1983.

49. Solitons and particles (C. Rebbi and G. Soliani, Eds), World Scientific, Singapore, 1984.

50. *Y. Matsuno*, Bilinear transformation method, Academic Press, 1984.

51. Nonlinear and Turbulent Processes in Physics, Proc. 2nd Intern. Workshop, Kiev, 1983 (R.Z. Sagdeev, Ed.), Gordon and Breach, New York, 1984.

52. A.C. Newell, Solitons in Mathematics and Physics, SIAM, Philadelphia, 1985.

53. A.N. Leznov and M.V. Saveliev, Group methods for the integration of nonlinear dynamical systems, Nauka, Moskva, 1985.

54. V.A. Dubrovin, I.M. Krichever and S.P. Novikov, Integrable systems, I, Sovremenniye Probl. Matem., **4**, 179, VINITI, Moscow, 1985.

55. Solitons and coherent structures, Proc. Conf. on Solitons and coherent structures held at Santa Barbara, USA, January 1985 (D.K. Campbell, A.C. Newell, P.J. Schrieffer and H. Segur, Eds.), Physica, **18D**, N1-3 (1986).

56. L.D. Faddeev and L.A. Takhtajan, Hamiltonian methods in the theory of solitons, Nauka, Moscow, 1986; Springer, Berlin, 1987.

57. V.A. Marchenko, Nonlinear equations and operator algebras, Naukova Dumka, Kiev, 1986; Kluwer Acad. Publ., Netherlands, 1988.

58. Solitons, (S.E. Trullinger, V.E. Zakharov and V.L. Pokrovsky, Eds.), series "Modern problems in condensed matter sciences", **17**, North-Holland, Amsterdam, 1986.

59. Solitons and nonlinear systems, (D.K. Sinha and R. Ghose, Eds.), South Asian Publ., New Delhi, 1986.

60. Topics in soliton theory and exactly solvable nonlinear equations, (M. Ablowitz, B. Fuchssteiner and M. Kruskal, Eds.), Proc. Conf. on Nonlinear evolution equations, solitons and the inverse scattering transform, Oberwolfach, West Germany, 1986, World Scientific, Singapore, 1987.

61. B.A. Kuperschmidt, Elements of superintegrable systems, Basic techniques and results, D. Reidel P. C., The Netherlands, 1987.

62. Soliton theory, A survey of results (A.P. Fordy, Ed.), Manchester Univ. Press, Manchester, 1987.

63. B.G. Konopelchenko, Nonlinear integrable equations. Recursion operators, group theoretical and Hamiltonian structures of soliton equations, Lecture Notes in Physics, **270**, Springer, Berlin, 1987.

64. Yu.A. Mitropol'skij, N.N. Bogolyubov (Jr.), A.K. Prikarpatsky and V.G. Samoilenko, Integrable dynamical systems: spectral and differential-geometric aspects, Naukova Dumka, Kiev, 1987.

65. Nonlinear Evolutions (J.J.-P. Léon, Ed.), Proc. of IV Workshop on Nonlinear evolution equations and dynamical systems, Balaruc-les Bains, France, June 1987, World Scientific, Singapore, 1988.

66. Solitons: Introduction and application (M. Lakshmanan, Ed.), Springer, Berlin, 1988.

67. R. Beals, P. Deift and C. Tomei, Direct and inverse scattering on the line, AMS, Providence, 1988.

68. Integrability and kinetic equations for solitons (V.G. Bar'jakhtar, V.E. Zakharov and V.N. Chernousenko, Eds.), Naukova Dumka, Kiev, 1990.

69. N. Asano and Y. Kato, Algebraic and spectral methods for nonlinear wave equations, Longman, New York, 1990.

70. L.P. Nizhnik, Inverse scattering problems for hyperbolic equations, Naukova Dumka, Kiev, 1991 (in Russian).

71. O.I. Bogoyavlensky, Overlapping solitons, Nauka, Moscow, 1991.

72. L.A. Dickey, Soliton equations and Hamiltonian systems, World Scientific, Singapore, 1991.

73. What is Integrability? (V.E. Zakharov, Ed.), Springer-Verlag, Berlin, 1991.

74. V.B. Matveev, M.A. Salle, Darboux transformations and solitons, Springer-Verlag, Berlin, 1991.

75. R.W. Carroll, Topics in soliton theory, North-Holland, Amsterdam, 1991.

76. M.J. Ablowitz and P.A. Clarkson, Solitons, nonlinear evolution equations and inverse scattering, Cambridge University Press, Cambridge, 1991.

77. B.G. Konopelchenko, Introduction to multidimentional integrable equations, Plenum Press, New York, 1992.

78. V.E. Korepin, A.G. Isergin and N.M. Bogoliubov, Quantum inverse scattering method and correlation functions, Algebraic Bethe Ansatz, Cambridge University Press, Cambridge, 1992.

## References

79. E.D. Belokolos, A.I. Bobenko, V.Z. Enolsky, A.R. Its and V.B. Matveev, Algebraic-Geometrical Approach to Nonlinear Evolution equation, Springer-Verlag, Berlin, 1992.

80. N.I. Muskhelishvili, Singular integral equations, Noordholf, Groninger, 1953.

81. N.P. Vekua, Systems of singular integral equations, Gordon and Breach, 1967.

82. B.B. Kadomtsev and V.I. Petviashvili, On the stability of solitary waves in weakly dispersive media, DAN SSSR, **192**, 75 (1970).

83. V.S. Dryuma, Analytic solution of the two-dimensional Korteweg-de Vries equation, Pis'ma ZETF, **19**, 753 (1974); Soviet ZETP Lett., **19**, 387 (1974).

84. B.G. Konopelchenko, On the gauge-invariant description of the evolution equations integrable by Gelfand–Dikii spectral problems, Phys. Lett., **92A**, 323 (1982).

85. M. Jimbo and T. Miwa, Solitons and infinite dimensional Lie algebras, Publications of Res. Inst. Math. Sci., **19 (3)**, 943 (1983).

86. M.J. Ablowitz and R. Haberman, Nonlinear evolution equations — two and three dimensions, Phys. Rev. Lett., **35**, 1185 (1975).

87. V.E. Zakharov, The inverse scattering method, in: Solitons (R.K. Bullough and P.J. Caudrey, Eds.), Springer, Berlin, 1980, p.243.

88. A. Davey and K. Stewartson, On three-dimensional packets of surface waves, Proc. Roy. Soc. London, **A338**, 101 (1974).

89. V.E. Zakharov, Exact solutions to the problem of the parametric interaction of three-dimentional wave packets, DAN SSSR, **69**, 1651, (1975); Soviet Phys. Docl., **21**, 322 (1976).

90. H. Cornille, Solutions of the nonlinear 3-wave equations in three spatial dimensions, J. Math. Phys., **20**, 1653 (1979).

91. D.J. Kaup, The solution of the general initial value problem for the full three dimensional three-wave resonant interaction, Physica, **3D**, 374 (1981).

92. Y. Ishimori, Multi-vertex solutions of a two-dimensional nonlinear wave equation, Prog. Teor. Phys., **72**, 33 (1984).

93. *F. Calogero*, A method to generate solvable nonlinear evolution equations, Lett. Nuovo Cim., **14**, 443 (1975).

94. *V.E. Zakharov and S.V. Manakov*, On the generalization of the inverse problem method, Teor. Mat. Fyz., **27**, 283 (1976).

95. *V.E. Zakharov*, Integrable systems in multidimensional spaces, Lecture Notes in Physics, **153**, 190 (1982).

96. *O.I. Bogoyavlensky*, Overlapping solitons in a novel two-dimensional integrable equations, Izvestiya AN SSSR, Ser. mat., **53**, N2, 243 (1989).

97. *O.I. Bogoyavlensky*, Overlapping solitons, Izvestiya AN SSSR, Ser. mat., II, **53**, N4, 907 (1989). III, **54**, N1, 123 (1990).

98. *S.V. Manakov and V.E. Zakharov*, Three-dimensional model of relativistic-invariant field theory integrable by the inverse scattering transform, Lett. Math. Phys., **5**, 247 (1981).

99. *A.V. Mikhailov*, The reduction problem and the inverse scattering method, Physica, **D3**, 73 (1981).

100. *K. Chadan and P.C. Sabatier*, Inverse problems in quantum scattering theory, Springer, New York, 1977.

101. *S.V. Manakov*, Inverse scattering transform method and two-dimensional evolution equations, Usp. Mat. Nauk, **31**, 245 (1976).

102. *L.P. Nizhnik*, Integration of multidimentional nonlinear equations by the method of inverse problem, DAN SSSR, **254**, 332 (1980).

103. *A.P. Veselov and S.P. Novikov*, Finite-zone two-dimensional potential Schrödinger operators. Explicit formulae and evolution equations, DAN SSSR, **279**, 20 (1984).

104. *B.G. Konopelchenko*, Soliton eigenfunction equations: the IST integrability and some properties, Reviews in Math. Phys. **2**, 399 (1990).

105. *B.G. Konopelchenko and C. Rogers*, On (2+1)-dimensional nonlinear systems of Loewner-type, Phys. Lett., **158A**, 391 (1991).

106. *C.A. Loewner*, Generation of solutions of systems of partial differential equations by composition of infinitesimal Bäcklund transformations, J. Anal. Math., **2**, 219 (1952).

107. *P.J. Caudrey*, Discrete and periodic spectral transforms related to the Kadomtsev–Petviashvili equation, Inverse Problems, **2**, 281 (1986).

108. *V.E. Zakharov and S.V. Manakov*, Multidimensional integrable nonlinear systems and methods for constructing their solutions, Zapiski nauchn. semin. LOMI, **133**, 77 (1984), Leningrad.

109. *V.E. Zakharov and S.V. Manakov*, Construction of multidimensional integrable nonlinear systems and their solutions, Funk. Anal. Pril., **19**, N2, 11 (1985).

110. *B.G. Konopelchenko*, On spectral problems and compatibility conditions in multidimensions, J. Phys. A: Math. Gen., **20**, L1057 (1987).

111. *V.S. Dryuma*, On geometry of differential equations of second order and its applications, Proc. of conf. "Nonlinear phenomena" (K.V. Frolov, Ed.), Moscow, Nauka, 1991;
*V.S. Dryuma*, Three-dimensional exactly integrable system of nonlinear equations and its application, Mathematical Studies, vol. **124**, Shdinitza, Kishinev, 1992, p.56.

112. *B.G. Konopelchenko and C. Rogers*, On generalized Loewner systems: novel integrable equations in 2+1-dimensions, J. Math. Phys., 1992.

113. *B.G. Konopelchenko*, Integrable equations: from the Lax pair to "Lie algebra" type operator representation. Preprint PM/88-57, Montpellier, 1988.

114. *S.V. Manakov, V.E. Zakharov, L.A. Bordag, A.R. Its and V.B. Matveev*, Two-dimensional solitons of the Kadomtsev–Petviashvili equation and their interaction, Phys. Lett., **63A**, 205 (1977).

115. *M. Boiti, J. Léon, L. Martina and F. Pempinelli*, Scattering of localized solitons in the plane, Phys. Lett., **132A**, 432 (1988).

116. *A.S. Fokas and P.M. Santini*, Coherent structures in multidimensions, Phys. Rev. Lett., **63**, 1329 (1989).

117. *P.M. Santini and A.S. Fokas*, The initial value problem for the Davey–Stewartson 1 equation: how to generate and drive localized coherent structures in multidimensions, in: Partially integrable evolution equations in Physics (R. Conte and N. Boccara, Eds), p.223, Kluwer Acad. Publ., Dordrecht, 1990.

118. *S.V. Manakov*, The inverse scattering transform for the time-dependent Schrödinger equation and Kadomtsev–Petviashvili equation, Physica, **3D**, 420 (1981).

119. *M.J. Ablowitz, D. Bar Yaacov and A.S. Fokas*, On the inverse scattering transform for the Kadomtsev–Petviashvili equation, Stud. Appl. Manh., **69**, 135 (1983).

120. *L. Hörmander*, An introduction to complex analysis in several variables, D. van Nostrand C., Princeton, 1966.

121. *A.S. Fokas and M.J. Ablowitz*, The inverse scattering transform for multidimensional (2+1) problems, Lecture Notes in Physics, **189**, 137; Proc. CIFMO Mexico, 1982 (K.B. Wolf, Ed.), Springer, Berlin, 1983.

122. *R. Beals and R.R. Coifman*, Scattering transformations spectral, et equations d'evolutions nonlineaires I, II, Seminaire Goulaouic-Meyer-Schwartz, 1980–1981, exp. 22; 1981–1982.

123. *R. Beals and R.R. Coifman*, Inverse scattering and evolution equations, Comm. Pure Appl. Math., **38**, 28 (1985).

124. *R. Beals and R.R. Coifman*, Multidimensional inverse scattering and nonlinear P.D.E., Proc. Sym. Pure Math., **43**, 42 (1985).

125. *M.J. Ablowitz and A.I. Nachman*, Multidimensional nonlinear evolution equations and inverse scattering, Physica, **18D**, 223 (1986).

126. *R. Beals and R.R. Coifman*, The D-bar approach to inverse scattering and nonlinear evolutions, Physica, **180D**, 242 (1986).

127. *M.J. Ablowitz*, Exactly solvable multidimensional nonlinear equations and inverse scattering, in: Topics in soliton theory and exactly solvable nonlinear equations (M.J. Ablowitz, B. Fuchssteiner and M. Kruskal, Eds.), World Scientific, Singapore, 1987, p.20.

128. *A.S. Fokas and M.J. Ablowitz*, On the inverse scattering of the time dependent Schrödinger equation and the associated Kadomtsev–Petviashvili (I) equation, Stud. Appl. Math., **69**, 211 (1983).

129. *A.S. Fokas and M.J. Ablowitz*, On the inverse scattering transform of multidimensional nonlinear equations related to first order systems in the plane, J. Math. Phys., **25**, 2494 (1984).

130. P.G. Grinevich and S.V. Manakov, Inverse scattering problem for the two-dimensional Schrödinger operator, $\bar{\partial}$-method and nonlinear equations, Funk. Anal. Pril., **20** (7), 14 (1986).

131. R. Beals and R.R. Coifman, Linear spectral problems, non-linear equations and the $\bar{\partial}$-method, Inverse problems, **5**, 87 (1989).

132. M. Jaulent and M. Manna, The spatial transform method: $\bar{\partial}$-derivation of the AKNS hierarchy, Phys. Lett., **117A**, 62 (1986).

133. J.J.P. Léon, General evolution of the spectral transform from the $\bar{\partial}$-approach, Phys. Lett., **123A**, 65 (1987).

134. M. Jaulent and M. Manna, The spatial transform method for multidimensional (2+1) problems, Europhys. Lett., **2**, 891 (1986).

135. M. Jaulent, M. Manna and L. Martinez Alonso, $\bar{\partial}$-equations in the theory of integrable systems, Inverse Problems, **4**, 123 (1988).

136. P.G. Grinevich and S.P. Novikov, Inverse scattering problem for the two-dimensional Schrödinger operator at a fixed negative energy and generalized analytic functions, in: Proc. of Int. Workshop "Plasma theory and nonlinear and turbulent processes in physics", Kiev, April 1987 (V.G. Bar'yakhtar, V.M. Chernousenko, N.S. Erokhin, A.G. Sitenko and V.E. Zakharov, Eds), v.1, p.58, World Scientific, Singapore, 1988.

137. P.G. Grinevich and S.P. Novikov, Two-dimensional inverse scattering problem at negative energy and generalized analytic functions. I. Energy below a ground state, Func. Anal. Appl., **22** (1), 23 (1988).

138. L. Bers, Partial differential equations and generalized analytic functions, Proc. Math. Acad. Scien. USA, **37** (1), 42 (1951).

139. I.N. Vekua, Systems of the first order differential equations of elliptic type and boundary problems with applications to shell theory, Matem. Sbornik, **31**, 217 (1952).

140. L. Bers, Theory of pseudo-analytic functions, New York Univ., New York, 1953.

141. I.N. Vekua, Generalized analytic functions, Fizmatgiz, Moscow, 1959.

142. W.L. Wendland, Elliptic systems in the plane, Pitman, London, 1979.

143. L. Yu. Rodin, Generalised analytic functions on Riemann surfaces, Lecture Notes in Math., **1288**, Springer-Verlag, Heidelberg, 1987.

144. V.E. Zakharov, Commuting operators and nonlocal $\bar{\partial}$-problem, in: Nonlinear and turbulent processes in physics, Proc. of III Int. Workshop, Naukova Dumka, Kiev, 1988, v.1, p.152.

145. L.V. Bogdanov and S.V. Manakov, The nonlocal $\bar{\partial}$-problem and (2+1) dimensional soliton equations, J. Phys. A: Math. Gen., **21**, L537 (1988).

146. L.V. Bogdanov and S.V. Manakov, Nonlocal $\bar{\partial}$-problem and (2+1) dimensional soliton equations, in: Proc. of Int. Workshop on Plasma theory and nonlinear and turbulent processes in physics, Kiev, April 1987, World Scientific, Singapore, 1988, v.1, p.7.

147. V.E. Zakharov, On the dressing method, in Inverse Problems in Action (P.S. Sabatier, Ed.), Springer-Verlag, Berlin, 1990, p.602.

148. A.S. Fokas and V.E. Zakharov, The dressing method and nonlocal Riemann–Hilbert problems, Journal of Nonlinear Sciences, **2**, 109 (1992).

149. I.M. Gelfand, M.I. Graev and N.Ya. Vilenkin, Integral geometry and connected problems of representation theory, Fizmatgiz, Moscow, 1962.

150. W.V. Lovitt, Linear integral equations, McGraw-Hill Book C., New York, 1924.

151. F.C. Tricomi, Integral equations, Interscience Publ., New York, 1957.

152. B.G. Konopelchenko and B.T. Matkarimov, Inverse spectral transform for the nonlinear evolution equation generating the Davey–Stewartson and Ishimori equations, Stud. Appl. Math., **82**, 319 (1990).

153. M. Jaulent, M.A. Manna and L. Martinez Alonso, Scalar-bipolar asymptotic modules for solving (2+1)-dimensional nonlinear evolution equations with constraints, Phys. Lett., **132A**, 414 (1988).

154. M. Jaulent, M.A. Manna and L. Martinez Alonso, An integrable (2+1)-dimensional generalization of the Volterra model, J. Phys. A: Math. Gen., **21**, L719 (1988).

155. M. Jaulent, M.A. Manna and L. Martinez Alonso, Matrix-bipolar asymptotic modules for solving (2+1)-dimensional nonlinear evolution equations with constraints, J. Phys. A: Math. Gen., **21**, L1019 (1988).

156. M. Jaulent, M.A. Manna and L. Martinez Alonso, Nonlinear (2+1)-dimensional systems solvable through asymptotic modules, Phys. Lett., **145A**, 328 (1989).

157. M. Jaulent, M.A. Manna and L. Martinez Alonso, A solvable hierarchy of ($N$+1)-dimensional nonlinear evolution equations with constraints, J. Phys. A: Math. Gen., **22**, L13 (1989).

158. M. Jaulent, M.A. Manna and L. Martinez Alonso, Asymptotic modules for solving integrable models, Inverse Problems, **5**, 573 (1989).

159. M. Jaulent, M.A. Manna and L. Martinez Alonso, Multi-series Lie groups and asymptotic modules for characterizing and solving integrable models, J. Math. Phys., (1989).

160. E. Date, M. Kashiwara, M. Jimbo and T. Miwa, Transformation groups for soliton equations, in: Nonlinear integrable systems-classical theory and quantum theory, Proc. of RIMS Symposium (M. Jimbo and T. Miwa, Eds.), World Scientific, Singapore, 1983, p.39.

161. M. Jimbo and T. Miwa, Solitons and infinite dimensional Lie algebras, Publ. RIMS, Kyoto Univ., **19**, 943 (1983).

162. V.E. Zakharov, Shock wave propagating along solitons on the water surface, Izvestija VUZov, Radiophysica, **29**, 1073 (1986).

163. V.I. Petviashvili, Forming the three-dimensional Langmuir solitons by action of powerful radiowave in ionsphere, Fizika plasmi, **2**, 650 (1976).

164. I.M. Krichever, On the rational solutions of the Kadomtsev–Petviashvili equation and integrable systems of $n$ particles on a line, Funct. Anal. Pril., **12** (1), 76 (1978).

165. F. Calogero, Solution on the one-dimensional $N$-body problem with quadratic and/or inversely quadratic pair potential, J. Math. Phys., **12**, 419 (1971).

166. J. Moser, Three integrable Hamiltonian systems connected with isospectrum deformations, Adv. Math., **16**, 354 (1976).

167. I.M. Krichever and S.P. Novikov, Holomorphic bundles under Riemann surfaces and the Kadomtsev–Petviashvili equation (KD) I, Funct. Anal. Pril., **12** (4), 41 (1978).

168. *V.E. Zakharov*, Unstability and nonlinear oscillation of solitons, Pisma ZETF, **22**, 364 (1975).

169. *A.S. Fokas and M.J. Ablowitz*, On the inverse scattering and direct linearizing transforms for the Kadomtsev–Petviashvili equation, Phys. Lett., **94A**, 67 (1983).

170. *F.W. Nijhoff, H.W. Capel and G.L. Wiersma*, Integrable lattice systems in two and three dimensions, Lecture Notes in Physics, **239**, 263 (1985).

171. *F.W. Nijhoff*, Integrable hierarchies, Lagrangian structures and noncommuting flows, in: Topics in soliton theory and exactly solvable nonlinear equations (M. Ablowitz and M. Kruskal, Eds.), World Scientific, Singapore, 1987, p.150.

172. *F.W. Nijhoff*, Linear integrable transformation and hierarchies of integrable nonlinear evolution equation, Physica, **31D**, 339 (1988).

173. *G.L. Wiersma and H.W. Capel*, Lattice equations hierarchies and Hamiltonian structures. II. KP-type of hierarchies on 2D lattices, Physica A, **149**, 49 (1988); III. The 2D Toda and KP hierarchies, Physica A, **149**, 75 (1988).

174. *H.W. Capel, G.L. Wiersma and F.W. Nijhoff*, Linearizing integral transform for multicomponent lattice KP, Physica A, (1988).

175. *J.W. Miles*, Resonantly interacting solitary waves, J. Fluid Mech., **79**, 171 (1977).

176. *A.C. Newell and L.G. Redekopp*, Breakdown of Zakharov–Shabat theory and soliton creation, Phys. Rev. Lett., **38**, 377 (1977).

177. *R.S. Johnson and S. Thompson*, A solution of the inverse scattering problem for the Kadomtsev–Petviashvili equation by the method of separation of variables, Phys. Lett., **66A**, 279 (1978).

178. *J.J. Satsuma*, N-soliton solution of the two-dimensional Korteweg-de Vries equation, J. Phys. Soc. Japan, **40**, 286 (1976).

179. *M.J. Ablowitz and J. Satsuma*, Solitons and rational solutions of nonlinear evolution equations, J. Math. Phys., **19**, 2180 (1978).

180. *N.C. Freeman*, Soliton interactions in two dimensions, Adv. Appl. Mech., **20**, 1 (1980).

181. *S. Oishi*, A method of analysing soliton equations by bilinearization, J. Phys. Soc. Japan, **48**, 639 (1980).

182. A. Nakamura, Decay mode solution of the two-dimensional KdV equation and generalized Bäcklund transformation, J. Math. Phys., **22**, 2456 (1981).

183. K. Okhuma and M. Wadati, The Kadomtsev–Petviashvili equation, the trace method and soliton resonance, J. Phys. Soc. Japan, **52**, 749 (1983).

184. S.P. Burtsev, Damping of soliton oscillations in a medium with negative dispersion, ZETF, **88**, 461 (1985).

185. B.A. Dubrovin, The Kadomtsev–Petviashvili equation and the relations between the periods of holomorphic differentials on Riemann surfaces, Math. USSR Izv., **19**, 285 (1982).

186. H. Segur and A. Finkel, An analytic model of periodic waves in shallow water, Stud. Appl. Math., **73**, 183 (1985).

187. I.M. Krichever, Spectral theory of finite-zone nonstationary Schrödinger operators. Nonstationary Paierls's model, Funct. Anal. Pril., **20** (2), 42 (1986).

188. V.B. Matveev and A.O. Smirnov, On the Riemann Theta function of the trigonal curve and solutions of the Boussinesq and KP equations, Lett. Math. Phys., **14**, 25 (1987).

189. M. Schwartz, Periodic solutions of Kadomtsev–Petviashvili, Adv. Math., **66**, 217 (1987).

190. N. Sheffner, J. Hammack and H. Segur, The KP equation and biperiodic water waves, in: Nonlinear evolutions (J.J.-P. Léon, Ed.), World Scientific, Singapore, 1988, p.517.

191. I.M. Krichever, Periodic problem for Kadomtsev–Petviashvili equation, DAN SSSR, **298**, 802 (1988).

192. S.V. Manakov, P.M. Santini and L.A. Takhtajan, Asymptotic behaviour of the solutions of the Kadomtsev–Petviashvili equation (two-dimensional Korteweg-de Vries equation), Phys. Lett., **75A**, 451 (1980).

193. Z. Jiang, R.K. Bullough and S.V. Manakov, Complete integrability of the Kadomtsev–Petviashvili equation in 2+1 dimensions, Physica, **18D**, 305 (1986).

194. M.V. Wickerhauser, Inverse scattering for the heat operator and evolutions in 2+1 variables, Commun. Math. Phys., **108**, 67 (1987).

195. X. Zhou, Inverse scattering transform for the time dependent Schrödinger equation with applications to the KP-I equation, Commun. Math. Phys., **128**, 55 (1990).

196. A.S. Fokas and L.Y. Sung, On the solvability of the N-wave, the Davey–Stewartson and the Kadomtsev–Petviashvili equations, Inverse Problems, **8**, 673 (1992).

197. M. Boiti, J. Léon and F. Pempinelli, Spectral transform and orthogonality relations for the Kadomtsev–Petviashvili I equation, Phys. Lett., **141A**, 96 (1989).

198. M. Boiti. F. Pempinelli, A.K. Pogrebkov and M.C. Polivanov, Resolvent approach for the nonstationary Schrödinger equation, Inverse Problems **8**, 331 (1992).

199. B.G. Konopelchenko, V.G. Dubrovsky, Some new integrable nonlinear evolution equations in 2+1 dimensions, Phys. Lett., **102A**, 15 (1984).

200. B.G. Konopelchenko, V.G. Dubrovsky, Inverse spectral transform for the modified Kadomtsev–Petviashvili equation, Stud. Appl. Math., **86**, 219 (1992).

201. R. Hirota, Classical Boussinesq equation in a reduction of the modified KP equation, Journal of Phys. Soc. Japan, **54**, 2409 (1985).

202. S. Lie, Explicit solutions for the KP and mKP hierarchies, Proc. Symp. Pure Math., **49**, part I, 101 (1989).

203. V.G. Dubrovsky and B.G. Konopelchenko, On the interrelation between the solutions of the mKP and KP equations via the Miura transformation, J. Phys. A: Math. Gen., **24**, 4315 (1991).

204. B.G. Konopelchenko, Inverse spectral transform for the (2+1)-dimensional Gardner equation, Inverse Problems, **7**, 739 (1991).

205. D.J. Benney and G.J. Roskes, Wave instabilities, Stud. Appl. Math., **48**, 377 (1969).

206. V.E. Zakharov and E.A. Kuznetsov, Multi-scale expansion in the theory of systems integrable by the inverse scattering transform, Physica, **18D**, 455 (1986).

207. F. Calogero, Why are certain nonlinear PDEs both widely applicable and integrable, in: What is integrability? (V.E. Zakharov, Ed.), Springer-Verlag, Berlin, 1991, p.1.

208. D. Anker and N.C. Freeman, On the soliton solutions of the Davey–Stewartson equation for long waves, Proc. Roy. Soc. London, **360A**, 524 (1978).

209. J.J. Satsuma and M.J. Ablowitz, Two-dimensional lumps in nonlinear dispersive systems, J. Math. Phys., **20**, 1416 (1979).

210. P.M. Santini, Energy exchange of interacting solitons in multidimensions, Physica, **41D**, 26 (1990).

211. M. Boiti, J. Léon, L. Martina and F. Pempinelli, Localized solitons in the plane, in: Nonlinear evolution equations: integrability and spectral methods (A. Degasperis, A.P. Fordy, M. Lakshmanan, Eds.), p.249, Manchester University Press, Manchester, 1990.

212. M. Boiti, J. Léon, L. Martina and F. Pempinelli, Solitons in two dimensions, in: Integrable systems and applications (M. Balabane, P. Lochak, C. Sulem, Eds.), Lecture Notes in Physics, **342**, p.31, Springer-Verlag, Berlin, 1989.

213. M. Boiti, J. Léon and F. Pempinelli, Multidimensional solitons and their spectral transforms, Journal of Math. Phys., **31**, 2612 (1990).

214. M. Boiti, J. Léon, L. Martina and F. Pempinelli, Exponentially localized solitons in 2+1 dimensions, in: Nonlinear evolution equations and Dynamical Systems (S. Carillo, O. Ragnisco, Eds.), p.26, Reports in Physics, Springer-Verlag, Berlin, 1990.

215. F. Pempinelli, M. Boiti, J. Léon and L. Martina, Bäcklund transformations and localized solitons in multidimensions, in: Lie theory, differential equations and representation theory (V. Hussin, Ed.), p.337, Les Publications CRM, Montreal, 1990.

216. J. Léon, M. Boiti and F. Pempinelli, Spectral characterization of 2-d nonlinear coherent structures (M. Barthes and J. Léon, Eds.), Lecture Notes in Physics, **353**, p.241, Springer-Verlag, Berlin (1990).

217. F. Pempinelli, M. Boiti, J. Léon and L. Martina, Exponentially localized solitons in 2+1 dimensions, in: Nonlinear World (V.G. Bar'yakhtar et al., Eds.), p.285, World Scientific, Singapore, 1990.

218. M. Boiti, J. Léon and F. Pempinelli, Solitons and spectral transform for DSI equation, in: Nonlinear World (V.G. Bar'yakhtar et al., Eds), p.56, World Scientific, Singapore, 1990.

219. M. Boiti, J. Léon and F. Pempinelli, On the spectral theory of the Davey-Stewartson equation, in: Inverse methods in action (P.C. Sabatier, Ed.), p.544, Springer-Verlag, Berlin, 1990.

220. M. Boiti, J. Léon and F. Pempinelli, Bifurcation of solitons, Inverse Problems, **6**, 715 (1990).

221. M. Boiti, J. Léon and F. Pempinelli, Waves in the Davey-Stewartson equation, Inverse Problems, **7**, 175 (1991).

222. M. Boiti, J. Léon, L. Martina, F. Pempinelli and D. Perrone, Asymptotic bifurcation of multidimensional solitons, in: Nonlinear evolution equations and dynamical systems – NEEDS'90 (V. Makhankov and O.K. Pashaev, Eds.), p.47, Springer-Verlag, Berlin, 1991.

223. M. Boiti, L. Martina, O.K. Pashaev and F. Pempinelli, Dynamics of multidimensional solitons, Phys. Lett., **A160**, 55 (1991).

224. A.S. Fokas and P.M. Santini, Coherent structures and a boundary value problem for the Davey–Stewartson I equation, Physica, **44D**, 99 (1990).

225. J. Hietarinta and R. Hirota, Multidimension solutions to the Davey–Stewartson equation, Phys. Lett., **A145**, 237 (1990).

226. J. Hietarinta, One-dromion solutions for generic classes of equations, Phys. Lett., **A149**, 113 (1990).

227. J. van der Linden, Solutions of the Davey–Stewartson equation with non-zero boundary condition, preprint INLO-PUB-5/91, Leiden (1991).

228. M. Jaulent, M. Manna and L. Martinez Alonso, Fermionic analysis of Davey-Stewartson dromions, Phys. Lett. A, **151**, 303 (1990);
L. Martinez Alonso and E. Medina Reus, Localized coherent structures of the Davey–Stewartson equation in the bilinear formalism, J. Math. Phys., **33**, 2947 (1992).

229. A. Degasperis, The Davey–Stewartson I equation: A class of explicit solutions including the special case of dromions, in: Inverse Problems in Action (P.C. Sabatier, Ed.), Springer-Verlag, Berlin, 1991, p.505.

230. L.D. Landau and E.M. Lifshitz, Quantum mechanics, Nauka, Moscow, 1974.

231. H. Cornille, Solutions of the generalized Schrödinger equation in two spatial dimensions, J. Math. Phys., **20**, 199 (1979).

232. H. *Cornille*, GNLS (or Davey–Stewartson I) equation, with associated Marchenko-like equations, revisited, Note interne CEA SPLT/89-014, Saslay, 1989.

233. A.S. *Fokas*, On the inverse scattering of first order systems in the plane related to nonlinear multidimensional equations, Phys. Rev. Lett., **51**, 3 (1983).

234. A.S. *Fokas and M.J. Ablowits*, Method of solution for a class of multidimensional nonlinear evolution equations, Phys. Rev. Lett., **51**, 7 (1983).

235. L.P. *Nizhnik*, Nonstationary inverse scattering problem for Dirac equation, Ukrain. Matem. Zurnal, **24** (1), 110 (1972).

236. Fam Loi *Wu*, Inverse scattering problem for Dirac system on a whole axis, Ukrain. Matem. Zurnal, **24**, 666 (1973).

237. L.P. *Nizhnik*, Nonstationary inverse scattering problem, Naukova Dumka, Kiev, 1973.

238. E.K. *Sklyanin*, On the class of potentials for nonstationary Dirac equation, Zapiski nauchn. sem. LOMI, **77**, 214 (1978).

239. L.P. *Nizhnik and M.D. Pochinaiko*, Integration of the spatially two-dimensional nonlinear Schrödinger equation by the inverse problem method, Funct. Anal. Pril., **16** (1), 80 (1982).

240. L.P. *Nizhnik and V.G. Tarasov*, Nonstationary inverse scattering problem for the hyperbolic system of equations, DAN SSSR, **233**, 300 (1977).

241. M. *Boiti, J. Léon and F. Pempinelli*, A new spectral transform for the Davey–Stewartson I equation, Phys. Lett., **A141**, 101 (1989).

242. L.Y. *Sung and A.S. Fokas*, Inverse problem of $N \times N$ hyperbolic systems on the plane and $N$-wave interactions, Commun. Pure Appl. Math., (1992).

243. V.A. *Arkadiev, A.K. Pogrebkov and M.C. Polyvanov*, Inverse scattering transform method and soliton solutions for Davey–Stewartson II equation, Physica, **36D**, 189 (1989).

244. A. *Soyeur*, The Cauchy problem for the Ishimori equation, preprint Orsay, 1990.

245. V.G. *Dubrovsky and B.G. Konopelchenko*, Coherent structures for the Ishimori equations. I. Localized solitons with stationary boundaries, Physica, **48D**, 367 (1991).

246. V.G. Dubrovsky and B.G. Konopelchenko, Coherent structures for the Ishimori equations. II. Time-dependent boundaries, Physica, **55D**, 1 (1992).

247. B.G. Konopelchenko and B.T. Matkarimov, Inverse spectral transform for the Ishimori equation. I. Initial value problem, J. Math. Phys., **31**, 2737 (1990).

248. B.G. Konopelchenko and B.T. Matkarimov, On the inverse scattering transform for the Ishimori equation, Phys. Lett., **A135**, 183 (1989).

249. R. Beals and R.R. Coifman, The spectral problem for the Davey–Stewartson and Ishimori hierarchies, in: Nonlinear evolution equations: integrability and spectral methods, (A. Degasperis, A.P. Fordy and M. Lakshmanan, Eds.), Manchester Univ. Press, Manchester, 1990, p.15.

250. I.T. Khabibulin, Teor. Mat. Fiz. (1992).

251. V.D. Lipovsky and A.V. Shirokov, An example of gauge equivalence of multidimensional integrable equations, Funct. Anal. and its Appl., **23**, N3. 65 (1989).

252. R.A. Leo, L. Martina and G. Soliani, A noncompact spin model in (2+1) dimensions, Phys. Lett., **247B**, 562 (1990).

253. R.A. Leo, L. Martina and G. Soliani, Gauge equivalence theory of the noncompact Ishimori model and the Davey–Stewartson equation, J. Math. Phys., **33**, 1515 (1992).

254. G.L. Terng, A higher dimensional generalization of the sine-Gordon equation and its soliton theory, Ann. of Math., **111**, 491 (1980).

255. R. Beals and K. Tenenblat, Inverse scattering and the Bäcklund transformations for the generalized wave and sine-Gordon equations, Stud. Appl. Math., **78**, 227 (1988).

256. M. Boiti, J. Léon and F. Pempinelli, Integrable two-dimensional generalization of the sine- and sinh-Gordon equations, Inverse Problems, **3**, 37 (1987).

257. B.G. Konopelchenko and V.G. Dubrovsky, The 2+1-dimensional integrable generalization of the sine-Gordon equation. I. $\bar{\partial} - \partial$-dressing and initial value problem, Stud. Appl. Math. (to be published); preprint INP 92-1, Novosibirsk, 1992.

258. V.G. Dubrovsky and B.G. Konopelchenko, The 2+1-dimensional integrable generalization of the sine-Gordon equation. II. Localized solutions, Inverse problems (to be published).

259. *L.V. Bogdanov*, Veselov-Novikov equation as a natural two-dimensional generalization of the KdV equation, Teor. Mat. Fiz., **70**, 309 (1987).

260. *L.V. Bogdanov*, On two-dimensional Zakharov-Shabat problem, Teor. Mat. Fiz., **72**, 155 (1987).

261. *A.Yu. Orlov*, Hamiltonian formalism and Gelfand-Dikii identity for 2+1 D integrable systems, in: Nonlinear and turbulent processes in physics, Proc. of III Int. Workshop, Naukova Dumka, Kiev, 1988, v.1, p.126.

262. *B.G. Konopelchenko*, Symmetries of generating problems and integrable equations, Inverse Problems (1992).

263. *J.J.-P. Léon*, Nonlinear evolutions with singular dispersion law and forced systems, Phys. Lett., **144A**, 444 (1990).

264. *A.S. Fokas and S.V. Manakov*, The dressing method, symmetries and invariant solutions, Phys. Lett., **150A**, 369 (1990).

265. *H. Flaschka, A.C. Newell and T. Ratiu*, Kac-Moody algebras and soliton equations II, Physica, **9D**, 300 (1983).

266. *S.V. Manakov*, unpublished.

267. *C. Rogers*, The Harry Dym equation in 2+1 dimensions: a reciprocal link with Kadomtsev-Petviashvili equation, Phys. Lett., **120A**, 15 (1987).

268. *A.A. Belavin and V.E. Zakharov*, Yang-Mills equations as inverse scattering problem, Phys. Lett., **73B**, 53 (1978).

269. *A.M. Perelomov*, Nonlinear evolution equations that leave the spectrum of multidimensional Schrödinger equation invariant do not exist, Lett. Math. Phys., **1**, 175 (1976).

270. *P.M. Santini*, Old and new results on the algebraic properties of integrable systems in multidimensions, in: Some topics on inverse problems (P.C. Sabatier, Ed.), World Scientific, Singapore, 1988, p.231.

271. *B.G. Konopelchenko*, On the multidimensional evolution equations connected with multidimensional scattering problems, Phys. Lett., **93A**, 442 (1983),

272. *B.G. Konopelchenko*, On the integrable equations and generate dispersion laws in multidimensional spaces, J. Phys. A: Math. Gen., **16**, L311 (1983).

273. *A.I. Nachman and M.J. Ablowitz*, Multidimensional inverse scattering for first-order systems, Stud. Appl. Math., **71**, 251 (1984).

274. *A.S. Fokas*, Inverse scattering and integrability in multidimensions, Phys. Rev. Lett., **57**, 159 (1986).

275. *A.B. Zamolodchikov*, Tetrahedron equations and integrable systems in three dimensions, ZETF, **79**, 641 (1980).

276. *A.B. Zamolodchikov*, Tetrahedron equations and relativistic S-matrix of straightlines in 2+1 dimensions, Commun. Math. Phys., **79**, 489 (1981).

277. *R.S. Ward*, Complete solvable gauge-field equations in dimension greater than four, Nucl. Phys., **B236**, 381 (1984).

278. *M.J. Ablowitz, D.G. Kosta and K. Tenenblat*, Solutions of multidimensional extension of the anti-self-dual Yang–Mills equations, Stud. Appl. Math., **77**, 37 (1987).

279. *M.V. Saveliev*, Multidimensional nonlinear systems, Teor. Mat. Fiz., **69**, 411 (1986).

280. *M.V. Saveliev*, Integrable graded manifolds and nonlinear equations, Commun. Math. Phys., **95**, 199 (1984).

281. *G.L. Rcheushvili and M.V. Saveliev*, Multidimensional nonlinear systems related to Grassman's manifolds BI, DI, Funct. Anal. Pril., **21** (4), 83 (1987).

282. *J.-M. Maillet and F. Nijhoff*, Multidimensional lattice integrability and the simplex equations, preprint Clarkson Univ., INS 102 (1988).

283. *P.F. Dhooghe*, The KP and more dimensional KdV equations on $A_2^{(1)}$ and $A_3^{(1)}$, J. Phys. A: Math. Gen., **21**, 379 (1988).

284. *M. Boiti. J.J.-P. Léon, L. Martina and F. Pempinelli*, On the recursion operator for the KP hierarchy in two and three spatial variables, Phys. Lett., **123A**, 340 (1987).

285. *A. Degasperis and P.C. Sabatier*, Localized solutions of $(N+1)$-dimensional evolution equations, Phys. Lett., **150A**, 380 (1990).

286. *S. Friedlander and M.M. Vishik*, Lax pair formalism for the Euler equation, Phys. Lett., **148A**, 313 (1990).

# References

287. J. *Léon*, Nonlinear evolutions, spectral transforms and solitons in 3+1 dimensions, Phys. Lett., **156A**, 277 (1991).

288. L. *Martinez Alonso, E. Medina Reuz and R. Hernandez Heredero*, Multidimensional localized coherent structures in the bilinear formalism of integrable systems. Inverse Problems, **7**, L25 (1991).

289. K. *Takasaki,*, Integrable systems as deformations of D-modules, Proc. Symp. Pure Mathematics, **49**, part I, 143 (1989)

290. Y. *Ohyama*, Self-quality and integrable systems, Publ. RIMS, Kyoto Univ., **26**, 701 (1990).

291. A. *Nakayashiki*, Structure of Baker–Achiezer modules of principally polarized Abelian varieties, commuting partial differential operators and associated integrable systems, Duke Math. Journal, **62**, 315 (1991).

292. J. *Hietarinta and J. Satsuma*, The trilinear equations as a (2+2)-dimensional extension of the (1+1)-dimensional relativistic Toda lattice, Phys. Lett., **161A**, 267 (1991).

293. P.C. *Sabatier*, Spectral transform for nonlinear evolutions in $N$-dimensional space, Inverse Problems, **8**, 263 (1992).

294. P.C. *Sabatier*, Spectral transform for nonlinear evolution equations with $N$ space dimensions, Phys. Lett., **161A**, 345 (1992).

295. P.V. *Santini*, Integrable nonlinear evolution equations with constraints, Inverse Problems, **8**, 285 (1992).

296. D.V. *Chudnovsky*, The generalized Riemann–Hilbert problem and the spectral interpretation, Lecture Notes in Physics, **120**, 103, Springer-Verlag, Heidelberg, 1980.

297. B.G. *Konopelchenko and W. Strampp*, The AKNS hierarchy as symmetry constraint of the KP hierarchy, Inverse Problems, **7**, L417 (1991).

298. B.G. *Konopelchenko, Yu. Sidorenko and W. Strampp*, (1+1)-dimensional integrable systems as symmetry constraints of (2+1)-dimensional systems, Phys. Lett., **157A**, 17 (1991).

299. B.G. *Konopelchenko and W. Strampp*, New reductions of the KP and 2DTL hierarchies via symmetry constraints, J. Math. Phys., **33**, 3676 (1992); preprint SPL-T/91-162, Saclay, 1991.

300. *B.G. Konopelchenko and W. Strampp*, Reductions of 2+1-dimensional integrable systems via mixed potential-eigenfunction constraints, J. Phys. A: Math. Gen., **25**, 4399 (1992).

301. *Yi Cheng and Yi-shen Li*, The constraint of the Kadomtsev–Petviashvili equation and its special solutions, Phys. Lett., **157A**, 22 (1991).

302. *Z. Yunbo*, The constraint on potential and decomposition for (2+1)-dimensional integrable systems, J. Phys. A: Math. Gen., **24**, L1065 (1991).

303. *A.Yu. Orlov*, Symmetries for unifying different soliton systems into a single integrable hierarchy, preprint IINS/OCE-04/03, Moscow, 1991.

304. *Yi Cheng and Yi-shen Li*, Constraints of the 2+1-dimensional integrable soliton systems, J. Phys. A: Math. Gen., **25**, 419 (1992).

305. *M. Boiti, F. Pempinelli and P.C. Sabatier*, First and second order nonlinear evolution equations from an inverse spectral method, preprint Montpellier PM/92-27 (1992).

306. *R. Beals and B. Konopelchenko*, The $\bar{\partial}$-problem on the torus and multidimensional Landau–Lifshitz equation, (to be published).